Robert E. Dinnebier, Andreas Leineweber, John S.O. Evans
Rietveld Refinement

Also of interest

Highlights in Applied Mineralogy.
Soraya Heuss-Aßbichler, Georg Amthauer, Melanie John (Eds.), 2017
ISBN 978-3-11-049122-7, e-ISBN (PDF) 978-3-11-049734-2,
e-ISBN (EPUB) 978-3-11-049508-9

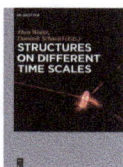

Structures on Different Time Scales
Theo Woike, Dominik Schaniel (Eds.), 2018
ISBN 978-3-11-044209-0, e-ISBN (PDF) 978-3-11-043392-0,
e-ISBN (EPUB) 978-3-11-043390-6

Multi-Component Crystals. Synthesis, Concepts, Function
Edward R. T. Tiekink, Julio Zukerman-Schpector (Eds.), 2017
ISBN 978-3-11-046365-1, e-ISBN (PDF) 978-3-11-046495-5,
e-ISBN (EPUB) 978-3-11-046379-8

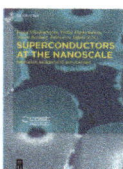

Superconductors at the Nanoscale.
From Basic Research to Applications
Roger Wördenweber, Victor Moshchalkov,
Simon Bending, Francesco Tafuri (Eds.), 2017
ISBN 978-3-11-045620-2, e-ISBN (PDF) 978-3-11-045680-6,
e-ISBN (EPUB) 978-3-11-045624-0

Electrochemical Storage Materials.
From Crystallography to Manufacturing Technology
Dirk C. Meyer, Tilmann Leisegang, Matthias Zschorner,
Hartmut Stöcker (Eds.), 2018
ISBN 978-3-11-049137-1, e-ISBN (PDF) 978-3-11-049398-6,
e-ISBN (EPUB) 978-3-11-049187-6

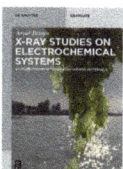

X-ray Studies on Electrochemical Systems.
Synchrotron Methods for Energy Materials
Artur Braun, 2017
ISBN 978-3-11-043750-8, e-ISBN (PDF) 978-3-11-042788-2,
e-ISBN (EPUB) 978-3-11-042796-7

Robert E. Dinnebier, Andreas Leineweber,
John S.O. Evans

Rietveld Refinement

Practical Powder Diffraction Pattern Analysis using
TOPAS

DE GRUYTER

Physics and Astronomy Classification Scheme 2010
Primary: 61.05.-a, 61.05.cp, 61.05.fn; Secondary: 61.72.-y, 61.72.Dd, 61.72.Nn

Authors
Prof. Robert E. Dinnebier
Max Planck Institute for
Solid State Research
Heisenbergstr. 1
70569 Stuttgart, Germany
r.dinnebier@fkf.mpg.de

Prof. Andreas Leineweber
TU Bergakademie Freiberg
Institute of Materials Science
Gustav-Zeuner-Str. 5
09599 Freiberg, Germany
Andreas.Leineweber@iww.tu-freiberg.de

Prof. John S.O. Evans
Durham University
Department of Chemistry
South Road
Durham DH1 3LE, UK
john.evans@durham.ac.uk

ISBN 978-3-11-045621-9
e-ISBN (PDF) 978-3-11-046138-1
e-ISBN (EPUB) 978-3-11-046140-4

Library of Congress Control Number: 2018954391

Bibliographic information published by the Deutsche Nationalbibliothek
The Deutsche Nationalbibliothek lists this publication in the Deutsche Nationalbibliografie; detailed
bibliographic data are available on the Internet at http://dnb.dnb.de.

© 2019 Walter de Gruyter GmbH, Berlin/Boston
Typesetting: Integra Software Services Pvt. Ltd.
Printing and binding: CPI books GmbH, Leck.
Cover image: Robert E. Dinnebier

www.degruyter.com

Preface

A detailed insight into atomic-level structure is of crucial importance in understanding, designing and optimizing structural and functional materials. The technique of choice for structure determination is usually single crystal diffraction. Unfortunately, single crystals of suitable size and shape are simply not available for many materials, and one has to rely on powder diffraction methods. These are often viewed as information poor, but actually contain a wealth of information which is not available in a routine single crystal study. Some of this is summarized schematically in Figure 1: Bragg peak intensities give detailed information about a sample's phase make-up and the composition and structure of phases; peak shapes give information about its microstructure; and the diffuse intensity between the Bragg peaks gives information on various aspects of local structure or disorder. The technique can also be more easily adapted to perform *in situ* or *operando* studies than single crystal methods, allowing us to follow materials in action. The information content of a powder pattern is huge, but sophisticated analysis methods are often needed to harvest the maximum amount of information from the data (Dinnebier & Billinge, 2008). This book discusses how this can be done.

Although the powder diffraction method was developed as early as 1916 by Peter Debye and Paul Scherrer (1916), for the first 50 years its use in crystallography was almost exclusively limited to qualitative and semiquantitative phase analysis. In materials science residual stress, texture and microstructure were also heavily investigated. What held the field back, especially in the area of determination of the atomic crystal structure, was the often-cited problem of powder diffraction: the accidental and systematic peak overlap caused by a projection of three-dimensional reciprocal space on to the one-dimensional 2θ axis of a powder pattern. This leads to a strongly reduced information content compared to a single crystal data set. The introduction of what has become known as the Rietveld method marked a turning point (Rietveld, 1967, 1969). Hugo Rietveld created a program to sidestep the overlap problem by modeling the whole powder pattern with a set of parameters that can be refined by minimizing the difference between the calculated and the measured powder pattern. One then finds that there is sufficient information in the one-dimensional data set to reconstruct the three-dimensional structure. The early development of different methods and Rietveld codes has been discussed by various authors (Young 1993, van Laar 2018). A great feature of the Rietveld method is the possibility to separate the main constituents of a powder pattern (background, lattice, crystal structure and microstructure – see Figure 1), allowing different aspects of a material's properties to be probed separately. We cover the fundamentals of powder diffraction in Chapter 1 and describe the Rietveld whole pattern fitting method in Chapter 2.

In its original implementation, the Rietveld method was used to refine a structural model from neutron diffraction data of single-phase samples. The method has been significantly extended over the last 50 years, and a variety of different whole-pattern fitting techniques are now used. One set is the LeBail (1988) and Pawley (1981) methods

https://doi.org/10.1515/9783110461381-201

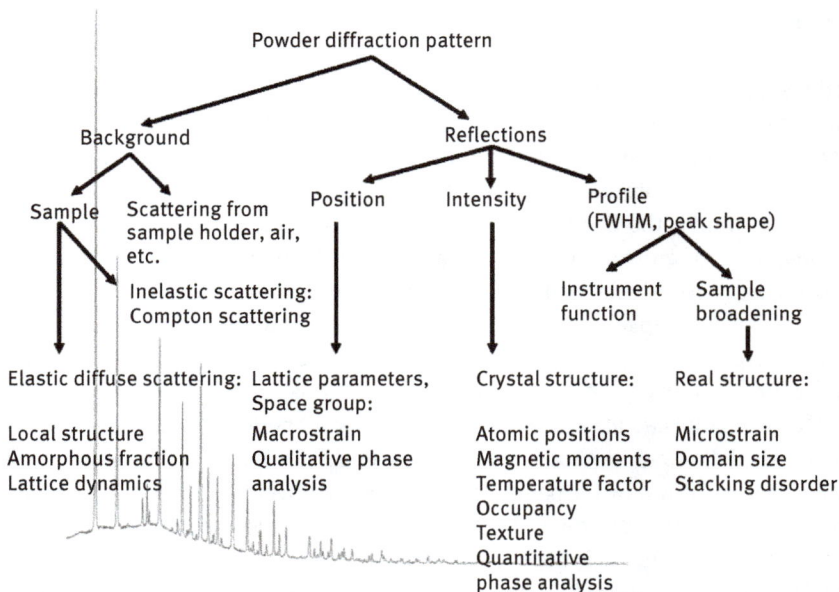

Figure 1: Schematic view of the different parts of a powder pattern and their physical meaning (adapted from Dinnebier & Billinge (2008) with permission from The Royal Society of Chemistry).

that are structure-independent whole powder pattern fitting approaches for deriving reflection intensities for structure determination, testing lattice parameters and space groups and for deriving microstructural parameters without the need for a crystal structure. The uses of these and related methods are discussed in Chapter 3.

In the earliest Rietveld codes, the shapes of peaks in a powder pattern were fitted with analytical functions such as Gaussian, Lorentzian or pseudo-Voigt functions, and the 2θ-dependence of peak widths was described using simple empirical functions. With increasing computational power it has become possible to develop more sophisticated approaches in which the contributions to the peak shape from the instrument and sample are treated separately giving access to exquisite detail about the microstructure of a material (size, strain and so on). We explore various approaches for this in Chapter 4. Recently, faster and more sophisticated algorithms and much faster computers have allowed Rietveld refinement using large supercells. This allows the description of complex stacking faults as structural contributions to peak shapes during Rietveld refinement (Coelho et al., 2016), and is discussed in Chapter 10.

The extension of the Rietveld method to X-ray powder diffraction data was published in 1977 (Malmros & Thomas, 1977) and has had enormous impact in many different areas. Most obviously, it allows the structural analysis of samples using laboratory-derived data. It has also opened up the possibility of routine quantitative analysis of multiphase mixtures (including amorphous content) by the Rietveld method. This is now a major application of the Rietveld method in industry (Bish

& Howard, 1988). We discuss the quantitative phase analysis of crystalline and amorphous samples in Chapter 5.

Chapter 6 introduces some of the ideas that are needed to refine particularly complex structures, where the information content in the powder diffraction data alone may not be sufficient. In these cases, extra information ("chemical knowledge") can be included in the form of constraints, restraints and rigid bodies. These can help stabilize refinements, and even small protein structures can now be refined using synchrotron powder diffraction data (Von Dreele, 1999). Similar information is also needed if one tries to solve structures *ab initio* using the Rietveld method coupled with simulated annealing (Newsam et al., 1992). We discuss some of the tricks needed to do this in Chapter 7, but refer the reader to more specialized texts (David et al., 2006) for discussion of other structure solution methods.

Chapter 8 discusses another approach that can help reduce the number of parameters needed in Rietveld refinement of samples that undergo symmetry-lowering phase transitions – so-called symmetry-adapted distortion mode refinements. The method can allow efficient analysis of otherwise complex problems and can help rapidly identify the key structural changes occurring at the transition. The symmetry ideas underlying this approach can be reapplied to a number of related problems such as understanding spin arrangements in magnetically ordered materials, which are covered in Chapter 9. We discuss here how the interaction of neutrons with magnetic moments allows the determination of magnetic structure and some of the different approaches to magnetic Rietveld refinement.

Turning to the bottom left corner of Figure 1, elastic diffuse scattering is measured during any powder diffraction experiment, but is usually ignored during Rietveld refinement and treated as part of the general "background" of the pattern. It is, however, possible to extract information about the local structure of a material from this scattering. With modern X-ray and neutron sources, the powder pattern can be measured over a large range of reciprocal space with good counting statistics. One can then take the Fourier transform of the normalized pattern to produce the real space pair distribution function (PDF) in the so-called total scattering approach. In Chapter 11, we introduce the real-space Rietveld-type approaches to analyzing such data and modeling local structure.

Finally, with modern instrumentation it is now possible to collect a vast number of powder diffraction patterns of a sample as a function of external variables with a time resolution in the second or even sub-second regime. This allows the detailed *in situ* or *operando* study of many important processes. In Chapter 12, we describe some of the ways to simulate multiple powder patterns to help plan such experiments. We also discuss how to automate data analysis to allow fitting of large numbers of patterns using either a sequential approach, where each data set is analyzed independently, or a parametric approach, where parameters are refined across multiple data sets (Stinton & Evans, 2007). There are many as yet unexplored applications of these methods.

Over 25 years have passed since the last textbook exclusively devoted to Rietveld refinement (Young, 1993) was published, and there have been significant developments since that time. We therefore decided to write a text that combines the fundamentals of the method with practical details of its implementation. Several powerful public domain Rietveld codes have been developed such as GSAS (Larson & Von Dreele, 1986), FULLPROF (Rodriguez-Carvajal, 1993), RIETAN (Izumi, 1989) and JANA (Petricek et al., 2014), to name just a few. When preparing this book, we decided to focus on the academic version of the program TOPAS (Coelho, 2018). The reasons are manifold. First of all, TOPAS is extremely fast and extremely robust, making it a great tool for beginners. It also tackles the different types of refinement problems (X-ray, neutron, fixed wavelength, time of flight, energy dispersive, structure solution, distortion modes, magnetic refinement, PDF analysis, multiple data sets and so on) that we think are most important. Our number one reason, however, is related to the unique scripting language of TOPAS, which allows experienced users to implement new developments in Rietveld refinement themselves. One example of this is the symmetry mode approach described in Chapter 8, which can be programmed without changing the TOPAS code (Campbell et al., 2007). The macro language of TOPAS also allows the implementation of new ideas by the user that can be rapidly shared with the entire user community (Scardi & Dinnebier, 2010). In this spirit we provide TOPAS code/macros for many of the corrections and algorithms described throughout the book. Many of these are available online at http://topas.dur.ac. uk/topaswiki/doku.php?id=book, so the reader can test the ideas. Most of the examples will work with any version of TOPAS, but some of the more advanced topics may only work with version 6 and above (Chapters 9–11, some of Chapter 12). Finally, the provision of standard macros via the topas.inc file acts as a wonderful "dictionary" detailing the mathematics behind many TOPAS commands.

It should be noted that we expect the reader to be familiar with the basics of crystallography and diffraction that are covered in many excellent introductory texts (e.g., Giacovazzo et al., 2011; Pecharsky & Zavalij, 2009; Gilmore et al. 2019). We have also kept the mathematical detail in the body of the text at the minimum level, which is necessary for understanding. Where more explanation is required we refer the reader to a stand-alone Chapter 13 that covers most of the higher mathematics needed to understand Rietveld refinement.

We would like to acknowledge Prof. Jörg Ihringer who gave permission for using his excellent script about diffraction for Chapter 13. Special thanks goes to Dr. Sebastian Bette for writing a major fraction of the chapter on "stacking faults." We would also like to thank Prof. Branton Campbell and Dr Phil Chater for educating some of us through on-going collaborations in the areas of Chapters 8, 9 and 11.

Robert Dinnebier
Andreas Leineweber
John Evans

The book is dedicated to Alan Coelho (Brisbane) and Arnt Kern (Karlsruhe) for developing the unique TOPAS program and making it available to the powder diffraction community.

References

Bish, D., Howard, S.A. (1988): *Quantitative phase analysis using the Rietveld method.* J. Appl. Cryst. 21, 86–91.

Campbell, B. J., Evans, J.S.O., Perselli, F., Stokes, H.T. (2007): *Rietveld refinement of structural distortion-mode amplitudes.* IUCr commission on crystallographic computing newsletter No. 8, 81–95.

Coelho, A.A. (2018): *TOPAS and TOPAS-Academic: an optimization program integrating computer algebra and crystallographic objects written in C++.* J. Appl. Cryst. 51,210–218.

Coelho, A.A., Evans, J.S.O., Lewis, J.W. (2016): *Averaging the intensity of many-layered structures for accurate stacking-fault analysis using Rietveld refinement.* J. Appl. Cryst. 49, 1740–1749.

Debye, P., Scherrer, P. (1916): *Interferenzen an regellos orientierten Teilchen im Röntgenlicht.* Nachrichten Kgl. Gesell. Wiss. Göttingen I. 1–15; II. 16–26.

David, W.I.F., Shankland, K., McCusker L.B., Baerlocher, Ch. (2006): *Structure determination from powder diffraction data,* IUCr Monographs on Crystallography, Oxford University Press, 337 pages.

Dinnebier, R.E., Billinge, S.J.L. (eds.) (2008): *Powder diffraction – theory and practice,* The Royal Society of Chemistry (RCS), Cambridge UK, 574 pages.

Giacovazzo, C., Monaco, H.L., Artioli, G., Viterbo, D., Milanesio, M., Gilli, G., Gilli, P., Zanotti, G., Ferraris, G. (2011): *Fundamentals of Crystallography (International Union of Crystallography Monographs on Crystallography),* 3rd edition, Oxford University Press, 872 pages.

Gilmore, C.J., Kaduk, J.A., Schenk H. (eds.): *International tables for crystallography, volume H, powder diffraction.* Wiley, New York (US), to be published in 2019.

Izumi, F. (1989): *RIETAN: a software package for the Rietveld analysis of X-ray and neutron diffraction patterns.* The Rigaku J. Vol. 6 10–20.

Larson, A.C., Von Dreele, R.B. (1986): *GSAS – general structure analysis system,* Report LAUR 86–748, Los Alamos National Laboratory, New Mexico, 13 Pages.

Le Bail, A., Duroy, H., Fourquet, J.L. (1988): *Ab-initio structure determination of LiSbWO$_6$ by X-ray powder diffraction.* Mat. Res. Bull. 23, 447–452.

Malmros, G., Thomas, J.O. (1977): *Least squares structure refinement based on profile analysis of powder film intensity data measured on an automatic microdensitometer.* J. Appl. Cryst. 10, 7–11.

Newsam, J.M., Deem, M.W., Freeman, C.M. (1992): Accuracy in powder diffraction II, NIST Special Publ. No. 846, 80–91.

Pawley, G. S. (1981): *Unit-cell refinement from powder diffraction scans.* J. Appl. Cryst. 14, 357–361.

Pecharsky, V., Zavalij, P. (2009): *Fundamentals of powder diffraction and structural characterization of materials,* 2nd edition, Springer US, 744 pages.

Petricek, V., Dusek, M., Palatinus, L. (2014): *Crystallographic computing system JANA2006: general features.* Z. Kristallogr. 229, 345–352.

Rietveld, H.M. (1967): *Line profiles of neutron powder-diffraction peaks for structure refinement.* Acta Cryst. 22, 151–152.

Rietveld, H.M. (1969): *A profile refinement method for nuclear and magnetic structures.* J. Appl. Cryst. 2, 65–71.

Rodriguez-Carvajal, J. (1993): *Recent advances in magnetic structure determination by neutron powder diffraction.* Physica B 192, 55–69.

Scardi, P., Dinnebier, R.E. (eds.) (2010): *Extending the reach of powder diffraction modelling by user defined macros,* Materials Science Forum 561, Trans Tech Publications Ltd., 219 pages.

Stinton, G.W., Evans, J.S.O. (2007): *Parametric Rietveld refinement.* J. Appl. Cryst. 40, 87–95.

Van Laar, B., Schenk, H. (2018): *The development of powder profile refinement at the Reactor Centre Netherlands at Petten*. Acta Cryst. A74, 88–92.

Von Dreele, R.B. (1999): *Combined Rietveld and stereochemical restraint refinement of a protein crystal structure*. J Appl. Cryst. 32, 1084–1089.

Young, R.A. (ed.) (1993): *The Rietveld method*, Oxford University Press, New York (US), Oxford, 298 pages.

Contents

1 The powder diffraction method

Powder diffraction is one of the most powerful techniques for studying the atomic structure of real materials. In this chapter we assume the reader has a knowledge of basic crystallography and refer them to one of the excellent texts on the topic for more advanced information (e.g., Giacovazzo et al., 2011; Egami & Billinge, 2012). We will restrict ourselves to giving sufficient information that we can discuss diffraction phenomena that are particularly important for powder diffraction. We describe the basic effects of diffraction by crystalline and noncrystalline samples. We also discuss the origins of peak broadening in powder diffraction patterns and quantify how peak overlap leads to information loss relative to single crystal experiments. We refer readers to specialist texts (e.g., Pecharsky & Zavalij, 2009; Dinnebier & Billinge, 2008) for detailed information on experimental aspects of powder diffraction.

1.1 Diffraction by crystallites

When radiation with a wavelength comparable to interatomic distances is incident on a crystallite, two types of coherent elastic scattering[1] occur: Bragg scattering that is restricted to the lattice points in reciprocal space and elastic diffuse scattering that can be found over wide regions of reciprocal space if a sample lacks translational order. In this section we discuss the basic concepts for understanding scattering by ordered crystallites.

An ideal crystal is an infinite three-dimensional periodic arrangement of structural motifs. These motifs can consist of any number of atoms. If the motifs are replaced by points, the points form a three-dimensional lattice in real space. The empty lattice can be represented by a three-dimensional sum of equidistant δ-functions (see also Chapter 13 – Appendix: Mathematical Basics) called the lattice function g:

$$g(\mathbf{x}) = \sum_{v_a = -N_a}^{N_a} \delta(\mathbf{x} - \mathbf{a}\, v_a) \sum_{v_b = -N_b}^{N_b} \delta(\mathbf{x} - \mathbf{b}\, v_b) \sum_{v_c = -N_c}^{N_c} \delta(\mathbf{x} - \mathbf{c}\, v_c), \qquad (1.1)$$

with the origin in the center of a lattice with basis vectors $\mathbf{a}, \mathbf{b}$ and $\mathbf{c}$ and a number of unit cells in the direction of each lattice vector of $2N_a,\ 2N_b,\ 2N_c$. For an ideal crystal the number of unit cells is infinite. In Fourier (= reciprocal/diffraction) space, the

---

1 Coherent means a set phase relationship between the scattered radiation from two objects such that interference occurs giving structural information. Elastic means no change in wavelength between incident and scattered beams.

https://doi.org/10.1515/9783110461381-001

amplitude scattered by the δ-functions at the lattice points in a direction represented by scattering vector $\mathbf{s}$ is:

$$G(\mathbf{s}) = \sum_{v_a = -N_a}^{N_a} \sum_{v_b = -N_b}^{N_b} \sum_{v_c = -N_c}^{N_c} e^{2\pi i \left(v_a \mathbf{a} \cdot \mathbf{s} + v_b \mathbf{b} \cdot \mathbf{s} + v_c \mathbf{c} \cdot \mathbf{s} \right)}. \tag{1.2}$$

The scattering vector can be expressed in reciprocal coordinates:

$$\mathbf{a} \cdot \mathbf{s} = h_1; \quad \mathbf{b} \cdot \mathbf{s} = h_2; \quad \mathbf{c} \cdot \mathbf{s} = h_3, \tag{1.3}$$

which results in the so-called Laue function which when squared gives the distribution of diffracted intensity in reciprocal space:

$$G(\mathbf{s})^2 = G(h_1, h_2, h_3)^2 = \frac{\sin^2(\pi N_a h_1)}{\sin^2(\pi a h_1)} \frac{\sin^2(\pi N_b h_2)}{\sin^2(\pi b h_2)} \frac{\sin^2(\pi N_c h_3)}{\sin^2(\pi c h_3)}. \tag{1.4}$$

The intensity of $G(\mathbf{s})^2$ is highest in directions where the phases of all summed scattered waves differ by multiples of 2π. $G(\mathbf{s})^2$ is plotted in two dimensions in Figure 1.1, and we see that the maxima occur at integer coordinates of the scattering

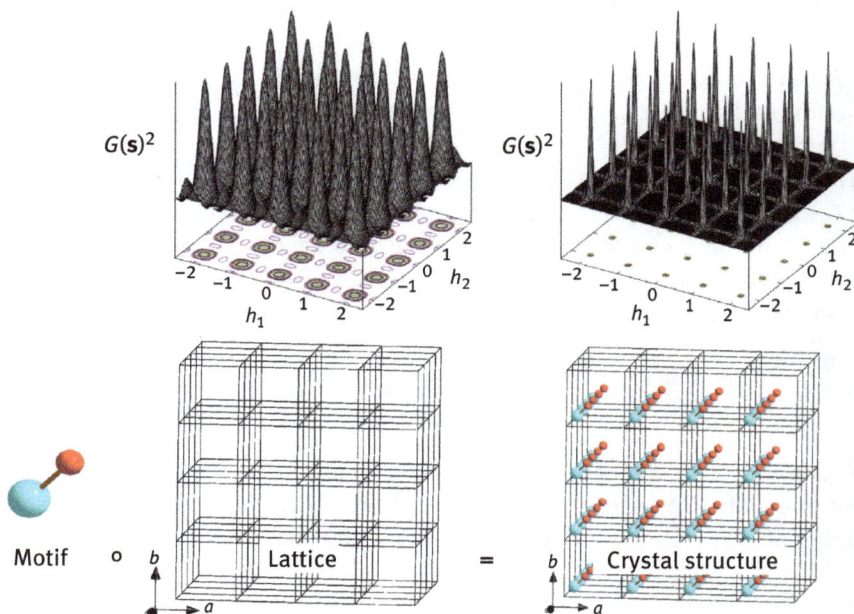

Figure 1.1: Top: The $G(\mathbf{s})^2$ function of eq. (1.4) plotted for $N_a = N_b = 3$ and $N_a = N_b = 10$; note how the peaks sharpen to become more like δ-functions as N increases. Bottom: Mathematical convolution of a structural motif (consisting of two atoms) and a three-dimensional lattice (consisting of mathematical points) to give a crystal structure.

vectors. Geometrically, these are at the lattice points of the reciprocal lattice.[2] This can be expressed mathematically as:

$$\mathbf{a} \cdot \mathbf{s} = h; \quad \mathbf{b} \cdot \mathbf{s} = k; \quad \mathbf{c} \cdot \mathbf{s} = l, \tag{1.5}$$

where h, k and l are integers. These are the three well-known Laue equations. They tell us how the direction of the scattered beam is related to the real or reciprocal lattices. Dividing the integer triple (h, k, l) by the largest common integer leads to the coprime Miller indices.

We can also see from Figure 1.1 that $G(\mathbf{s})^2$ becomes increasingly sharp at the reciprocal lattice points as the number of unit cells increases.

The motifs associated with the crystal lattice can be represented by a distribution of density (electron or nuclear density) ϱ_{cell}. X-rays interact with the electrons while neutrons interact with the nuclei (apart from magnetic scattering, see Chapter 9). The content of each unit cell can be expressed as the sum of individual objects (atoms). If symmetry elements in addition to the identity element (one-fold axis) exist, only the atoms of the asymmetric unit must be given, as the remaining atoms in the unit cell are created by the symmetry operators, while the entire crystal is built by the translational symmetry of the lattice.

Each object is represented by the convolution of its density and a δ-function defining its position $\mathbf{x}_j$ in the unit cell:

$$\varrho_{\text{cell}}(\mathbf{x}) = \sum_{j=1}^{n} \varrho_j(\mathbf{x}) \circ \delta(\mathbf{x} - \mathbf{x}_j). \tag{1.6}$$

In Fourier space the amplitude scattered by the unit cell at scattering vector $\mathbf{s}$ is:

$$F_{\text{cell}}(\mathbf{s}) = \sum_{j=1}^{n} \int_{-\infty}^{\infty} \int_{-\infty}^{\infty} \int_{-\infty}^{\infty} \varrho_j(\mathbf{x}) e^{2\pi i \, \mathbf{s} \cdot \mathbf{x}} d\mathbf{x} \int_{-\infty}^{\infty} \int_{-\infty}^{\infty} \int_{-\infty}^{\infty} \delta(\mathbf{x} - \mathbf{x}_j) e^{2\pi i \, \mathbf{s} \cdot \mathbf{x}} d\mathbf{x}. \tag{1.7}$$

The Fourier transform of a single atom depends on its form and scattering power and is called the *atomic form factor* in the case of X-rays and the *scattering length* in the case of neutrons:

$$f_j(\mathbf{s}) = \int_{-\infty}^{\infty} \int_{-\infty}^{\infty} \int_{-\infty}^{\infty} \varrho_j(\mathbf{x}) e^{2\pi i \, \mathbf{s} \cdot \mathbf{x}} d\mathbf{x} \tag{1.8}$$

and is usually assumed to show a spherically symmetric distribution of scattering power, implying that $f_j(\mathbf{s}) = f_j(s)$. The structure factor then greatly simplifies to:

---

2 See Section 13.10 for a definition of the reciprocal lattice.

$$F_{\text{cell}}(\mathbf{s}) = \sum_{j=0}^{n} f_j(s) \, e^{2\pi i \, \mathbf{s} \cdot \mathbf{x}_j}. \tag{1.9}$$

Finally, a crystal structure can be viewed as a *convolution* of the structural motif (content of one unit cell) and the lattice (Figure 1.1) such that:

$$\varrho_{\text{crystal}}(\mathbf{x}) = \varrho_{\text{cell}}(\mathbf{x}) \circ g(\mathbf{x}) \tag{1.10}$$

represents the density distribution of the crystal. In Fourier space the amplitude scattered by the crystal with scattering vector $\mathbf{s}$ reads:

$$F_{\text{crystal}}(\mathbf{s}) = F_{\text{cell}}(\mathbf{s}) \cdot G(\mathbf{s}). \tag{1.11}$$

We therefore find that a single crystal gives rise to scattering in certain discrete directions determined by the cell dimensions (through $G(\mathbf{s})$) with intensities determined by the cell contents (through $F_{\text{cell}}(\mathbf{s})$). These scattered or diffracted beams would appear as sharp spots on an X-ray film or a two-dimensional detector.

If we now consider a microcrystalline powder, there will be a large number of crystallites in different orientations. Ideally, the individual crystallite size will be of the order of 1 μm or less and all orientations will be equally probable. A key property of any Fourier transformation is that the intensity distribution of the diffracted image is invariant against translation of the object, which means that all diffracting crystallites in a

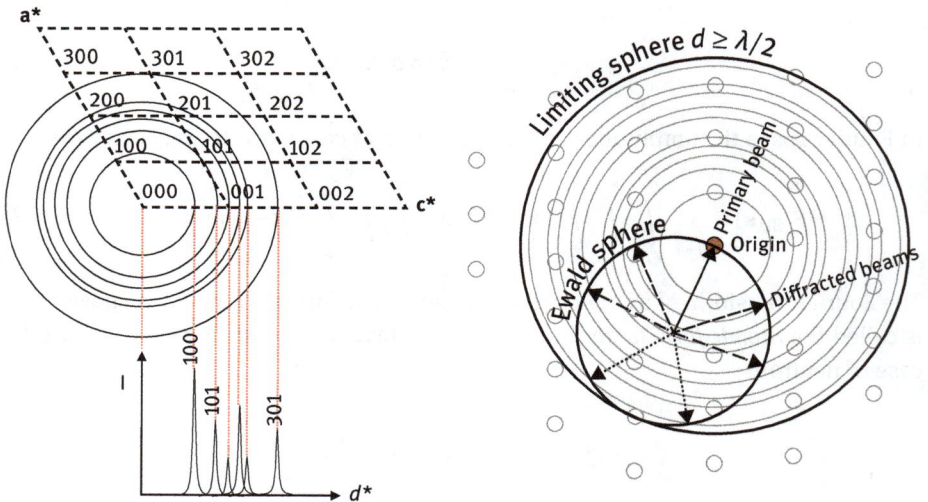

Figure 1.2: Left: Schematic illustration of how in-plane rotation of the projection of the reciprocal **a*c***-plane around its origin produces the one-dimensional powder pattern. Right: Illustration of the region of reciprocal space that is accessible in a powder measurement (outer circle). The smaller circle represents the Ewald sphere (adapted from Dinnebier & Billinge (2008) with permission from The Royal Society of Chemistry).

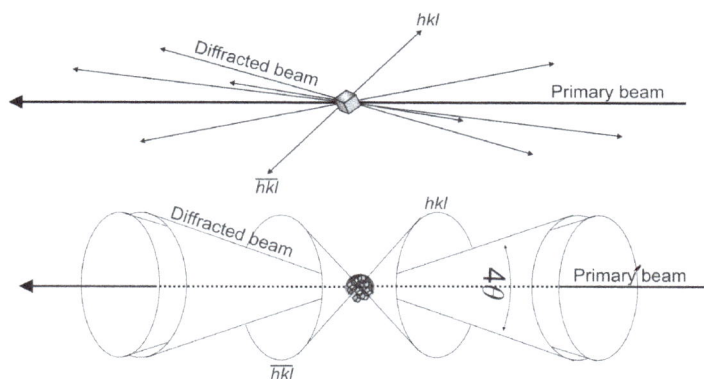

Figure 1.3: Comparison between the diffracted beams originating from a single crystal (top) and a powder (bottom). For the latter, some Debye–Scherrer cones are drawn (adapted from Dinnebier & Billinge (2008) with permission from The Royal Society of Chemistry).

powder sample, which have identical orientation, contribute to the same diffraction spots. Due to the different orientations of the crystallites, a single diffraction spot becomes smeared out on the surface of a sphere with a radius given by the length of the reciprocal lattice vector d^*. Therefore, the orientation of the vector $\mathbf{d^\star}$ (= $\mathbf{s}$) gets lost, or, in other words, the three-dimensional reciprocal space is projected onto the d^*-axis (Figure 1.2, left).

The circular intersection of the smeared reciprocal lattice with the Ewald sphere (Figure 1.2, right) results in the diffracted X-rays of the reflection hkl forming coaxial cones, the so-called Debye–Scherrer cones (Figure 1.3). The smearing of reciprocal space makes the measurement easier but results in a loss of information. Reflections from lattice planes whose vectors lie in different directions but have the same d-spacing will overlap and cannot be resolved or separated in the measurement. Some of these overlaps are dictated by symmetry (systematic overlaps) and others are accidental. Systematic overlap is not a serious issue for symmetry equivalent reflections (e.g., the six Bragg peaks (100), ($\bar{1}$00), (010), (0$\bar{1}$0), (001), (00$\bar{1}$) of a cubic sample) since all will have the same intensity and multiplicity (6). The problem is more serious for nonsymmetry-related peaks (e.g., (300) and (221) of any cubic material, or (210) and (120) of a cubic material with $m\bar{3}$ Laue symmetry) or accidental overlap.

The two most important formulas for understanding powder diffraction, Bragg's law (W. L. Bragg, 1912) and the Debye equation (Debye, 1915), are derived below.

1.2 The Bragg equation

The most famous way of understanding the peak positions in a powder diffraction pattern is via the Bragg equation (W. L. Bragg, 1912) that describes the principle of X-ray diffraction as what appears to be a reflection of X-rays by sets of parallel lattice planes,

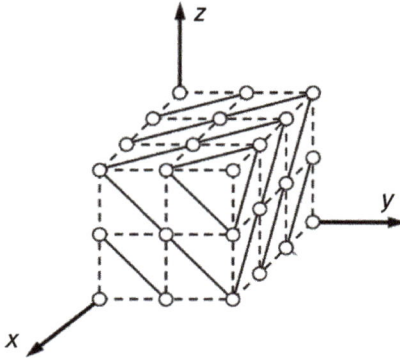

Figure 1.4: Schematic drawing of a set of parallel lattice planes (111) passing through all points of the cubic lattice.

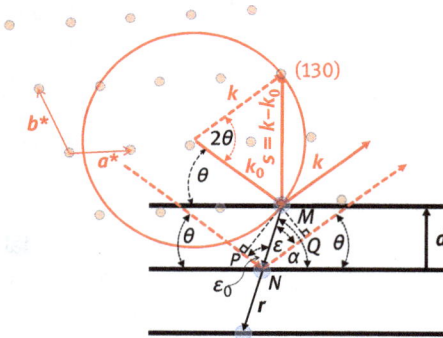

Figure 1.5: Schematic illustration of the geometry required to derive Bragg's law.

characterized by the Miller indices hkl. All planes in a set are identical and separated by a distance d_{hkl} (Figure 1.4).[3] Although this isn't a good picture of what is happening physically, it can be related to the Ewald approach given above using Figure 1.5.

The different path lengths of X-rays/neutrons scattered from atoms in different planes lead to a phase shift Δ, which can immediately be deduced from Figure 1.5 using standard trigonometric relationships[4]:

---

3 Note that X-rays and neutrons are scattered by the electrons and/or nuclei of atoms. The view of diffraction as reflection by lattice planes that may or may not contain atoms is a convenient mathematical description but doesn't represent the true physical process occurring.

4 $\cos(\alpha + \theta) = \cos\alpha\cos\theta - \sin\alpha\sin\theta$; $\cos(\alpha - \theta) = \cos\alpha\cos\theta + \sin\alpha\sin\theta$

$$\Delta = PN + NQ$$
$$= (MN)\cos\varepsilon_0 + (MN)\cos\varepsilon = (MN)[-\cos(\alpha + \theta) + \cos(\alpha + \theta)] \quad (1.12)$$
$$= (MN)[2\sin\alpha\sin\theta]$$

with

$$d = (MN)\sin\alpha \quad (1.13)$$

For constructive interference, Δ must be a multiple n of the wavelength λ, which immediately leads to the Bragg equation:

$$n\lambda = 2d\sin\theta. \quad (1.14)$$

For convenience, d is usually divided by n leading to:

$$\lambda = 2\frac{d}{n}\sin\theta = 2d_{hkl}\sin\theta. \quad (1.15)$$

Alternatively, Bragg's law can be written in vector notation (Figure 1.5). The scattering vector $\mathbf{s}$ is always perpendicular to the scattering plane, and is the difference between the wave vectors of the incoming and outgoing beams given by $\mathbf{k}_0$ and $\mathbf{k}$, respectively:

$$\mathbf{s} = \mathbf{k} - \mathbf{k}_0. \quad (1.16)$$

Setting the magnitude of $\mathbf{k}_0$ and $\mathbf{k}$ to $1/\lambda$ (the radius of the Ewald sphere) leads to the Bragg equation in terms of the magnitude of the scattering vector $\mathbf{s}$:

$$\mathbf{s} = \mathbf{d}_{hkl}^*. \quad (1.17)$$

Bragg's law results in sharp spots of high intensity that emerge from the crystallite in specific directions given by the Bragg equation, and there is a one-to-one correspondence between these Bragg spots and each set of crystallographic planes. Each Bragg spot is therefore labeled with the same set of Miller indices, hkl, as the set of planes that gave rise to it. The Debye–Scherrer cone created by the orientational averaging becomes a sharp peak in a one-dimensional powder pattern.

Peaks in powder patterns of real samples will invariably be broadened. We can understand one source of broadening by calculating the total derivative of the Bragg equation with the d-spacing as the subject. Applying the chain rule, this is:

$$dd = \frac{\partial d}{\partial\theta}d\theta + \frac{\partial d}{\partial\lambda}d\lambda = \frac{n\lambda}{2\sin\theta}\frac{\cos\theta}{\sin\theta}d\theta + \frac{n}{2\sin\theta}d\lambda, \quad (1.18)$$

which simplifies to:

$$\frac{dd}{d} = -\frac{d\theta}{\tan\theta} + \frac{d\lambda}{\lambda}. \quad (1.19)$$

The dimensionless quantity dd/d can, for example, be interpreted as a microscopic strain. Any variation of the d-spacing within a crystallite will then give rise to a strain peak broadening $d\theta$ as discussed in Chapter 4.

Finite crystallite domain size is a second source of broadening. If we assume an infinite stack of lattice planes, the Bragg equation gives the position of δ-function Bragg peaks. If the angle between the incoming beam and the lattice plane θ is different by an amount ε from the Bragg condition, there will always be a lattice plane inside the crystallite for which the accumulated extra path length produces a phase shift of $(n+1/2)\lambda$ causing destructive interference (Figure 1.6). For a thick crystal this will occur for an arbitrarily small ε, which explains why Bragg reflections are sharp.

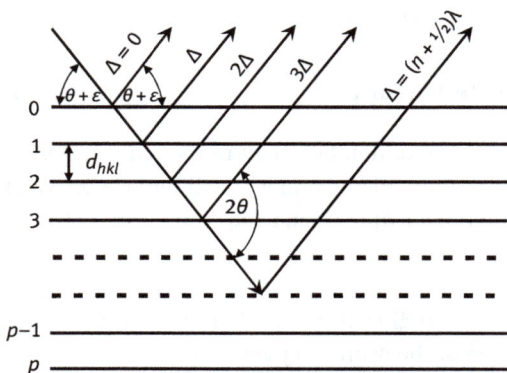

Figure 1.6: Path length difference of the scattered ray versus the depth of the lattice plane in the crystal (adapted from Dinnebier & Billinge (2008) with permission from The Royal Society of Chemistry).

The situation changes if the size of the crystallites is finite. In this case, for small ε the plane for which $\Delta = (n+1/2)\lambda$ holds may not be present in the crystal, thus leading to an intensity distribution over some small angular range in θ. This is called size broadening and is described by the Scherrer equation (Klug and Alexander, 1974). We can derive the Scherrer equation approximately by considering the crystallite in Figure 1.6, which will have a thickness in the direction perpendicular to $p+1$ lattice planes of separation d_{hkl} (Figure 1.6) of:

$$L_{hkl} = p d_{hkl}. \tag{1.20}$$

The additional beam path Δ between consecutive lattice planes at the angle $\theta + \varepsilon$ is[5]:

---

5 Note that for small θ, $\sin\theta \approx \theta$ and $\cos\theta \approx 1$.

$$\Delta = 2d\sin(\theta + \varepsilon) = 2d(\sin\theta\cos\varepsilon + \cos\theta\sin\varepsilon)$$
$$= n\lambda\cos\varepsilon + 2d\cos\theta\sin\varepsilon \approx n\lambda + 2d\ \varepsilon\cos\theta. \tag{1.21}$$

The phase difference is then:

$$\varphi = 2\pi\frac{\Delta}{\lambda} = 2\pi n + \frac{4\pi\varepsilon d\cos\theta}{\lambda}. \tag{1.22}$$

The phase difference φ_L between the top and the bottom layer p is then:

$$\varphi_L = p\frac{4\pi\varepsilon d\cos\theta}{\lambda} = \frac{4\pi\varepsilon L_{hkl}\cos\theta}{\lambda}. \tag{1.23}$$

If a is the amplitude of a wave diffracted at a single lattice plane, then according to Klug and Alexander (1974), the resulting amplitude of all diffracted waves is:

$$A = a\ p\frac{\sin(\varphi_L/2)}{(\varphi_L/2)}, \tag{1.24}$$

with the maximum amplitude $A_0 = a\ p$ at $\varepsilon = 0$. Half maximum intensity is then:

$$\frac{A^2}{A_0^2} = \frac{1}{2} = \frac{\sin^2(\varphi_L/2)}{(\varphi_L/2)^2}, \tag{1.25}$$

which is satisfied when:

$$\frac{\varphi_L}{2} = 1.4. \tag{1.26}$$

From this we can approximate the misalignment angle at half maximum intensity using eq. (1.23):

$$\varepsilon_{1/2} = \frac{2.8\lambda}{4\pi L_{hkl}\cos\theta}. \tag{1.27}$$

The measured angular width $fwhm_{hkl}$ between the two points of half maximum intensity on a 2θ scale leads to the Scherrer equation:

$$fwhm_{hkl} = 4\varepsilon_{1/2} = \frac{0.89\lambda}{L_{hkl}\cos\theta}, \tag{1.28}$$

which gives a measure of the peak width *in radians* due to the finite particle size. The prefactor depends on the shape of the grains (e.g., it is 0.89 for perfect spheres and 0.94 for cubic-shaped grains) but is always close to unity. Note that this equation is not valid for crystallites that are too large or too small. For very large crystallites the peak width is governed by the coherence of the incident beam and not by particle size. For nanoscale crystallites, Bragg's law fails and needs to be replaced by the Debye equation discussed in the next section.

1.3 The Debye equation

In general, decreasing the size of a crystal leads to an increase in the width of its Bragg peaks as we have seen in the previous section. When the size of the crystallite becomes very small, the width of the Bragg peaks is so large that they merge and overlap, and it no longer makes sense to use δ-function Bragg peaks as the starting point for the analysis. Even though the coherent diffraction is diffuse in nature and distributed throughout reciprocal space at this point, it still contains useful structural information.

Under the assumption that only elastic scattering occurs and each photon is scattered only once, the structure amplitude for an ensemble of n atoms at fixed positions without periodicity is given by[6]:

$$A(\mathbf{s}) = \sum_{j=1}^{n} f_j(s) e^{2\pi i \, \mathbf{s} \cdot \mathbf{x}_j} \tag{1.29}$$

with the conjugate complex:

$$A^*(\mathbf{s}) = \sum_{j=1}^{n} f_j(s) e^{-2\pi i \, \mathbf{s} \cdot \mathbf{x}_j}. \tag{1.30}$$

The intensity can be calculated as:

$$I(\mathbf{s}) = |A(\mathbf{s})|^2 = A(\mathbf{s})A^*(\mathbf{s}) = \sum_{j=1}^{n}\sum_{k=1}^{n} f_j(s) f_k(s) e^{-2\pi i \, \mathbf{s} \cdot \mathbf{r}_{jk}}, \tag{1.31}$$

where $\mathbf{r}_{jk} = \mathbf{x}_k - \mathbf{x}_j$ is the vector between the atoms j and k. For a simple diatomic molecule (Figure 1.7), this would be[7]:

$$I(\mathbf{s}) = f_1(s)^2 + f_2(s)^2 + f_1(s)f_2(s) e^{-2\pi i \mathbf{s} \cdot \mathbf{r}_{12}} + f_1(s)f_2(s) e^{2\pi i \mathbf{s} \cdot \mathbf{r}_{12}}$$

$$= f_1(s)^2 + f_2(s)^2 + 2f_1(s)f_2(s)\cos(2\pi \mathbf{s} \cdot \mathbf{r}_{12}) \tag{1.32}$$

Eq. (1.32) simplifies further for a homonuclear diatomic molecule (e.g., oxygen O_2) with $f_1(s) = f_2(s)$ to:

$$I(\mathbf{s}) = 2f_1(s)^2(1 + \cos(2\pi \mathbf{s} \cdot \mathbf{r}_{12})). \tag{1.33}$$

For O_2 gas, scattering from an individual molecule will be coherent but that from different molecules will be incoherent. In this case we will see a sum of

---

6 For simplicity, only the real part of the form factor is taken into account. See Section 2.2.1 for more detail.

7 Using $\cos \varphi = \frac{1}{2}\left(e^{i\varphi} + e^{-i\varphi}\right)$.

scattering from all the molecules present, which will have every orientation relative to the beam with equal probability. It is therefore necessary to take an orientational average of the scattering leading to spherically averaged intensity. The result is the equation known as the *Debye equation*, which is derived below (Debye, 1915).

Analogous to the first step of eq. (1.32), eq. (1.31) can be rewritten as:

$$I(\mathbf{s}) = \sum_{k=1}^{n} f_j(s)^2 + \sum_{j=1}^{n} \sum_{k=1, k \neq j}^{n} f_j(s) f_k(s) e^{2\pi i \mathbf{s} \cdot \mathbf{r}_{jk}}. \tag{1.34}$$

The double sum in eq. (1.34) contains $n(n-1)$ addends, which can be grouped into $\frac{1}{2}n(n-1)$ pairs of addends describing the same atom–atom pair with the distance vectors $\mathbf{r}_{jk}$ and $\mathbf{r}_{kj} = -\mathbf{r}_{jk}$. Hence the corresponding pairs of addends are complex conjugates. Decomposition of each addend according to eq. (13.1) (see Chapter 13) leads to cancellation of the imaginary terms. Thus, eq. (1.34) can be rewritten as:

$$I(\mathbf{s}) = \sum_{k=1}^{n} f_j(s)^2 + \sum_{j=1}^{n} \sum_{k=1, k \neq j}^{n} f_j(s) f_k(s) \cos(2\pi \mathbf{s} \cdot \mathbf{r}_{jk}) \tag{1.35}$$

corresponding to the second step in eq. (1.32). In order to determine the direction average of eq. (1.35), let's consider an arbitrary addend of the double sum, and assume that φ is the angle between the scattering vector $\mathbf{s}$ and the direction vector $\mathbf{r}_{jk}$ between two atoms j and k (Figure 1.7):

$$\mathbf{s} \cdot \mathbf{r}_{jk} = s \, r_{jk} \cos \varphi. \tag{1.36}$$

The direction average of the cosine of that particular addend then becomes the integral over the surface of the sphere of radius r_{jk} (Figure 1.8) divided by the area $4\pi r_{jk}^2$ of that sphere:

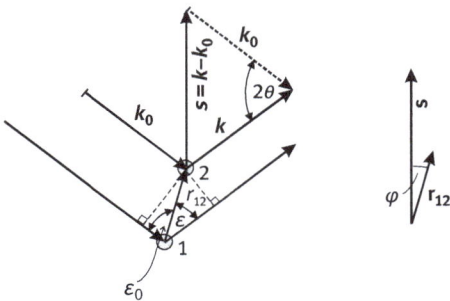

Figure 1.7: Scattering by an object consisting of two scatterers 1 and 2, separated by the vector $\mathbf{r}_{12}$.

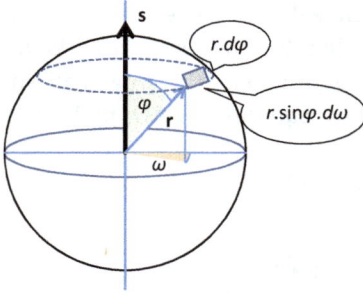

Figure 1.8: Definition of a volume element in spherical coordinates.

$$\cos(2\pi \mathbf{s} \cdot \mathbf{r}_{jk}) = \frac{1}{4\pi r_{jk}^2} \int_{\text{surface of sphere}} \cos(2\pi s \, r_{jk} \cos\varphi) d\mathbf{S}. \tag{1.37}$$

As illustrated in Figure 1.8, the area element $d\mathbf{S}$ in eq. (1.37) is calculated by:

$$d\mathbf{S} = r_{jk}^2 \sin\varphi d\varphi d\omega. \tag{1.38}$$

Using appropriate limits of integration, this leads to:

$$\cos(2\pi \mathbf{s} \cdot \mathbf{r}_{jk}) = \frac{1}{4\pi r_{jk}^2} \int_0^\pi \int_0^{2\pi} \cos(2\pi s \, r_{jk} \cos\varphi) r_{jk}^2 \sin\varphi d\omega d\varphi. \tag{1.39}$$

The integration over ω simply yields 2π. The second integration over φ can be accomplished by substitution using $z = \cos\varphi$, $d\varphi = -dz/\sin\varphi$ and adaption of the integration boundaries to give:

$$\cos(2\pi \mathbf{s} \cdot \mathbf{r}_{jk}) = \frac{\sin(2\pi s \, r_{jk})}{2\pi s \, r_{jk}}. \tag{1.40}$$

The direction average of eq. (1.35) is then obtained by using eq. (1.40), leading to the Debye equation:

$$I(s) = \sum_{k=1}^n f_j(s)^2 + \sum_{j=1}^n \sum_{k=1, k \neq j}^n f_j(s) f_k(s) \frac{\sin(2\pi s \, r_{jk})}{2\pi s \, r_{jk}}. \tag{1.41}$$

The intensity must be multiplied by the number of particles N (molecules, nanocrystalline grains and so on) that contribute to the scattering. An example of using the Debye equation to describe the scattering of CCl_4 gas is shown in Figure 1.9. For clusters of atoms such as small nanoparticles that are intermediate in size between a diatomic molecule and a small chunk of crystal, the Debye equation is exact and may be used to calculate the scattering intensity. We find that as the length scale of order

Intensity [a.u]

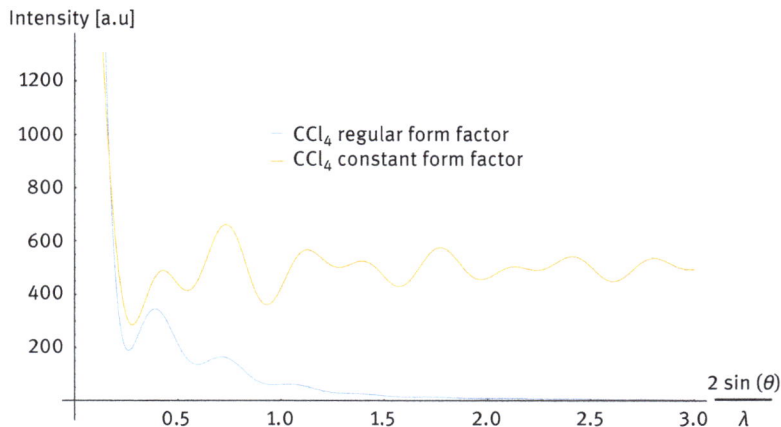

Figure 1.9: Qualitative intensity distribution from elastic diffuse scattering (Debye equation) of CCl_4 gas taking the form factors of carbon and chlorine into account (lower blue line) and with a constant form factor (upper orange line).

increases, the diffracted intensity gradually evolves from having broad features as a function of $s = 2\sin\theta/\lambda$ (as in Figure 1.9) to the sharp peaks we associate with microcrystalline powders.

1.4 Information loss in a powder pattern

The loss of information due to peak overlap in a powder pattern can be quantified. According to David et al. (2006), the average number of reflections within a shell of width Δd^* at a radius $d^* = 1/d$ (=s) in reciprocal space can be approximated for a triclinic material with unit cell volume V by:

$$\Delta N(d^*) \sim 2\pi V d^{*2} \Delta d^*. \tag{1.42}$$

The value steadily increases with a constant Δd^*. If the expression is converted to equidistant intervals in °2θ using the Bragg equation (eq. (1.15)), a maximum occurs as illustrated in Figure 1.10 (left), which shows the average number of reflections in 0.1° 2θ intervals as a function of the scattering angle for Cu-$K_{\alpha1}$ radiation for triclinic unit cell volumes of 1000 and 1500 Å^3.

All peaks with a separation that is larger than the experimentally resolvable separation can be counted as distinguishable reflections. Choosing the resolvable separation as a constant strain variation with the Bragg peak resolution $\Delta d/d$, the number of distinguishable reflections for a triclinic system can be calculated as (David & Shankland, 2008):

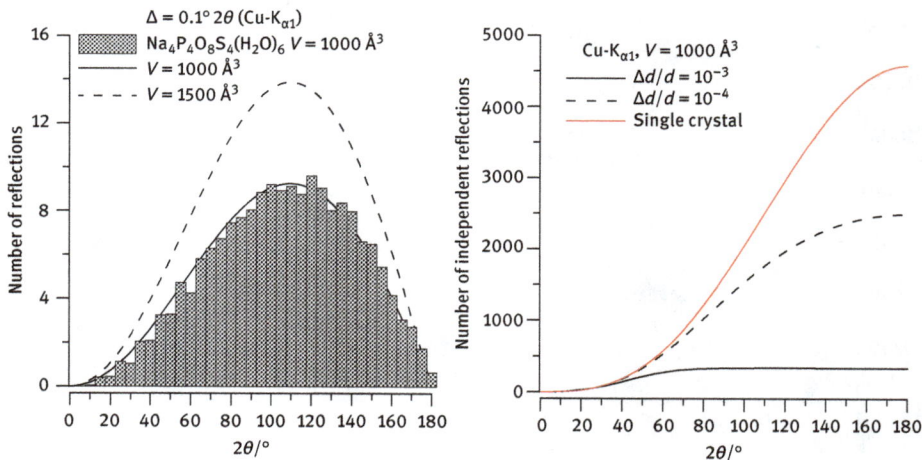

Figure 1.10: Left: Number of Bragg reflections within 0.1° 2θ intervals for Cu-K$_{\alpha 1}$ radiation for different triclinic unit cell volumes over the entire diffraction range. The bar chart contains binned data from a simulated powder pattern of Na$_4$P$_4$O$_8$S$_4$(H$_2$O)$_6$ that has a unit cell volume of 1000.04 Å^3 (a = 8.815 Å, b = 9.313 Å, c = 14.259 Å, α = 78.33°, β = 95.07°, γ = 119.26° (Ilyukhin et al., 1982)). Right: Number of distinguishable Bragg reflections for a triclinic unit cell volume of 1000 Å^3 for different high-resolution powder and single crystal experiments over the entire diffraction range.

$$N_{ind}\left(d^{*}_{max}\right) \sim \frac{1 - \exp\left(- 2\pi V \frac{\Delta d}{d} {d^{*}_{max}}^{3}\right)}{3\frac{\Delta d}{d}}.\tag{1.43}$$

Figure 1.10 (right) shows the effect of increasing resolution on the number of distinguishable reflections for a unit cell volume of 1000 Å^3 as a function of the scattering angle for Cu-K$_{\alpha 1}$ radiation. Even for the highest instrument resolution of $\Delta d/d = 10^{-4}$, as typical for third generation synchrotron sources, the number of reflections that can be resolved (even assuming no sample broadening) quickly falls below what is possible in single crystal diffraction as 2θ increases.

References

Bragg, W.L. (1912): *The diffraction of short electromagnetic waves by a crystal*. Proc. Camb. Phys. Soc. 17, 43–57.

David, W.I F.,Shankland, K. (2008): *Structure determination from powder diffraction data*. Acta Cryt. A64, 52–64.

David, W.I.F., Shankland, K., McCusker L.B., Baerlocher, Ch. (2006): *Structure determination from powder diffraction data*, IUCr Monographs on Crystallography, Oxford University Press, New York (US), 337 pages.

Debye, P. (1915): *Zerstreuung von Röntgenstrahlen*. Annalen der Physik, 351, 809–823.

Dinnebier, R.E., Billinge, S.J.L. (eds) (2008): *Powder diffraction – theory and practice*, The Royal Society of Chemistry (RCS), 574 pages.

Egami, T., Billinge, S. (2012): *Underneath the Bragg peaks - structural analysis of complex materials*, Volume 16, 2nd edition, Pergamon, 422 pages.

Giacovazzo, C., Monaco, H.L., Artioli, G., Viterbo, D., Milanesio, M. Gilli, G., Gilli, P., Zanotti, G., Ferraris, G. (2011): *Fundamentals of crystallography (International Union of Crystallography Monographs on Crystallography)*, 3rd edition, Oxford University Press, New York (US), 872 pages.

Ilyukhin, V.V., Kalinin, V.P., Kuvshinova, T.B., Tananaev, I.V. (1982): *The crystal structure of $Na_4P_4O_8S_4(H_2O)_6$*. Dokl. Akad. Nauk SSSR 266, 1387–1391.

Klug, H.P., Alexander, L.E. (1974): *X-ray diffraction procedures*, 2nd edition, John Wiley & Sons, New York, 966 pages.

Pecharsky, V., Zavalij, P. (2009): *Fundamentals of powder diffraction and structural characterization of materials*, 2nd Edition, Springer US, 744 pages.

2 The Rietveld method

As we discussed at the end of Chapter 1, one of the key features of powder diffraction is the accidental and/or systematic overlap of some Bragg reflections. The degree of overlap of individual reflections depends on their separation in $\Delta d/d$ and their peak widths. Ultimately, the intensity information content in the powder pattern will be determined by these factors (Figure 2.1).

A powder pattern consists of a set of N consecutively measured intensities $y_{obs,i}$, where the running index $i \in [1, .., N]$ represents the measured position $X_{obs,i}$ as the diffraction angle 2θ (°), energy E (keV) or time-of-flight TOF (ms).[1] The steps do not have to be equidistant.

Most frequently, a powder pattern is measured at constant wavelength as a function of the diffraction angle 2θ with equidistant angular steps of width $\Delta 2\theta$. In such a case, the index i represents the angular position in the powder pattern according to $2\theta_i = 2\theta_{start} + (i-1)\Delta 2\theta$, with the starting angle $2\theta_{start}$.

In the case of angular dispersive constant wavelength X-ray or neutron data (Figure 2.2), the relationship between the measured angular space and the d-spacing scale follows directly from the Bragg equation[2]:

$$2\theta/rad = 2\arcsin\left(\frac{\lambda}{2d}\right) \tag{2.1}$$

For energy dispersive X-ray data (Figure 2.4), the relationship between the measured energy scale and the d-spacing scale is given by:

$$E/keV = \frac{6.2}{(d/\text{Å}) \cdot \sin\theta_{fixed}} \tag{2.2}$$

with θ_{fixed} the fixed detector angle used for data collection. The conversion factor depends on the chosen units.

For TOF data (Figure 2.5), the relationship between the measured TOF scale and the d-spacing scale is given by:

$$TOF = t_0 + t_1 d + t_2 d^2 \tag{2.3}$$

where t_0, t_1 and t_2 are diffractometer constants characteristic of a given detector bank on a TOF powder diffractometer. TOF is typically given in milli- or microseconds.

The fundamental idea behind the Rietveld method is simple: Instead of analyzing the integrated peak intensities from a powder pattern in a single crystal-like fashion, the entire information content of a powder pattern (Figure 2.2) available in step-scanned intensity data is fitted with a model whose parameters are refined using a

---

[1] TOPAS uses X for all the scales. We adopt this practice where needed.
[2] Note the need in TOPAS to multiply by $360/2\pi$ to convert to degrees.

https://doi.org/10.1515/9783110461381-002

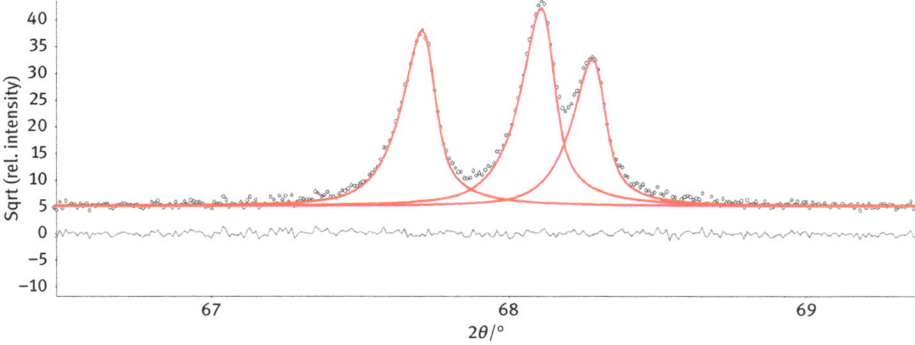

Figure 2.1: Zoomed-in part of the powder pattern of quartz showing the measured pattern (black circles), three single peaks (red lines) fitted to the individual Bragg reflections exhibiting different degrees of overlap, and the difference curve below. The square root of the intensity is displayed.

least squares procedure to optimize the fit. This procedure intrinsically accounts for peak overlap. According to the method of least squares, the squared sum of differences between the N observed $y_{obs,i}$ and calculated $y_{calc,i}$ step-scanned intensities is subjected to minimization (Figures 2.2–2.5):

$$\sum_i \left(w_i (y_{obs,i} - y_{calc,i})^2 \right) \rightarrow Min \qquad (2.4)$$

The weight w_i is usually derived from the variance of $y_{obs,i}$ as $1/\sigma^2(y_{obs,i})$ while all covariances between different $y_{obs,i}$ are assumed to be zero.

The calculated intensity $y_{calc,i}$ is expressed by combinations of mostly nonlinear analytic or nonanalytic functions as:

$$y_{calc,i} = \sum_p \left(S_p \sum_{\mathbf{s}(p)} \left(|F_{calc,\mathbf{s},p}|^2 \Phi_{\mathbf{s},p,i} Corr_{\mathbf{s},p,i} \right) \right) + Bkg_i \qquad (2.5)$$

The outer sum runs over all crystalline phases p with Bragg peaks in the powder pattern, while the inner sum runs over all Bragg reflections $\mathbf{s} = (hkl)$ of a phase p, which contribute to the position i in the powder pattern.[3] A scaling factor S_p, which is proportional to the weight fraction of phase p, is applied to the reflection intensities of each phase. $Corr_{\mathbf{s},p,i}$ represents the product of various correction factors that need to be applied to the reflection intensities $|F_{calc,\mathbf{s},p}|^2$ that may depend on the diffraction geometry and/or individual reflection indices. The contributions to the factor $Corr$ are discussed in more detail in Section 2.3. The value of the profile function $\Phi_{\mathbf{s},p,i}$ is given for the profile point i

---

3 The components of $\mathbf{s}$, the Laue indices h, k and l, refer to the basis vectors $\mathbf{a^\star}$, $\mathbf{b^\star}$ and $\mathbf{c^\star}$ of the reciprocal lattice.

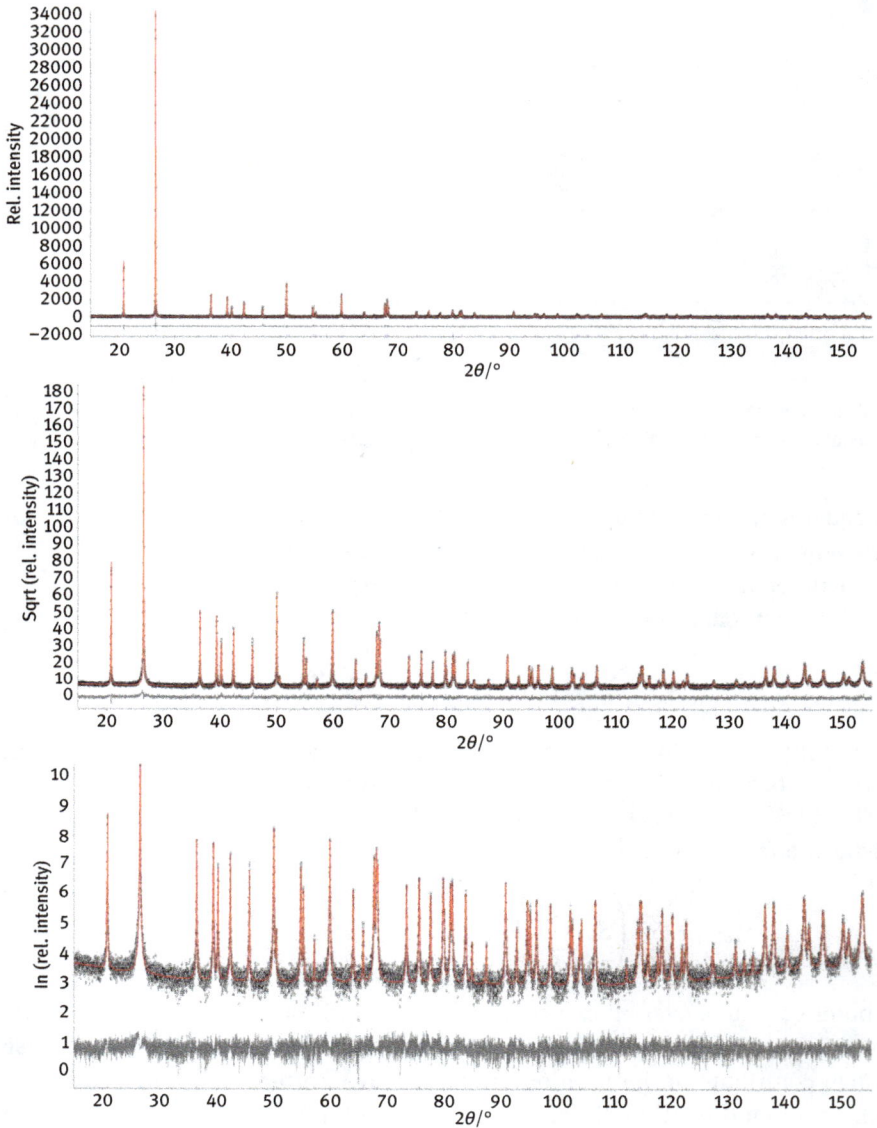

Figure 2.2: Typical single phase Rietveld plot showing the observed angular dispersive powder pattern (black circles) of quartz, the calculated powder pattern (red line), the difference curve (grey line) and the markers of the reflection positions below. Top: linear intensity; middle: square root of the intensity; bottom: logarithmic intensity.

Note: For better visibility of lower intensity reflections, the square root representation of the measured intensity is typically used throughout the book.

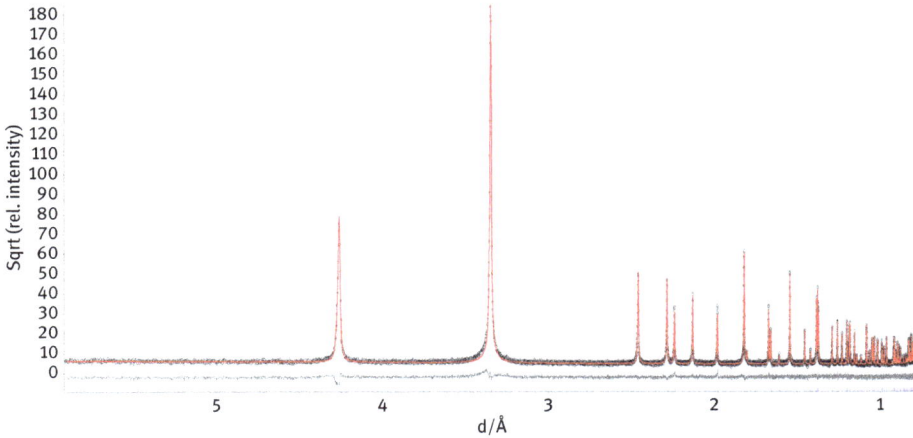

Figure 2.3: Rietveld plot of Figure 2.2 converted to d-spacing (in Å units) with the square root of the intensity.

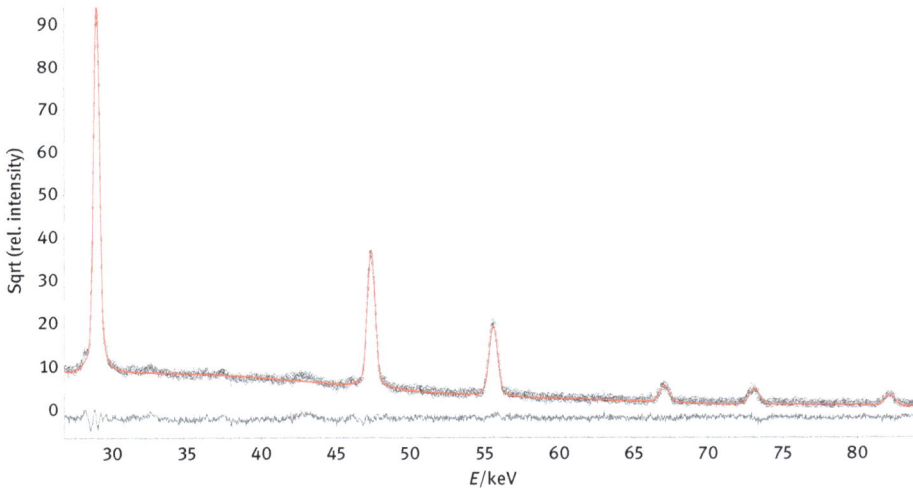

Figure 2.4: Typical Rietveld plot showing the energy dispersive powder pattern of silicon recorded with $\theta_{fixed}= 7.77°$.

relative to the position $s = |\mathbf{s}| = 2\sin\theta/\lambda$ of the Bragg reflection $\mathbf{s}$. Therefore, the peak profile depends only on the peak position given by the scalar s, and not on hkl (i.e., $\mathbf{s}$). This restriction is lifted in the case of anisotropic line broadening, where an explicit hkl dependent $\Phi_{\mathbf{s},p,i}$ is considered (see Chapter 4). The observed background coming from thermal diffuse scattering, incoherent

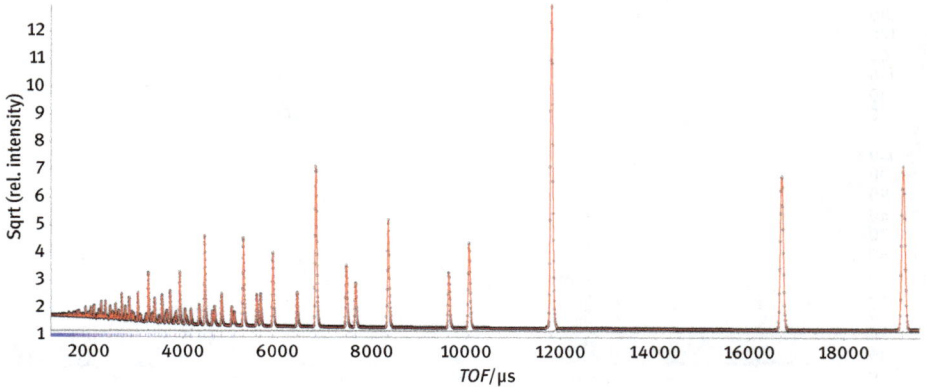

Figure 2.5: Rietveld fit to *TOF* powder pattern of CeO_2.

scattering, inelastic scattering, sample environment and so on at position i in the powder pattern is denoted as Bkg_i.

This approach requires modeling of the *entire* powder pattern. To simplify and understand this complex task, the information content of the powder pattern can be divided into several parts, allowing the separation of groups of parameters with respect to their origin:

- Peak position s that is geometrically determined by the crystallographic lattice, space-group symmetry and instrumental factors.
- Integrated peak intensity $|F_{calc,\,s,\,p}|^2 Corr_{s,\,p}$ that is determined by the time- and space-averaged crystal structure and geometrical contributions. $Corr_{s,\,p}$ is the value of $Corr$ at the peak position s for phase p.
- Peak profile $\Phi_{s,\,p,\,i}$ that is determined by the instrument profile and microstructural parameters of the sample.
- Background Bkg_i.

Each part has contributions from both the sample *and* the instrument. Since Rietveld refinement requires starting values of all parameters within a (relatively narrow) range of convergence, it can be useful to consider different aspects of the pattern separately according to different empirical, phenomenological or physical models.

In the following sections, the factors affecting these different parameter groups are described. The main focus will be on *angular dispersive* powder diffraction as this is the standard method in most laboratories. Where useful, the corresponding code in the TOPAS scripting (INP) language is given between border lines.[4] Table 2.1 summarizes

---

[4] In general, we quote fewer significant figures than would be carried within the TOPAS script for clarity of presentation.

Table 2.1: Selection of important instrumental and sample contributions affecting position (as a shift) and/or intensity (as a correction factor) and/or shape (as a convolution) of the Bragg reflections.

Correction	Position (shift)	Intensity (factor)	Profile (convolution)	Section
Instrument				
Zero shift	x			2.1.1
Specimen displacement	x			2.1.1
Equatorial divergence (fixed slit)			x	2.5.8
Equatorial divergence (variable slit)		x	x	2.3.6, 2.5.8
Size of source in the equatorial plane			x	2.5.1
Specimen tilt			x	2.5.1
Receiving slit length in the axial plane			x	2.5.6
Receiving slit width in the equatorial plane			x	2.5.1
Emission profile		x	x	2.5.9
Tube tails			x	2.5.1, 2.5.9
Axial divergence (prim., sec. Soller etc.)	x		x	2.5.6
Lorentz-polarization		x		2.3.2
Sample				
Linear absorption/transparency	x	x	x	2.1.1, 2.3.3, 2.5.7
Surface roughness		x		2.3.4
(An)isotropic microstrain			x	2.5.2–2.5.4
(An)isotropic crystallite size			x	2.5.2–2.5.4
Preferred orientation		x		2.3.7
Extinction		x		2.3.5
Overspill		x		2.3.6
Multiplicity		x		2.3.1
Displacement parameter		x		2.2.3

common aberrations/contributions affecting position and/or intensity and/or shape (as a convolution) of the Bragg reflections in a Rietveld refinement.

2.1 The peak position

For an angular dispersive powder pattern, the observed scattering angle $2\theta_s$ of a Bragg reflection $\mathbf{s} = (hkl)$ can be calculated from the corresponding d-spacing by the Bragg equation corrected by aberrations $\Delta 2\theta_{corr}$ due to misalignment of the diffractometer or the sample, or due to transparency, axial divergence (or similar) effects:

$$2\theta_\mathbf{s} = 2\arcsin\left(\frac{\lambda}{2}\frac{1}{d_\mathbf{s}}\right) + \Delta 2\theta_{corr}. \tag{2.6}$$

Given a set of lattice parameters (a, b, c, α, β, γ) or their reciprocal counterparts (a^*, b^*, c^*, α^*, β^*, γ^*) and the unit cell volume V, the positions for all possible reflections **s** can be calculated according to:

$$\frac{1}{d_\mathbf{s}} = \frac{1}{V}\sqrt{\begin{array}{c} h^2b^2c^2\sin^2\alpha + k^2a^2c^2\sin^2\beta + l^2a^2b^2\sin^2\gamma + \\ 2hkabc^2(\cos\alpha\cos\beta - \cos\gamma) \\ +2kla^2bc(\cos\beta\cos\gamma - \cos\alpha)+2hlab^2c(\cos\alpha\cos\gamma - \cos\beta)\end{array}}$$
$$= \sqrt{h^2a^{*2} + k^2b^{*2} + l^2c^{*2} + 2hka^*b^*\cos\gamma^* + 2hla^*c^*\cos\beta^* + 2klb^*c^*\cos\alpha^*} \tag{2.7}$$

for the triclinic case. The equation simplifies considerably with increasing lattice symmetry (Giacovazzo et al., 2011); for example, for orthorhombic, tetragonal and cubic systems to:

$$\frac{1}{d_\mathbf{s}} = \sqrt{\frac{h^2}{a^2} + \frac{k^2}{b^2} + \frac{l^2}{c^2}}. \tag{2.8}$$

2.1.1 Corrections to the peak position

The position of the Bragg reflection can be affected in a linear or nonlinear manner by a series of contributions coming from the sample and from the instrument. For constant wavelength data, the absolute error in the interplanar spacing Δd as a function of the measured diffraction angle can be easily calculated by evaluating the exact differential of the Bragg equation (eq. (1.15)) (Chapter 1):

$$\mathrm{d}d = \left(\frac{\lambda\cos\theta}{2\sin^2\theta}\right)\mathrm{d}\theta + 2d\cos\theta\mathrm{d}\lambda. \tag{2.9}$$

Neglecting the error in the wavelength,[5] eq. (2.9) immediately shows the strong nonlinear increase of Δd at low diffraction angles even for small constant errors in 2θ (Figure. 2.6).

The simplest angular correction is a constant shift ("zero error"), which in TOPAS is defined as[6]:

```
prm c 0.0 min -0.1 max 0.1
th2_offset = c;
```

---

5 This would not be a valid assumption with, for example, wavelength (energy) dispersive data.
6 We'll see later that the parameter (*prm*) c is allowed to refine in this example. A fixed parameter would be preceded by ! giving !c. In the early code snippets of the book we will often omit the ! of fixed parameter names for ease of reading.

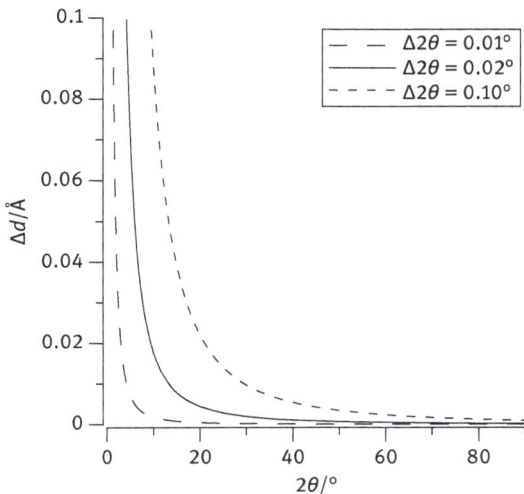

Figure 2.6: Absolute error in the interplanar spacing Δd versus diffraction angle 2θ for different constant errors in the measured diffraction angle using monochromatic Cu-K$_{\alpha 1}$ radiation.

or as predefined macro:

```
Zero_Error(@, 0)
```

For *TOF* data, any constant shift is included in the parameter t_0 of the transformation function between the *TOF* scale and d-spacing. The relevant TOPAS code is:

```
prm !t0     -16.50
prm !t1    48262.60
prm !t2     -8.35
pk_xo = t0 + t1 D_spacing + t2 D_spacing^2;
```

or as predefined macro:

```
TOF_x_axis_calibration( !t0, -16.50, !t1, 48262.60, !t2, -8.35)
```

t_1 is the most important factor and is related to the source to detector flight path. t_0 is usually a small contribution from electronic factors and t_2 describes minor d^2-dependent aberrations. Typically, all three parameters are refined from a reference pattern and are fixed in later refinement.

A constant shift in energy dispersive data is also included in the transformation of the measured energy scale into a d-spacing scale. The corresponding TOPAS code is:

```
prm zero -0.16068
prm !detector_angle_in_radians = 7.77 Deg_on_2;
prm wavelength = 2 D_spacing Sin(detector_angle_in_radians);
prm energy_in_eV = 10^5 / (8.065541 wavelength);
pk_xo = 10^-3 energy_in_eV + zero;
```

A common nonlinear correction for angular dispersive Bragg–Brentano geometry is the $\cos\theta$-dependent peak shift caused by a flat sample whose surface deviates from the focusing circle (Figure 2.7). This is the so-called height error c in mm:

$$\Delta 2\theta_{corr}/^\circ = -2\left(\frac{180^\circ}{\pi}\right)\frac{\cos\theta}{R_{DS}}c \tag{2.10}$$

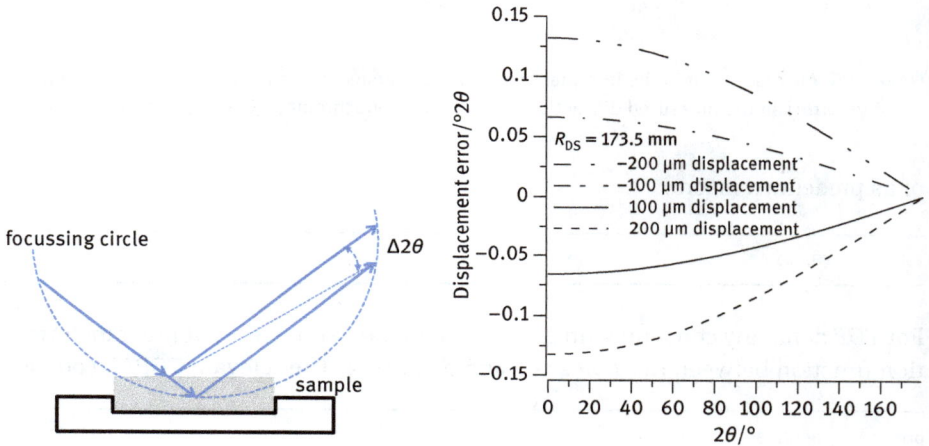

Figure 2.7: Angular shift due to specimen displacement in Bragg–Brentano geometry. Schematic drawing (left) and dependence on diffraction angle for different displacements (right).

with R_{DS} the distance between sample and detector in mm. In TOPAS language[7]:

```
prm c 0.0 min -0.1 max 0.1
th2_offset = -2 Rad c Cos(Th) / Rs;
```

---

[7] Note that the reserved parameter name *Th* will take all the different values of *s* for each reflection **s** = (*hkl*) of a phase occurring in the diffraction data. In most equations, however, *s* will be simply written as θ. Similar things are valid for the parameter *D_spacing*.

The corresponding predefined macro in TOPAS is:

```
Specimen_Displacement(@, 0.0)
```

Note that due to the nature of the cosine function, there is a high correlation between the constant zero shift and specimen height displacement, particularly if the powder pattern covers only a small angular range. It's also worth noting that sample height corrections are particularly problematic when indexing powder data. Equation. (2.10) shows that even a 15 μm sample height error causes sufficient peaks shifts (e.g., 0.01° 2θ at 10° 2θ for Cu-K$_\alpha$ radiation with a diffractometer of radius 173 mm) that indexing could be difficult.

A displacement of a capillary away from the center of the goniometer in Debye–Scherrer geometry also causes a nonlinear shift in the angular position. A suitable correction function is:

$$\Delta 2\theta_{corr} = \arcsin\left(\frac{d_L}{R_{DS}}\sin(2\theta)\right) - \arcsin\left(\frac{d_V}{R_{DS}}\cos(2\theta)\right) \tag{2.11}$$

with the displacement of the capillary in the direction of the beam d_L and perpendicular to the beam d_V (Gozzo et al., 2010). Figure 2.8 shows that these effects are much less significant than in Bragg–Brentano geometry.

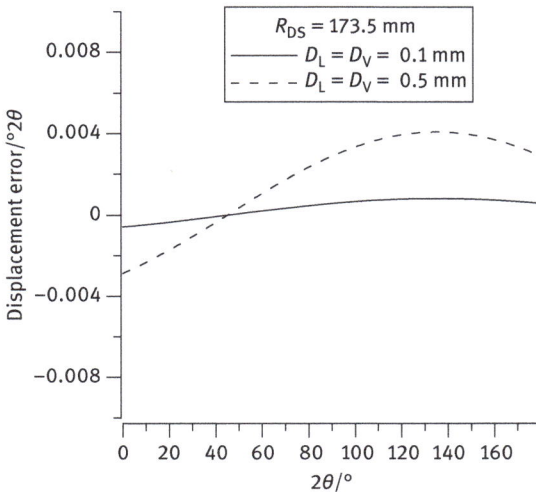

Figure 2.8: Angular shift as a function of diffraction angle caused by the displacement of a capillary with respect to the center of the goniometer.

In TOPAS this correction can be realized by:

```
prm dv 0
prm dl 0
th2_offset =(Rad ArcSin(dv Cos(2 Th))/Rs))+(Rad ArcSin(dl Sin(2 Th))/Rs));
```

Another type of peak shift for capillary samples in angular dispersive powder diffraction experiments is caused by θ-dependent absorption. An empirical expression was given by Sabine (1988) (Figure 2.9):

$$\Delta 2\theta_{corr}/° = 2A\theta^B(90 - \theta)^C$$
$$A = 0.000033\mu_{eff}R$$
$$B = 1.168 - 0.22\mu_{eff}R + 0.0168\left(\mu_{eff}R\right)^2 \qquad (2.12)$$
$$C = 1.155 + 0.2054\mu_{eff}R - 0.0224\left(\mu_{eff}R\right)^2$$

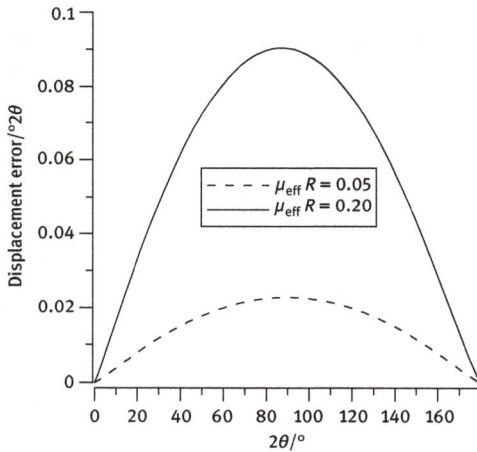

Figure 2.9: Peak shift of capillary samples in angular dispersive powder diffraction experiments caused by θ-dependent absorption as a function of diffraction angle for different values of $\mu_{eff}R$.

with the effective (taking packing effects into account) linear absorption of the sample μ_{eff} in mm^{-1} and the radius R of the capillary in mm. This correction can be applied in TOPAS as:

```
prm c 0.15 ' = µeff R
th2_offset = 0.000033 c (Th Rad)^(1.168 - 0.22 c + 0.0168 c^2)
             (90 - Th Rad)^(1.155 + 0.2054 c - 0.0224 c^2);
```

or by using the predefined macro:

```
Cylindrical_2Th_Correction(@, 0.15)
```

Other factors, which affect the position of Bragg reflections are axial divergence of the incident beam and transparency/absorption. The appropriate corrections are normally built into the peak shape functions. Due to high correlation, it is important to keep the number of correction factors in a refinement low.

2.2 The intensity of a Bragg reflection

The (integrated) intensity of a Bragg reflection is proportional to the squared complex structure factor that itself is the vector sum of complex atomic form factors (for X-rays) or coherent atomic scattering lengths (for neutrons) weighted by additional complex phase factors. The following sections discuss factors that influence peak intensities.

2.2.1 The atomic form factor

The atomic form factor describes the scattering power of an atom or ion as a function of the scattering vector length s. In the case of X-rays, the form factor f_j depends strongly on s with a marked decrease at higher values. Note that most explicit parameterizations of f_j and others are formulated as a function of $\tilde{s} = s/2 = \sin\theta/\lambda$ and not of s.

The value at $\tilde{s} = 0$ is normalized to the number of electrons of the scatterer (atom or ion). The form factor consists of a wavelength independent (normal scattering) and a complex wavelength-dependent part (anomalous scattering):

$$f_j(\tilde{s}) = f_j^0(\tilde{s}) + \Delta f_j'(\lambda) + \sqrt{-1}\Delta f_j''^i(\lambda). \tag{2.13}$$

The real part of the anomalous scattering factor has a phase shift of 180° with respect to the normal scattering factor, thus directly reducing the scattering power, while the complex part has a phase shift of 90° (Figure 2.10).

Anomalous scattering effects are often disregarded for simplicity, but become extremely important if the wavelength used is in the vicinity of an absorption edge of an atomic species in the sample. For a strong scatterer, the change in scattering power can amount to the equivalent of several electrons and so-called anomalous dispersion measurements can be used to give extra element-specific information on structures. By default, TOPAS uses dispersion coefficients $\Delta f_j'(\lambda)$ and $\Delta f_j''^i(\lambda)$ (from *http://henke.lbl.gov/optical_constants/*) that cover the energy range from 10 eV to 30 keV (Figure 2.11).

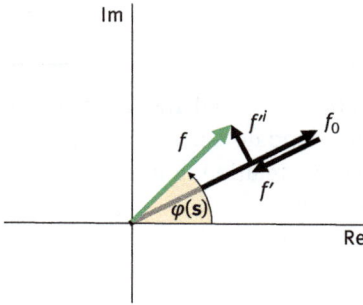

Figure 2.10: Vector (pointer) representation of the complex atomic scattering factor f with normal scattering (f_0) and real (f') and complex (f'') parts of anomalous scattering.

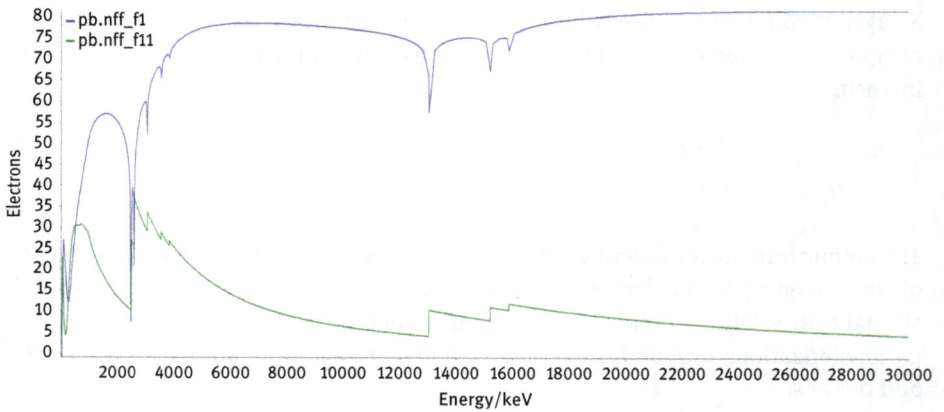

Figure 2.11: TOPAS screenshot showing anomalous scattering factors $Z - \Delta f'_j(\lambda)$ (nff_f1) and $\Delta f''_j(\lambda)$ (nff_f11) for Pb as a function of energy ($\lambda/\text{Å} = 12.3985/(E/\text{keV})$). Data were taken from http://henke.lbl.gov/optical_constants/.

The functional dependence of the form factors for all common atoms and ions has been parameterized by an empirical linear combination of four Gaussian functions:

$$f_j^0(\tilde{s}) = c_0^j + \sum_{k=1}^{4} a_k^j e^{-b_k^j \tilde{s}^2} \tag{2.14}$$

with the nine parameters a_1, a_2, a_3, a_4, b_1, b_2, b_3, b_4 and c_0 tabulated, for example, in Tables for X-Ray Crystallography (1995) Vol C. The obtained form factors are valid in the range $0 \le \tilde{s} \le 2$, which is sufficient for most cases (Figure 2.12).

For more precise approximations covering the range $0 \le \tilde{s} \le 6$ (Waasmaier & Kirfel, 1995), the normal form factor is approximated by an empirical linear combination of five Gaussian functions, leading to 11 parameters. This is the default in TOPAS (Figure 2.13):

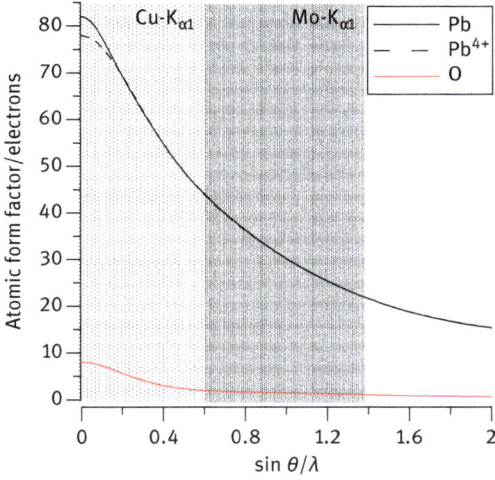

Figure 2.12: Normal atomic form factors of Pb, Pb^{4+} and O calculated using the nine-parameter approximation function as a function of $(\tilde{s})$. The theoretical maximum ranges $(0 \leq 2\theta \leq 180°)$ for Cu-$K_{\alpha 1}$ and Mo-$K_{\alpha 1}$ radiations are shaded.

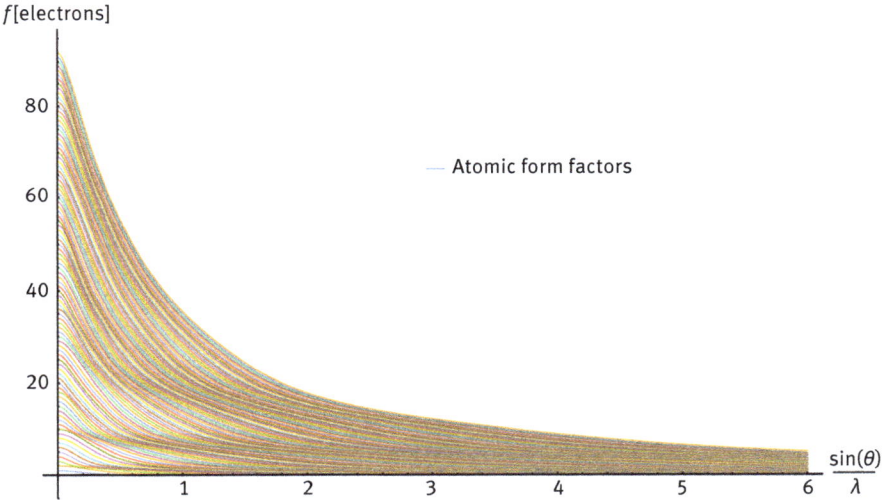

Figure 2.13: Atomic form factors of all atoms with atomic numbers ranging from 1 (H) to 92 (U) calculated using the 11-parameter approximation function as a function of $(\tilde{s})$.

$$f_j^0(\tilde{s}) = c_0^j + \sum_{k=1}^{5} a_k^j e^{-b_k^j \tilde{s}^2}. \tag{2.15}$$

TOPAS allows manual input and also refinement of $f_j^0(\tilde{s})$, $\Delta f_j'(\lambda)$ and $\Delta f_j''(\lambda)$, but high correlations exist between $\Delta f_j'(\lambda)$, $\Delta f_j''(\lambda)$, scale factor and displacement parameters (see Section 2.2.3 for latter). The syntax uses the *f0_f1_f11_atom* keyword ($f0 \equiv f_j^0(\tilde{s})$, $f1 \equiv \Delta f_j'(\lambda)$ and $f11 \equiv \Delta f_j''(\lambda)$). Defaults are used when either *f1* or *f11* are not defined. The following example defines *f1* and *f11* for Pb^{2+}, S and O^{2-} and the refinement flag (@) is set for Pb^{2+}:[8]

```
'  Default values from NFF files
load f0_f1_f11_atom f1 f11
{
    Pb+2 @ -3.7275704 @ 8.93529065
    S   0.335135942 0.550512703
    O-2  0.0523209357  0.0337069703
}
```

In the following example, the *a1* parameter of *f0* and *f11* of Pb are refined:

```
' Values for Pb from atmscat.cpp
' Pb+2 27.392647 16.496822 19.984501 6.813923  5.233910  4.065623 1.058874  0.106305
  6.708123 24.395554 1.058874

prm a1 25 min -50 max 50
load f0_f1_f11_atom f0 f11 {
    Pb+2
    = a1          Exp(1.058874    (-0.25) / D_spacing^2) +
      16.496822  Exp(0.106305    (-0.25) / D_spacing^2) +
      19.984501  Exp(6.708123    (-0.25) / D_spacing^2) +
       6.813923  Exp(24.395554  (-0.25) / D_spacing^2) +
       5.233910  Exp(1.058874    (-0.25) / D_spacing^2) +
       4.065623; ' this is f0 for Pb
    @ 5           ' this is f11 for Pb
}
```

2.2.2 The coherent atomic scattering length

For neutron diffraction data, $\Delta f_j'(\lambda) = \Delta f_j''(\lambda) = 0$, and $f_j^0(\tilde{s})$ is replaced by the bound coherent scattering length, *b*, which is independent of the scattering angle (Figure 2.14). Neutron scattering length data in TOPAS are from Sears (1992) and stored in the file *NEUTSCAT.CPP*.

---

8 The "load { }" keyword in TOPAS is used to simplify user input. It allows the loading of keywords of the same type by typing the keywords once.

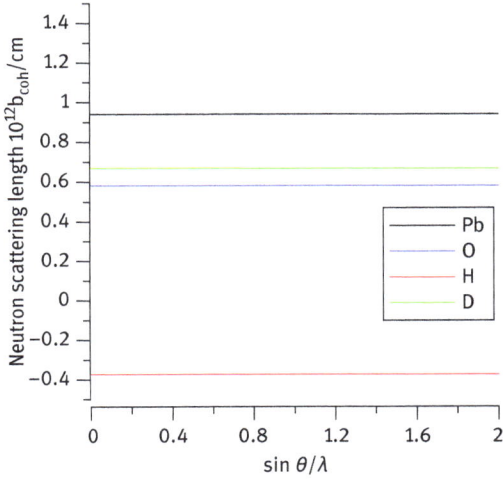

Figure 2.14: Neutron coherent scattering lengths for Pb, O, H and D.

2.2.3 Displacement parameter

At any temperature, atoms vibrate about their equilibrium position. Moreover, static local atomic displacements may exist in disordered structures like solid solutions. The corresponding displacements lead to a decrease in peak intensities, which can be described by multiplying the atomic form factor with a correction factor. One can distinguish between isotropic and anisotropic displacements. For the isotropic case the displacement factor (Debye–Waller factor) for the entire crystal structure, groups of atoms or an individual atom is defined as:

$$t = e^{-B\bar{s}^2}, \tag{2.16}$$

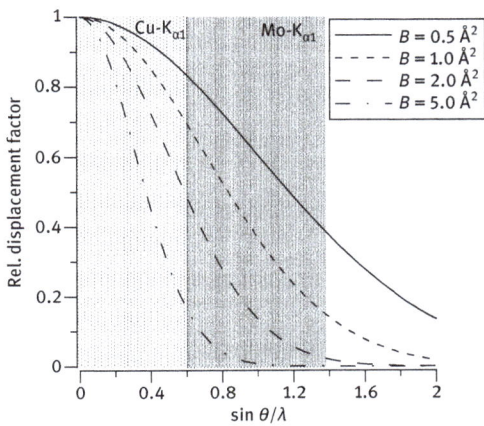

Figure 2.15: Intensity reduction t as a function of $\bar{s}$ for a series of normalized displacement parameters B.

where B is the isotropic displacement parameter (usually, and also in TOPAS, in Å^2). t is shown in Figure 2.15 and used in the structure factor calculation of eq. (2.21).

The physical meaning of B is given by:

$$B = 8\pi^2 u^2, \tag{2.17}$$

where u^2 is the mean square deviation from the equilibrium position of the atom or atomic group. A range of $0.1 \text{ Å}^2 \leq B \leq 1.5 \text{ Å}^2$ is considered normal for inorganic compounds, while for coordination compounds $B \leq 3 \text{ Å}^2$ is usually acceptable. Larger values usually indicate errors or severe disorder in the crystal structure. Negative values often indicate systematic errors in the intensities due to, for example, absorption or surface roughness or misassignment of an atom type. TOPAS uses B as the isotropic displacement parameter.

In order to determine individual displacement parameters with good precision from powder data, a large range of $\tilde{s}$ must be covered. In addition, the $\tilde{s}$-dependence of the intensity reduction (Figure 2.15) is similar to that of many other correction factors that are often poorly treated. For this reason, the displacement factor is frequently (harshly!) referred to as a "physical trash collector."

With high-quality powder diffraction data measured over an extended $\tilde{s}$ range, it may be possible to refine anisotropic displacement parameters for the strong scatterers in the crystal structure.[9] The anisotropic displacement parameter can be geometrically described as a three-axis ellipsoid and can be described in terms of a second-rank tensor symmetric about its principal diagonal:

$$\mathbf{u} = \begin{pmatrix} u_{11} & u_{12} & u_{13} \\ u_{12} & u_{22} & u_{23} \\ u_{13} & u_{23} & u_{33} \end{pmatrix}, \tag{2.18}$$

which leads to a direction $\mathbf{s} = (hkl)$ dependent displacement factor of:

$$t = e^{-2\pi^2 \left(u_{11}h^2 a^{*2} + u_{22}k^2 b^{*2} + u_{33}l^2 c^{*2} + 2u_{12}hka^* b^* + 2u_{13}hla^* c^* + 2u_{23}klb^* c^* \right)}. \tag{2.19}$$

In order to prevent physically meaningless results, the $\mathbf{u}_{ij}$ matrix must be kept positive definite, which can be achieved with the following boundary conditions:

$$\begin{aligned} u_{ii} &> 0 \\ u_{ii} \cdot u_{jj} &> u_{ij}^2 \\ u_{11} \cdot u_{22} \cdot u_{33} + u_{12}^2 \cdot u_{13}^2 \cdot u_{23}^2 &> u_{11} \cdot u_{23}^2 + u_{22}^2 \cdot u_{13}^2 + u_{33}^2 \cdot u_{12}^2. \end{aligned} \tag{2.20}$$

---

9 Especially for neutron data where the scattering factor doesn't fall off with $\tilde{s}$.

In TOPAS, the site-dependent macro *"ADPs_Keep_PD"* does this job.[10] TOPAS uses u_{ij} as anisotropic displacement parameters, which are activated by the *"adps"* keyword:

```
site Cl1   x @ 0.03947`  y @ 0.27566`  z @ 0.25331`  occ Cl  1 adps
```

that leads to (in the order u_{11}, u_{22}, u_{33}, u_{12}, u_{13}, u_{23}):

```
site Cl1 x @ 0.03947`  y @ 0.27566`  z @ 0.25331`  occ Cl  1
    ADPs { @  0.04367` @  0.06117` @  0.08074` @  0.00592` @  0.01280` @  0.01262` }
```

TOPAS automatically places appropriate symmetry restrictions on individual u_{ij} components.

Higher levels of complexity using an anharmonic approximation of the atomic displacement parameters are not of relevance for powder diffraction, except in very special cases (e.g., Wahlberg et al., 2016).

2.2.4 The structure factor

Ignoring anomalous scattering, the structure factor of a Bragg reflection is defined as a complex sum over all atoms j in the unit cell (Figure 2.16):

$$F(\mathbf{s}) = \sum_{j} \left(t_j \, f_j(s) e^{2\pi i \mathbf{s} \cdot \mathbf{x}_j} \right) \tag{2.21}$$

with the positional vector $\mathbf{x}_j$ of an atom j in the unit cell defined by the fractional crystal coordinates:

$$\mathbf{x}_j = \begin{pmatrix} x \\ y \\ z \end{pmatrix}. \tag{2.22}$$

Equation (2.21) contains the displacement factor t_j for every atom j. This factor is omitted in the following for brevity.

Using the Euler identity, the real and complex parts of the structure factor can be separated (Figure 2.16):

$$F(\mathbf{s}) = \sum_{j} \left(f_j(s)\cos(2\pi\mathbf{s} \cdot \mathbf{x}_j) \right) + \sqrt{-1} \sum_{j} \left(f_j(s)\sin(2\pi\mathbf{s} \cdot \mathbf{x}_j) \right) = A(\mathbf{s}) + iB(\mathbf{s}). \tag{2.23}$$

---

10 This behavior can be turned off using the keyword *adp_no_limits*.

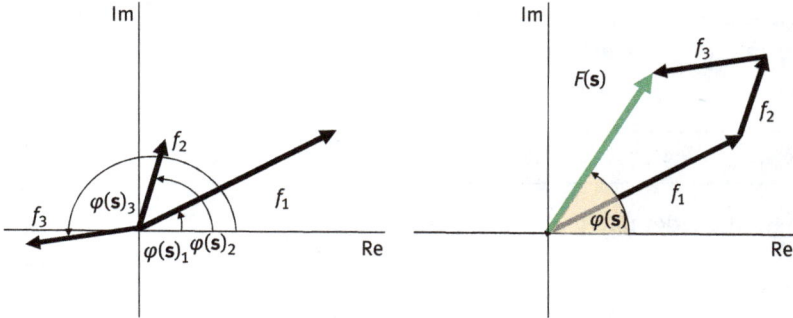

Figure 2.16: Graphical representation of the complex structure factor as vector sum (right) of the individual form factors (left).

If anomalous scattering is taken into account, the structure factor amplitude becomes:

$$F(\mathbf{s}) = \sum_j \left(\left(f_j^0(s) + \Delta f_j'(\lambda) + i \Delta f_j'^i(\lambda) \right) \cdot \cos(2\pi \mathbf{s} \cdot \mathbf{x}_j) \right)$$
$$+ i \sum_j \left(\left(f_j^0(s) + \Delta f_j'(\lambda) + i \Delta f_j'^i(\lambda) \right) \cdot \sin(2\pi \mathbf{s} \cdot \mathbf{x}_j) \right). \tag{2.24}$$

After separating the real and the imaginary parts, this turns into:

$$F(\mathbf{s}) = \left\{ \begin{array}{l} \sum_j \left(\left(f_j^0(s) + \Delta f_j'(\lambda) \right) \cdot \cos(2\pi \mathbf{s} \cdot \mathbf{x}_j) \right) \\ - \sum_j \left(\left(\Delta f_j'^i(\lambda) \right) \cdot \sin(2\pi \mathbf{s} \cdot \mathbf{x}_j) \right) \end{array} \right\}$$
$$+ i \left\{ \begin{array}{l} \sum_j \left(\left(f_j^0(s) + \Delta f_j'(\lambda) \right) \sin(2\pi \mathbf{s} \cdot \mathbf{x}_j) \right) \\ + \sum_j \left(\left(\Delta f_j'^i(\lambda) \right) \cdot \cos(2\pi \mathbf{s} \cdot \mathbf{x}_j) \right) \end{array} \right\} \tag{2.25}$$
$$= \{ A_{01}(\mathbf{s}) - B_{11}(\mathbf{s}) \} + i \{ B_{01}(\mathbf{s}) + A_{11}(\mathbf{s}) \} = A(\mathbf{s}) + i B(\mathbf{s}).$$

The intensity of a Bragg reflection $\mathbf{s}$ is proportional to the structure factor multiplied by its conjugate complex, which is equivalent to the squared absolute value of the structure factor amplitude $|F(\mathbf{s})|$:

$$I(\mathbf{s}) \propto F(\mathbf{s}) \, F^*(\mathbf{s}) = |F(\mathbf{s})|^2. \tag{2.26}$$

For practical purposes it is easier to separate the real and imaginary parts of the structure factor (Figure 2.17) leading to:

$$
\begin{aligned}
|F(\mathbf{s})|^2 &= A(\mathbf{s})^2 + B(\mathbf{s})^2 \\
&= [|A(\mathbf{s})| + \mathrm{i}|B(\mathbf{s})|][|A(\mathbf{s})| - \mathrm{i}|B(\mathbf{s})|] \\
&= A_{01}(\mathbf{s})^2 + B_{01}(\mathbf{s})^2 + A_{11}(\mathbf{s})^2 + B_{11}(\mathbf{s})^2 + 2B_{01}(\mathbf{s})A_{11}(\mathbf{s}) - 2A_{01}(\mathbf{s})B_{11}(\mathbf{s}).
\end{aligned}
\tag{2.27}
$$

The phase angle can be directly deduced from Figure 2.17 as:

$$
\varphi(\mathbf{s}) = \arctan\left(\frac{|B(\mathbf{s})|}{|A(\mathbf{s})|}\right).
\tag{2.28}
$$

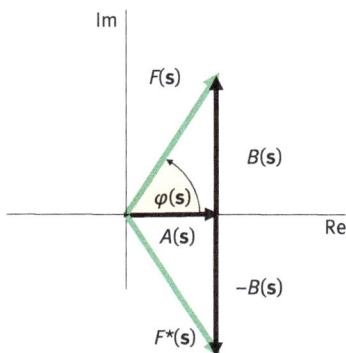

Figure 2.17: Vector (pointer) representation of the structure factor and its conjugate complex.

Due to the squaring of the structure factor amplitude, the relative contribution to the diffraction pattern of light elements in the presence of heavy elements is low if X-rays are used. As an example, the squared scattering power (squared form factors) of Pb and O differ by a factor of more than 100.

TOPAS gives direct access to all the quantities in the structure factor equation (eq. (2.27)). Even though it is not needed for regular work, this can sometimes be useful. As a simple example, the notation of atomic sites for structure factor calculation in a structure (*str*) phase of TOPAS is shown for LaB$_6$. Please note that the special positions should be given as equations.[11] The keywords *num_posns*, *beq* and *occ* define the number of unique equivalent positions generated from the space group, the isotropic displacement parameter B and the site occupancy factor, respectively:

```
site La num_posns 1 x =0; :   0  y =0;    :   0    z =0;   :  0    occ La  1 beq @ 0.38934
site B  num_posns 6 x @ 0.19986 y =1/2; :  0.5 z =1/2; :  0.5 occ B   1 beq @ 0.23502
```

---

11 Particularly for recurring numbers like 1/3 (use =1/3; not 0.33333).

The reflection intensities and structure factors can be written to the file *"reflec-tions.txt"* with[12]:

```
phase_out "reflections.txt" append
      load out_record out_fmt out_eqn
{
   "%4.0f"    = H;   "%4.0f" = K;   "%4.0f" = L; " %3.0f" = M;
   "%12.2f"   = I_no_scale_pks;            ' I without any scale_pks corrections applied
   "%12.2f"   = Iobs_no_scale_pks;         ' Iobs from Rietveld decomposition formula
   "%10.2f o" = Iobs_no_scale_pks_err;
   " %11.5f"  = A01; " %11.5f" = A11; " %11.5f" = B01; " %11.5f\n" = B11;
}
```

2.3 Intensity correction factors

To calculate the integrated reflection intensities observed experimentally in a powder diffraction pattern, a series of correction factors have to be applied to the squared structure factors, which depend on the scattering vector $\mathbf{s}$ or its length s. A list of the most common correction factors is given by the product:

$$Corr(\mathbf{s}) = M(\mathbf{s})LP(s)A(s)PO(\mathbf{s})E(\mathbf{s})\ldots, \tag{2.29}$$

which includes the multiplicity $M(\mathbf{s})$ of a reflection given by the lattice symmetry, an absorption correction $A(s)$, the (solely geometrical) Lorentz–polarization factor $LP(s)$, a preferred orientation correction $PO(\mathbf{s})$ and a correction for primary extinction $E(\mathbf{s})$, which is only relevant for highly crystalline materials. The absolute values of the individual correction factors are *not* important, as any constant gets absorbed in the scale factor.[13]

2.3.1 Multiplicity

Due to the overlap of Friedel pairs, the observed intensity is always doubled corresponding to a minimum value of two for the reflection multiplicity for all crystal systems. In addition, for symmetries higher than triclinic, symmetry-equivalent reflections of identical d-spacing have *identical intensity* and *overlap completely*. The total number of these reflections is called multiplicity and lies between 2 and

---

12 Expressions like " *%11.5f\n*" are formatting commands, for example writing real numbers (f) 11 characters wide with five decimal places followed by a new line (\n).

13 Note that all corrections are applied to the *calculated* intensities in order to match the observed intensities.

48. TOPAS takes care of the multiplicity automatically (Figure 2.18). The reserved keyword "*M*" is used to store all reflection multiplicities of a Pawley/LeBail (*hkl_Is*) (see Chapter 3) or Rietveld (*str*) phase. An example of how to save the reflection parameters including the multiplicity of a *hkl_Is* phase to the file "*reflections.txt*" is:

```
phase_out reflections.txt load out_record out_fmt out_eqn
{
     "%3.0f" = H; " %3.0f" = K; " %3.0f" = L; " %3.0f" = M;
    " %11.5f"   = D_spacing;
    " %11.5f"   = 2 Rad Th;
    " %11.5f\n" = I_no_scale_pks;
}
```

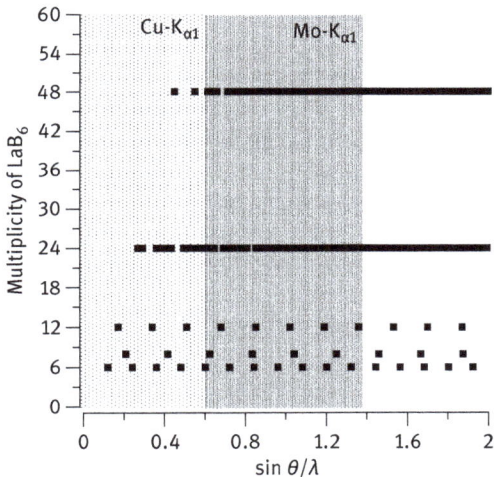

Figure 2.18: Multiplicities (6, 8, 12, 24, 48) of the Bragg reflections of LaB$_6$ (*Pm$\bar{3}$m*) as a function of $\bar{s}$.

2.3.2 Lorentz–polarization factor

The Lorentz and the polarization factors are purely geometric factors. The Lorentz factor has several contributions. One takes into account the relative time that a reciprocal lattice point moving with angular velocity ω spends passing through the finite thickness of the Ewald sphere. According to Figure 2.19, the component of the linear velocity v of a reciprocal lattice point along the radius of the Ewald sphere is:

$$v = \omega d^{*} \cos\theta = \omega \frac{2\sin\theta}{\lambda}\cos\theta \propto \sin\theta\cos\theta, \ L \propto \frac{1}{v}. \tag{2.30}$$

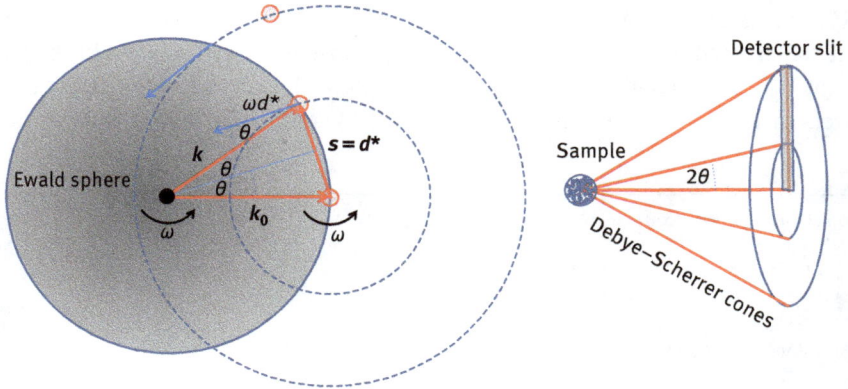

Figure 2.19: Schematic drawing of the geometrical origin of the Lorentz factor. Left: Reciprocal lattice points rotating with different speed through the Ewald sphere. Right: Effect of the opening angle of the Debye–Scherrer cones on the intensity distribution.

In the case of powder diffraction, an additional geometrical factor occurs, which normalizes the different radii of the Debye–Scherrer rings (Figure 2.19, right). The fraction of the diffraction cone that intersects the detector is highest at low angles and at very high angles (backscattering). The factor is proportional to $1/\sin\theta$. The typical form of the Lorentz factor for powders is then (Figure 2.20, left):

$$L = \frac{1}{\cos\theta\sin^2\theta} \propto \frac{1}{\sin\theta\sin(2\theta)} \tag{2.31}$$

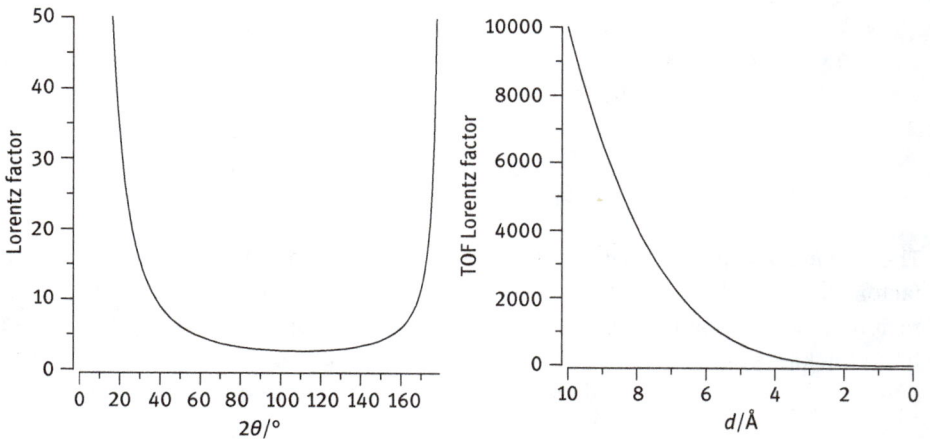

Figure 2.20: Lorentz factor for angle dispersive data as a function of the scattering angle (left) and for *TOF* data as a function of *d*-spacing (right).

where any constant factor gets absorbed by the overall scale factor. In the case of constant wavelength neutron data or fully polarized synchrotron radiation, the macro *Lorentz_Factor* can be used to describe these effects.

For neutron TOF data, the Lorentz factor calculates to (Figure 2.20, right):

$$L_{TOF} = d^4. \tag{2.32}$$

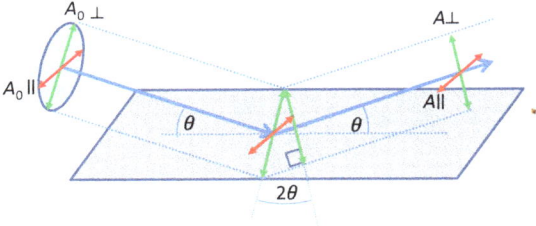

Figure 2.21: Schematic drawing of the effect of polarization for an unpolarized primary beam.

The polarization factor originates from partial polarization of the scattered electromagnetic wave. For the case depicted in Figure 2.21 with:

$$A_{0\perp} = A_{0\parallel}, \; I_0 = A_{0\perp}^2 + A_{0\parallel}^2 = 2A_{0\perp}^2,$$
$$A_\parallel = A_{0\parallel}, A_\perp = A_{0\perp}\cos 2\theta, \; I = A_\perp^2 + A_\parallel^2 = A_{0\perp}^2 \cos^2 2\theta + A_{0\perp}^2, \tag{2.33}$$

the intensity ratio between the diffracted and the primary beam follows:

$$P = \frac{I}{I_0} = \frac{1 + \cos^2 2\theta}{2}. \tag{2.34}$$

This equation is valid for unpolarized radiation from a laboratory X-ray tube. When a primary or secondary beam monochromator is present, a more general equation is used:

$$P = \frac{1 - \cos^2 2\theta \cdot \cos^2 2\theta_m}{2} \tag{2.35}$$

where $2\theta_m$ is the Bragg angle of the reflection from the monochromator (Table 2.2). For unpolarized radiation $2\theta_m$ can be set to $0°$ (e.g., X-ray diffractometers without any monochromator), for fully polarized radiation $2\theta_m$ can be set to $90°$ (e.g., synchrotron radiation or constant wavelength neutron diffraction) (Figure 2.22). In reality, synchrotron radiation is 95–97% polarized. In order to account for fractional polarization of the beam, a factor K can be introduced with $K = 0.5$ for circularly polarized X-rays (i.e., laboratory X-ray tubes), $K = 0$ for fully polarized X-rays (ideal synchrotron source) and $K \sim 0.05$ for a "real" synchrotron source:

Table 2.2: Bragg angle of different combinations of monochromators and radiation for angle-dispersive laboratory powder diffractometers.

Monochromator	Radiation	Bragg angle $2\theta_m/°$
None	Any	0
Ge (111)	Cu-$K_{\alpha 1}$	27.3
Ge (111)	Mo-$K_{\alpha 1}$	17.3
Ge (111)	Ag-$K_{\alpha 1}$	9.8
Ge (220)	Mo-$K_{\alpha 1}$	12.4
SiO$_2$ (101)	Cu-$K_{\alpha 1}$	26.4
Graphite	Cu-$K_{\alpha 1}$	26.6

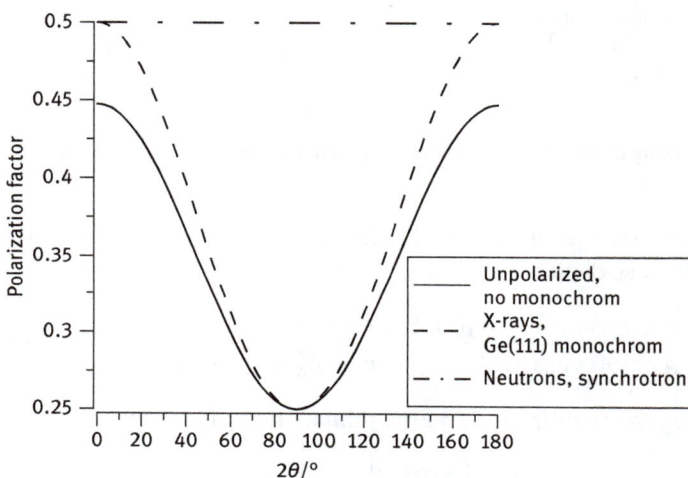

Figure 2.22: Polarization factor for unpolarized, partly polarized and polarized radiation as a function of diffraction angle.

$$P = \frac{(1 - K + K \cdot \cos^2 2\theta \cdot \cos^2 2\theta_m)}{2}. \tag{2.36}$$

For practical reasons, the Lorentz and the polarization factors are usually combined in a single Lorentz–polarization factor (LP factor) (Figure 2.23, left). The effect of the LP factor on the intensities of an X-ray powder pattern is enormous (Figure 2.23, right).

The corresponding correction macros in TOPAS are called *LP_Factor*($2\theta_m$) for laboratory X-ray data and *LP_Factor_Synchrotron(1 – K, $2\theta_m$)* for a "real" synchrotron. Although possible, one should not attempt to refine the K factor and/or the Bragg angle of the monochromator. To apply a general LP factor for angular dispersive data as defined above in TOPAS, the following code can be used:

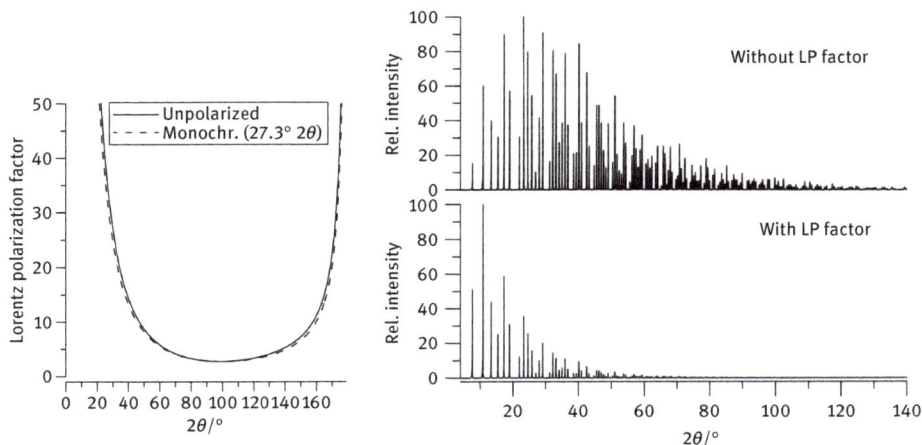

Figure 2.23: Effect of the Lorentz–polarization (LP) factor for angular dispersive data as a function of Bragg angle. Left: LP factor for unpolarized beam (solid line) and partly polarized beam by a Ge (111) primary beam monochromator with Cu-$K_{\alpha 1}$ radiation. Right: Simulated powder pattern of LaB_6 for Ag-$K_{\alpha 1}$ radiation with (bottom) and without (top) LP factor.

```
prm k    0.5 min 0.0 max  1.0
prm mono 0   min 0.0 max 90
scale_pks = (1/( Sin(Th)^2 Cos(Th))) (1 - k + k Cos(mono) Deg)^2 Cos(2 Th)^2) / 2;
```

For *TOF* neutron data, the following code is appropriate:

```
scale_pks = D_spacing^4;
```

Note that the keyword *scale_pks* is used for applying intensity corrections to phase peaks, while *scale_phase_X* scales calculated patterns point by point. Multiple definitions are allowed and each is applied to the peaks and pattern, respectively.

2.3.3 Absorption correction

For accurate powder diffraction work it is important to consider the effects of absorption on experimental intensities. For simple transmission through a solid material, the transmitted intensity I with respect to the initial intensity I_0 depends on the thickness x of the material and its linear absorption coefficient μ (Figure 2.24):

$$I = I_0 e^{-\mu x}. \tag{2.37}$$

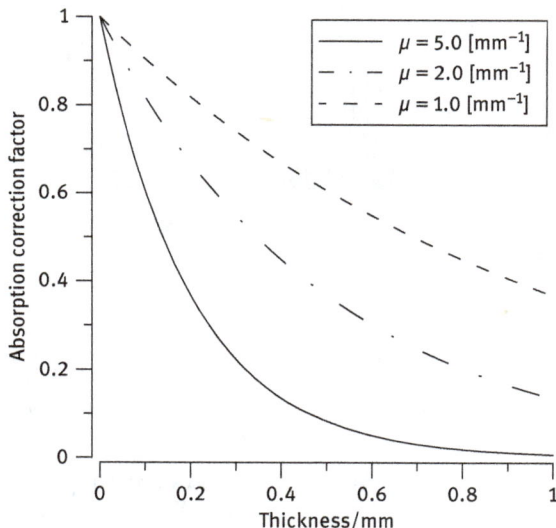

Figure 2.24: Intensity correction factor required for solid samples in transmission geometry with different absorption coefficients.

The appropriate absorption correction factor (transmission factor) for calculated intensities therefore is $A = I/I_0$. Absorption depends strongly on the energy (wavelength) of the radiation used and also changes rapidly close to the absorption edges discussed in Section 2.2.1. To minimize absorption effects it is therefore important to match the experimental wavelength to the system being studied.

One important consideration in Bragg–Brentano geometry is the requirement that the sample is "infinitely thick", meaning that a negligible fraction of the beam passes straight through the sample. One way to estimate the minimum sample thickness, t_{min}, required to meet this criterion is to calculate the sample depth required for the incident beam to be reduced to 1/1000th of its initial intensity. We set[14] $t_{min} = 2x \sin \theta_{max}$ and $I(surface)/I(tmin) = 1000$ in eq. (2.37), which simplifies to give:

$$t_{min} \cong \frac{3.45 \ \sin \theta_{max}}{\mu}. \tag{2.38}$$

For Ni powder $t_{min} \approx 0.013$ cm and for a typical organic sample $t_{min} \approx 1.3$ cm at $\theta = 90$ using Cu-K$_\alpha$ radiation. Note that it is practically unlikely that an organic material will meet the infinite thickness criterion.

In a Rietveld analysis, the effects of absorption on peak intensities must be taken into account. The corrections differ for different experimental geometries. Many of these have been reviewed and a collection of TOPAS macros given (Rowles &

---

14 The factor of 2 arises as both the incoming and the outgoing beams must be taken into account.

Buckley, 2017). To calculate the reduction of the diffracted intensity, one must take the total path l of the incident and the diffracted beams in the sample into account and integration must be performed over the entire volume V of the sample that contributes to scattering. Instead of the linear absorption coefficient μ, an *effective* linear absorption coefficient μ_{eff} should be used to account for the lower packing density of a loose powder.

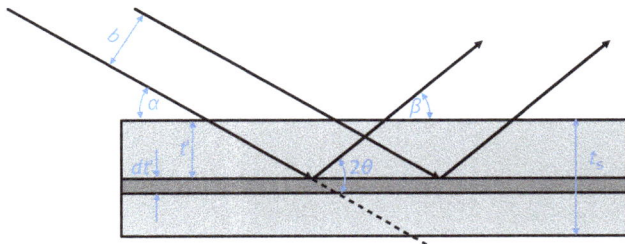

Figure 2.25: Schematic drawing of asymmetric flat plate reflection geometry.

For flat plate asymmetric reflection (Figure 2.25), where a (largely) parallel beam[15] is incident on the specimen surface at a fixed angle, the irradiated volume contributing to the diffracted intensity can be calculated (Egami & Billinge, 2003) by:

$$V = \frac{Area}{\sin\alpha} \int_0^{t_s} e^{-\mu_{eff}\, t'\left(\frac{1}{\sin\alpha} + \frac{1}{\sin(2\theta - \alpha)}\right)} dt'$$

$$= \frac{Area}{\mu}\left(1 + \frac{\sin\alpha}{\sin(2\theta - \alpha)}\right)^{-1}\left\{1 - e^{-\mu_{eff}\, t_s\left(\frac{1}{\sin\alpha} + \frac{1}{\sin(2\theta - \alpha)}\right)}\right\}, \tag{2.39}$$

where $Area$ is the area of the incident X-ray beam on the specimen surface, t_s is the specimen thickness, α is the angle between the incident beam and the specimen surface and $2\theta - \alpha\ (=\beta)$ is the angle between the diffracted beam and the specimen surface. In the case of an "infinitely thick" sample (exponential term in eq. (2.39) is less than 0.001), the equation simplifies to:

$$V_{Asym} = \frac{Area}{\mu_{eff}}\left(1 + \frac{\sin\alpha}{\sin(2\theta - \alpha)}\right)^{-1}. \tag{2.40}$$

For symmetric reflection $\alpha = \beta = \theta$ (Bragg–Brentano geometry), this volume further simplifies into a constant:

---

15 The tube to specimen distance in most experiments is sufficiently large that we can consider the beam to be parallel in these calculations.

$$V_{Bragg} = \frac{Area}{2\mu_{eff}}. \tag{2.41}$$

Consequently, for Bragg–Brentano geometry with an infinitely thick sample, the diffracting volume is constant, and no correction to the intensity is necessary.

For asymmetric reflection geometries a correction will be needed to account for the changing volume of diffracting material:

$$A = \frac{V_{Asym}}{V_{Bragg}} = 2\left(1 + \frac{\sin\alpha}{\sin(2\theta - \alpha)}\right)^{-1} \tag{2.42}$$

that, for samples fulfilling the criterion of infinite thickness, is independent of the absorption coefficient.

In the case when the specimen does not fulfill the criterion of "infinite thickness," the correction factor becomes:

$$A = 2\left(1 + \frac{\sin\alpha}{\sin(2\theta - \alpha)}\right)^{-1}\left\{1 - e^{-\mu_{eff}\,t_s\left(\frac{1}{\sin\alpha} + \frac{1}{\sin(2\theta - \alpha)}\right)}\right\}. \tag{2.43}$$

Figure 2.26 plots this correction factor for a specimen with different angles between the incident beam and specimen surface as a function of diffraction angle. For the case of a 10° incident angle, the effect of different values of the specimen thickness is shown.

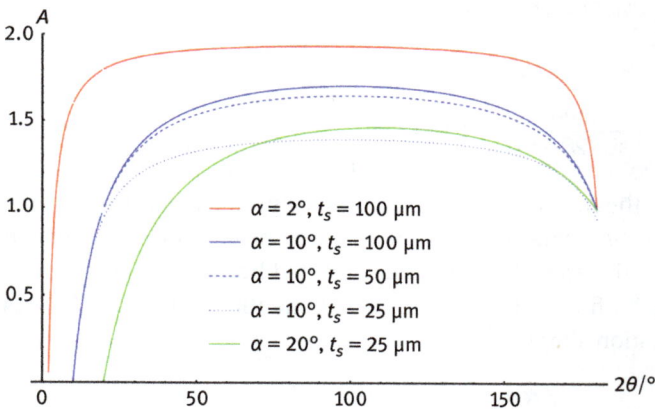

Figure 2.26: Intensity enhancement for asymmetric reflection geometry of specimen with different angle between incident beam and specimen surface. For a 10° angle, the effect of different values for the specimen thickness is shown using a linear absorption coefficient of 100 cm^{-1}.

Equation (2.43) simplifies significantly for symmetric Bragg reflection of noninfinitely thick samples to (Figure 2.27):

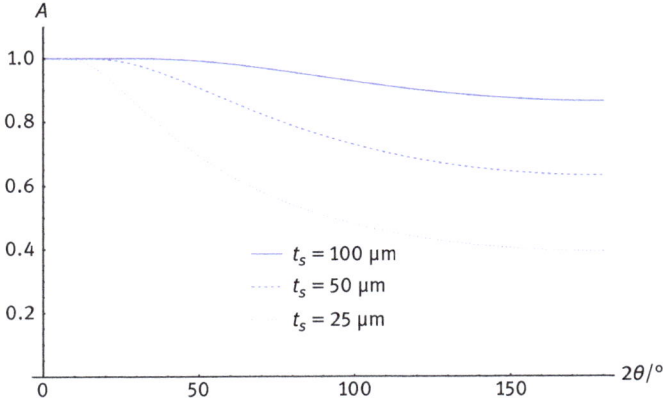

Figure 2.27: Intensity enhancement for symmetric reflection geometry (Bragg–Brentano) showing the effect of different values for the specimen thickness using a linear absorption coefficient of 100 cm^{-1}.

$$A = 1 - e^{-\mu_{eff}\, t_s \frac{2}{\sin\theta}}$$ (2.44)

In TOPAS scripting language this can be applied by:

```
prm uT 1.0 min 0.0001 max 12
scale_pks = 1-exp(-uT 2 / Sin(Th));
```

For cylindrical samples (Debye–Scherrer geometry), the beam must pass through the entire capillary at low angles. A reasonable approximation for an absorption correction factor has been given by Sabine et al., 1988:

$$A = A_L\cos^2\theta + A_B\sin^2\theta$$ (2.45)

where A_L and A_B are the absorption factors at the Laue condition ($\theta = 0°$) and the Bragg condition ($\theta = 90°$), respectively.

The absorption factors depend on $\mu_{eff}R$ with R the cylinder (capillary) radius, and are calculated using modified Bessel and Struve functions. A typical absorption correction factor for $\mu_{eff}R = 1.0$ or 2.5 is shown in Figure 2.28. The formula gives satisfactory results for $\mu_{eff}R < 10$.

In TOPAS scripting language such a cylindrical absorption correction can be realized by:

```
prm uR 1.0 min 0.0001 max 12
scale_pks = AL_Cyl_Corr(@, uR)) Cos(Th)^2 + AB_Cyl_Corr(@, uR) Sin(Th)^2;
```

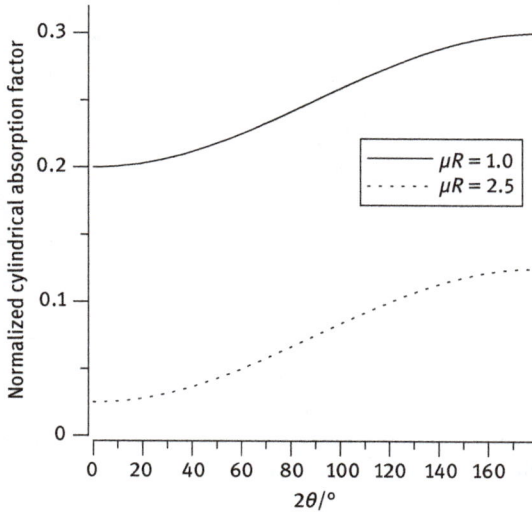

Figure 2.28: Absorption correction factor for cylindrical samples as a function of scattering angle for $\mu_{eff}R = 1.0$ and 2.5.

Alternatively a predefined macro with $\mu_{eff}R$ as argument can be used:

```
Cylindrical_I_Correction(@, 1.0)
```

A more rigorous treatment using radial symmetry for the calculation of cylindrical absorption coefficients taking the capillary loading into account was published by Kalifah (2015).

2.3.4 Surface roughness

If the packing density in Bragg–Brentano geometry varies with depth, thus creating a "rough surface," the so-called porosity effect reduces the intensity at low Bragg angles. This is also a kind of absorption effect. The two most common corrections are those by Pitschke et al. (1993) (Figure 2.29, left):

$$A = \frac{1 - a_1(1/\sin\theta - a_2/\sin^2\theta)}{1 - a_1(1 - a_2)} \tag{2.46}$$

and by Suortti (1972) (Figure 2.29, right):

$$A = \frac{a_1 + (1 - a_1)e^{-a_2/\sin\theta}}{a_1 + (1 - a_1)e^{-a_2}} \tag{2.47}$$

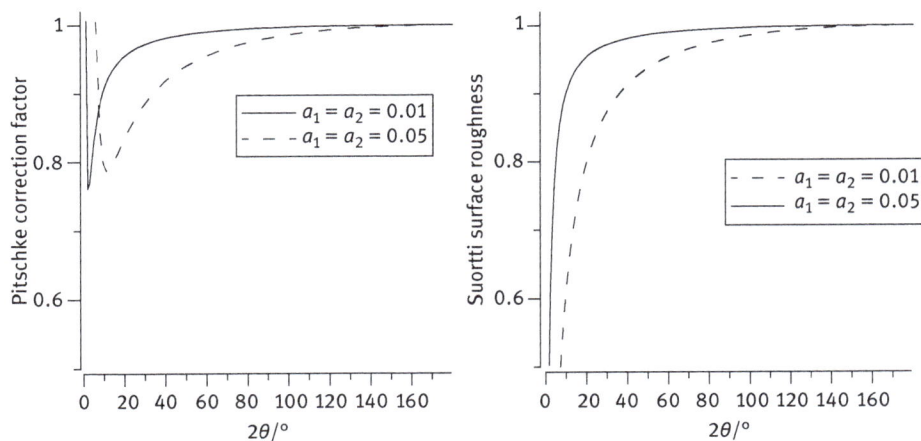

Figure 2.29: Correction factor for porosity effect in Bragg–Brentano geometry according to Pitschke et al. (1993) (left) and Suortti (1972) (right) as a function of diffraction angle.

where a_1 and a_2 are refinable parameters. The latter function works better on low-angle reflections. In TOPAS scripting language the surface roughness correction according to Pitschke can be realized by:

```
prm a1c min .0001 max 1
prm a2  min .0001 max 1
scale_pks = (1 - a1 (1 / Sin(Th) - a2) / Sin(Th)^2)) / (1 - a1 + a1 a2);
```

and according to Suortti by:

```
prm a1c min .0001 max 2
prm a2  min .0001 max 2
scale_pks = (a1 + (1 - a1) Exp( -a2 /Sin(Th))) / (a1 + (1 - a1) Exp( -a2));
```

The two corresponding macros in TOPAS are:

```
Surface_Roughness_Pitschke_et_al(@, a1v, @, a2v)
Surface_Roughness_Suortti(@, a1v, @, a2v)
```

2.3.5 Extinction

In rare cases, for nearly perfectly crystalline materials, primary extinction effects (within the same crystallite) can occur for powders. The observed intensity for strong reflections at low Bragg angle will decrease. The interested reader is referred to Sabine (1985) and Sabine et al. (1988).

2.3.6 Overspill effect

In many diffraction geometries it is important that the incident beam remains smaller than the sample area at all angles in order to ensure the constant illumination volume condition (in the case of an infinitely thick specimen). This is particularly important in Bragg–Brentano geometry. Nevertheless, at low angles it is common for the irradiated area to become greater than the area covered by the sample on the sample holder. This "overspilling" reduces the intensities up to the diffraction angle at which the two areas are identical (Figure 2.30).

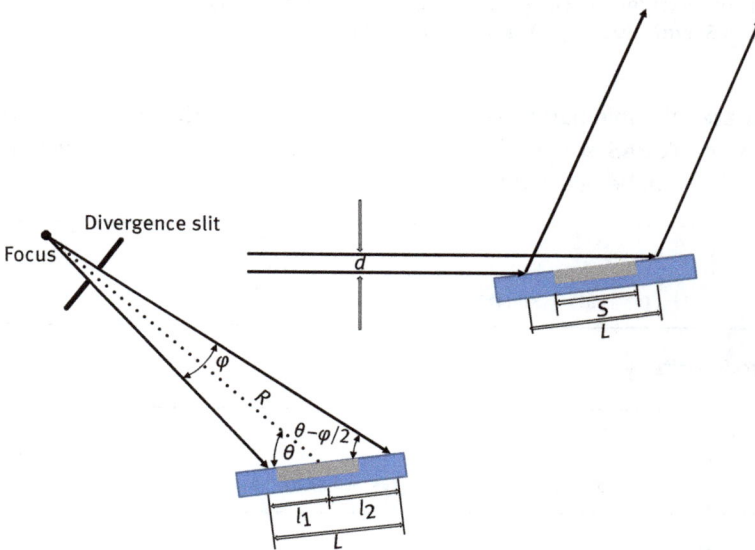

Figure 2.30: Irradiated length on the surface of a flat plate sample in Bragg–Brentano geometry for a divergent beam (left) and a perfectly parallel beam (right).

For divergent beam Bragg–Brentano geometries with a tube opening angle φ, which is determined by the divergence slit, the irradiated length calculates to:

$$L = l_1 + l_2 = \frac{R \sin \frac{\varphi}{2}}{\sin\left(\theta + \frac{\varphi}{2}\right)} + \frac{R \sin \frac{\varphi}{2}}{\sin\left(\theta - \frac{\varphi}{2}\right)} \simeq \frac{R\ \varphi[\text{rad}]}{\sin \theta} \tag{2.48}$$

with the goniometer radius R (Figure 2.30, left) (Fischer, 1996, Krüger & Fischer, 2004, Pecharsky & Zavalij, 2003). In the case of small divergence the beam can be regarded as quasi-parallel and the term $R\ \varphi$[rad] corresponds to the thickness of the beam d (Figure 2.30, right).

An intensity correction factor as a function of the diffraction angle can thus be calculated for a sample length S (Figure 2.31):

$$Ov = \frac{S}{L_D} \ for \ 0 \le 2\theta[\text{rad}] \le 2\arcsin\left(R\frac{\varphi}{S}\right). \tag{2.49}$$

In TOPAS this simple overspill correction for (quasi) parallel beam geometry can be realized by:[16]

```
prm !len 3 min 0.0001 'length of sample
prm !wid 1 min 0.0001 'beam width
scale_pks = if ( Sin(Th) < wid/len, len/(wid/sin(Th), 1);
```

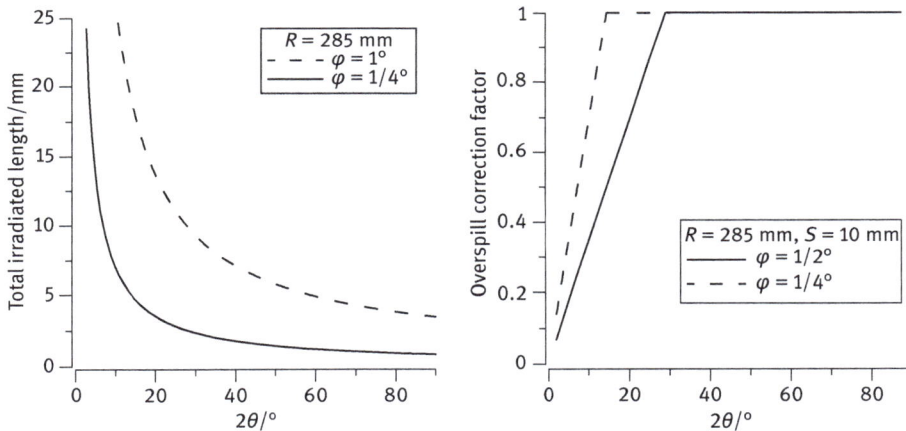

Figure 2.31: Left: Irradiated length on the surface of a flat plate sample in Bragg–Brentano geometry with a divergent beam for different opening angles φ. Right: Corresponding intensity correction function for the overspill effect for a sample length of 10 mm.

Alternatively, automatic variable divergence slits can be used during the measurement, which have a small opening at low 2θ and then widen as a function of 2θ. In TOPAS the correction of the peak intensity can be accomplished by scaling peaks by a θ-dependent sine function:

---

16 The Divergence_Sample_Length macro in TOPAS accommodates both overspill and the flat surface aberration. In other words the intensity, peak shift and peak shape is accounted for.

```
scale_pks = Sin(Th);
```

The corresponding TOPAS macro is:

```
Variable_Divergence_Intensity
```

We recommend not using variable slits in Rietveld work as any imprecision in slit opening will influence intensities, and there is a progressive deterioration of the parafocussing condition causing the resolution to continuously decrease with increasing 2θ. Moreover, the peak asymmetry changes (for a correction function see Section 2.5.8).

2.3.7 Preferred orientation

The idea of powder diffraction is based on the perfect randomness of the orientations of the crystallites as pointed out as early as in 1917 by Hull. Experimentally, this is only easily realized in the case of spherical crystallites. If needle or plate-like crystallites are prepared in flat plate sample holders for reflection geometry or between foils in transmission geometry, the crystallites tend to align themselves in one or more preferred orientation(s). If the corresponding lattice planes are in reflection condition, their intensities are strongly increased (Figure 2.32). A detailed introduction into the topic is given by Pecharsky and Zavalij (2009).

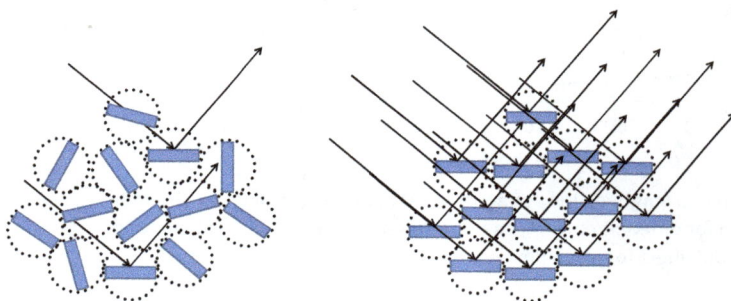

Figure 2.32: Schematic drawing of the effect of preferred orientation (in two dimensions). Left: Randomly oriented crystallites with a proportional number of crystallites in reflection condition. Right: preferentially oriented crystallites with a disproportional high number of crystallites in reflection condition.

In one simple approach where there is a single preferred orientation direction, the angle between the reciprocal lattice vector $\mathbf{s}$ of each Bragg reflection and the specific reciprocal lattice vector $\mathbf{s}_{pref}$ of the preferred orientation is calculated using the scalar product:

$$\cos\omega_{\mathbf{s}} = \frac{\mathbf{S}_{pref} \cdot \mathbf{S}}{|\mathbf{S}_{pref}||\mathbf{S}|} \tag{2.50}$$

A correction factor can be calculated by the March–Dollase function (March, 1932) according to which:

$$T_{\mathbf{s}} = \frac{1}{N}\sum_{i=1}^{N}\left(\tau^{2}\cos^{2}\omega_{\mathbf{s}}^{i} + \tau^{-1}\sin^{2}\omega_{\mathbf{s}}^{i}\right)^{-3/2}, \tag{2.51}$$

where the sum runs over all N symmetry equivalent reciprocal lattice points and τ is the refined preferred orientation parameter, which is defined as the ratio between the correction factors for Bragg peaks perpendicular and parallel to the direction of the preferred orientation.

In TOPAS the code for a single direction of preferred orientation (here 100) is:

```
str_hkl_angle ang1 1 0 0
prm tau min 0.0001 max 2
scale_pks = Multiplicities_Sum(((tau^2 Cos(ang1)^2 + Sin(ang1)^2 / tau)^(-1.5)));
```

Alternatively, the following macros for one or two preferred orientation directions can be used:

```
PO(@, c, ang, hkl)
PO_Two_Directions(c1, v1, ang1, hkl1, c2, v2, ang2, hkl2, w1c, w1v)
```

In a more general approach, a symmetry-adapted spherical harmonic expansion[17] for fixed sample orientation can be used:

$$T_{\mathbf{s}} = 1 + \sum_{l=2}^{L}\left(\frac{4\pi}{2l+1}\sum_{m=-l}^{l}C_{l}^{m}k_{l}^{m}(\mathbf{s})\right), \tag{2.52}$$

where C_{l}^{m} and $k_{l}^{m}(\mathbf{s})$ are the harmonic coefficients and factors (Järvinen, 1993), respectively, and L is the order of the spherical harmonic. Due to the inversion symmetry present in any powder pattern, only even orders need to be taken into account. L in TOPAS can be 2, 4, 6 or 8. A parameter, J, quantifying the magnitude of the preferred orientation can be calculated with:

---

17 See Chapter 13 (Mathematical basics).

$$J = 1 + \sum_{l=2}^{L} \left(\frac{1}{2l+1} \sum_{m=-l}^{l} |C_l^m|^2 \right) \tag{2.53}$$

with $J \geq 1$ (unity in case of random orientation).

In TOPAS, spherical harmonics functions to scale intensities can be applied using:

```
spherical_harmonics_hkl sh
sh_order 8
scale_pks = sh;
```

or alternatively by the predefined macro:

```
PO_Spherical_Harmonics(sh, order)
normals_plot = sh;normals_plot_min_d .25
```

where *sh* is the parameter and *order* the order of the spherical harmonics function. A three-dimensional plot of the spherical harmonics function can be performed by using the keyword *normals_plot* (Figure 2.33).

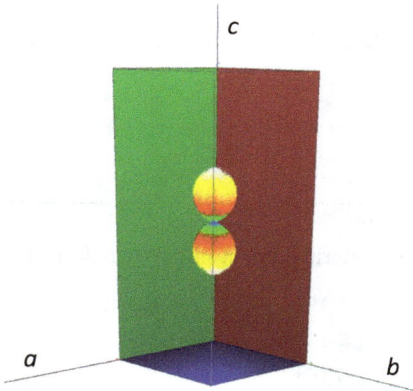

Figure 2.33: Three-dimensional second-order spherical harmonic representation of the preferred orientation correction of a 2H graphite sample measured in flat plate mode, showing strong preferred orientation along the *c*-direction.

Preferred orientation should be distinguished from graininess, where a small number of large oriented crystallites will lead to incorrectly measured intensities. There is no simple correction for graininess, except for collecting data on a better-prepared sample.

2.4 The scale factor

The scale factor is a linear, phase-specific factor that absorbs the constants of all intensity correction factors. It is specific for a particular instrumental configuration and depends on the incident intensity and measuring time. In the case of multi-phase Rietveld refinement using Bragg–Brentano geometry, the scale factor can be used for full standardless quantitative phase analysis, based on the following equation:

$$X_p = \frac{S_p (ZMV)_p \mu_m^*}{K} \tag{2.54}$$

where X_p is the relative weight fraction of the phase p in a mixture of several crystalline phases, Z is the number of formula units of phase p in the unit cell, M is the molecular mass of the formula unit of phase p, V is the volume of the unit cell in Å^3 of phase p and S_p is the scale factor of phase p. K is a scaling factor, which depends on the instrumental conditions and is independent of sample and phase related parameters. The mass absorption coefficient of the entire sample is μ_m^*. Quantitative Rietveld refinement can be performed without knowledge of K and μ_m^*, since the instrumental conditions and the absorption coefficient enter the equation as constants and are identical for all phases. Therefore, in the case of a multiphase mixture, the scale factor is directly related to the weight fraction X_α of the phase α and can be used for quantitative phase analysis according to:

$$X_\alpha = \frac{S_\alpha \rho_\alpha}{\sum_p \left(S_p \rho_p \right)} \tag{2.55}$$

with the density of a single phase ρ_α (in g/cm^3), which can easily be calculated according to:

$$\rho_\alpha = \frac{Z_\alpha M_\alpha \cdot 1.66055}{V_\alpha}. \tag{2.56}$$

Correction factors like the Brindley correction for spherical particles (Brindley, 1945) are often applied for mixtures with very different absorptions. They should be used with extreme caution. A more detailed analysis on the scale factor and its importance for quantitative phase analysis is given in Chapter 5.

2.5 The peak profile

In general, the profile $\Phi(X)$ of a Bragg reflection centered at the peak position x_0 can be regarded as a mathematical convolution[18] of contributions from the instrument, the so-called instrumental resolution function $IRF(X)$, and from the microstructure $MS(X)$ of the sample (Klug & Alexander, 1974):

---

18 See Chapter 13 (Mathematical basics).

$$\Phi(X) = (IRF \circ MS)(X) \tag{2.57}$$

Here $X = x - x_0$, where x_0 is the observed peak position on the scale (2θ, TOF, E) in which the data are recorded. The profile function is therefore described relative to the peak center $x_0(s)$ (i.e., what we called $2\theta_s$ in Section 2.1; more details in Chapter 4) for each phase p. In the present section, the indices s and p are omitted for simplicity.

The instrumental resolution function is split into contributions coming from the finite width of the X-ray source (X-ray tube or synchrotron), the so-called emission profile (EP), and a series of horizontal and vertical instrumental aberrations of the diffractometer. These include the angular acceptance function of the Soller slit(s) controlling the axial beam divergence, the angular acceptance function of the plug-in slit controlling the equatorial beam divergence, the angular acceptance function of the receiving slit and so on. For linear position sensitive detector (PSD) systems, the receiving slit aberration is replaced by functions describing the defocusing due to asymmetric diffraction, the parallax error and the point spread function of the detector.

The microstructure contribution of the sample contains contributions from effects like domain size, isotropic and/or anisotropic microstrain, dislocations, faulting and so on.

In the following section, several mathematical functions that are frequently used to approximate the effect of aberrations from the instrument and from the sample to the profile of a Bragg reflection are discussed in more detail. As most people use constant wavelength X-ray or neutron data, the following examples are given on a 2θ-scale and we explicitly write $\Phi(X) = f(2\theta - 2\theta_0)$.

All functions need to be normalized to unity:

$$\int_{-\infty}^{+\infty} \Phi(X)dX = 1 \tag{2.58}$$

in order not to alter the integrated intensity of the Bragg reflections. However, reflection-independent deviations from an integral of unity will only affect the value of the scale factor.

The different mathematical functions that contribute to the peak profile can be either of phenomenological or physical nature. When the parameters of these functions are directly related to geometrical properties of the diffraction experiments, they are called *fundamental parameters* (*FP*) and the convolution procedure the *fundamental parameter approach*. The convolution in TOPAS is either done numerically in direct space or by the product of the Fourier transforms in reciprocal space. In the following sections, a list of common functions is explained. A more detailed analysis on the influence of microstructural properties on the peak profile is given in Chapter 4 (Peak shapes: Instrument ∘ microstructure).

2.5.1 The box function

There are several aberrations that are commonly described by a box function. These include the size of the source in the equatorial plane, thickness of sample surface as projected onto the equatorial plane, width of the receiving slit in the equatorial plane, width of strips in position sensitive strip detectors, and so on. A box function of width a is defined as:

$$\text{box}(X) = \begin{cases} A \text{ for } -\frac{a}{2} < (X) < \frac{a}{2} \\ 0 \text{ for}(X) \leq -\frac{a}{2} \text{ and } (X) \geq \frac{a}{2} \end{cases} \tag{2.59}$$

with the normalization $A=1/a$ (Figure 2.34). The Fourier transform of a box function with the reciprocal variable h is calculated as:

$$\text{BOX}(h) = Aa\frac{\sin(\pi ha)}{\pi ha} \tag{2.60}$$

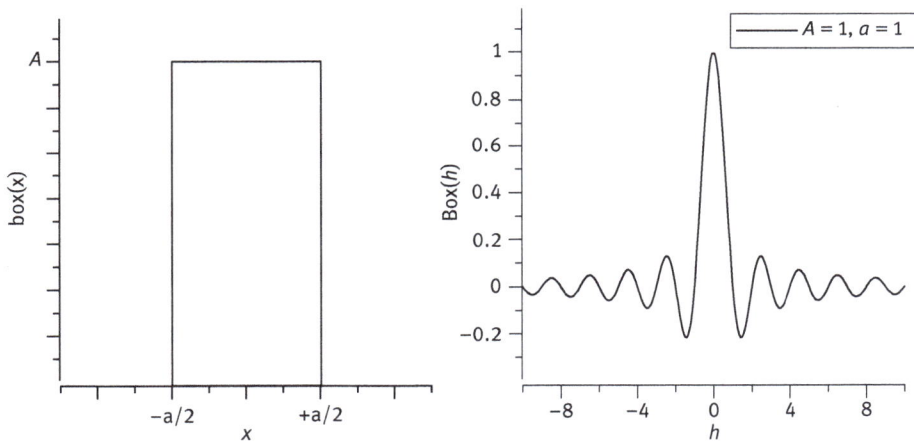

Figure 2.34: The box function (left) and its Fourier transform (right).

The following TOPAS code defines a box function (called *hat* in TOPAS) mimicking the transmittance of a rectangular slit with a width c of 0.1 mm:

```
prm !c 0.1 min 0.000001
hat = Rad c / Rs;
```

with the sample to detector distance Rs. The box function is then automatically convoluted with all Bragg reflections. The predefined equivalent macro is:

```
Slit_Width( , 0.1)
```

A delta function δ can be simulated by using a very small slit width of, for example, 10^{-5}.

2.5.2 Gaussian distribution

The normal (or Gaussian) distribution is a very common continuous probability distribution (Figure 2.35). Physical quantities that are expected to be the sum of many independent processes (such as measurement errors) often have distributions that are nearly normal. The expression for a normalized Gaussian distribution in terms of its full width at half maximum *fwhm* is:

$$\mathrm{gauss}(X) = \frac{2\sqrt{\ln(2)/\pi}}{fwhm} \mathrm{e}^{-4\ln(2)\left(\frac{X}{fwhm}\right)^2} \tag{2.61}$$

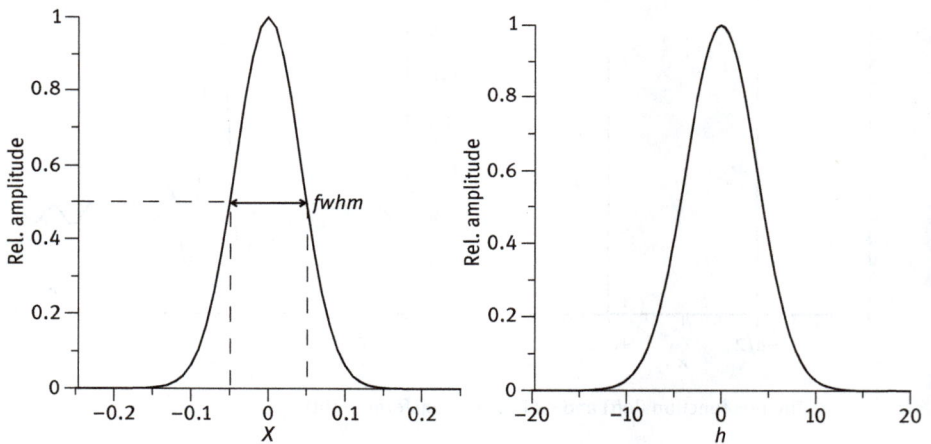

Figure 2.35: The Gaussian function (left) and its Fourier transform (right).

The Fourier transform of a Gaussian function is itself a Gaussian function:

$$\mathrm{GAUSS}(h) = \mathrm{e}^{-\frac{\pi^2 fwhm^2}{4\ln 2}h^2} \tag{2.62}$$

In TOPAS, a convolution of a Gaussian function with a constant *fwhm* of 0.1° 2θ into all Bragg reflections of the powder pattern can be achieved by:

```
prm fwhm 0.1
user_defined_convolution = (2 Sqrt(Ln(2)/Pi)/fwhm) Exp(-4 Ln(2)(X/fwhm)^2);
```

where X is the measured x-axis (usually 2θ). Since the Gaussian function in TOPAS is predefined, a simpler expression can be used:

```
prm c 0.1 min 0.00001
gauss_fwhm = c;
```

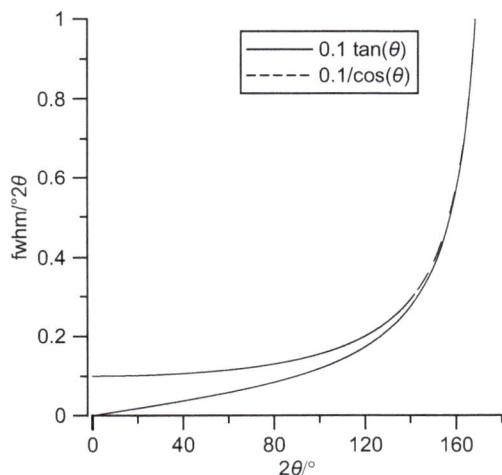

Figure 2.36: Angular dependence of the *fwhm* broadening according to microstrain (tan θ dependence) and crystallite size (1/cos θ dependence).

The dependence of peak widths on θ can be introduced in various ways depending on the origin of the broadening (Figure 2.36). In the case of microstrain the dependence is usually on tan θ:

```
prm c 0.1 min 0.00001
gauss_fwhm = c Tan(Th);
```

while for crystallite size it is $1/\cos\theta$:

```
prm c 0.1 min 0.00001
gauss_fwhm = c / Cos(Th);
```

In order to get physically meaningful values for crystallite size and microstrain, the parameter c must be appropriately scaled as explained in Chapter 4.

Alternatively, a Fourier transform (FT) of a response function (here a Gaussian function) can be convoluted on to peaks using a Fast Fourier Transform (FFT):

```
prm !fwhm 0.1 min 0.00001
ft_conv = Exp(-(Pi FT_K fwhm)^2 / (4 Ln(2)));
ft_min = 1e-8;
ft_x_axis_range = 40 fwhm;
```

FT_K is a reserved parameter name and it returns the Fourier transform variable divided by the *X*-axis range of the peak with extension *ft_x_axis_range*. *ft_min* defines the smallest value to which the transform is calculated.

2.5.3 Cauchy (Lorentz) distribution

The normalized Cauchy or Lorentz distribution (Figure 2.37) is defined as:

$$lorentz(X) = \frac{2\pi/fwhm}{1 + 4\left(\frac{X}{fwhm}\right)^2} \tag{2.63}$$

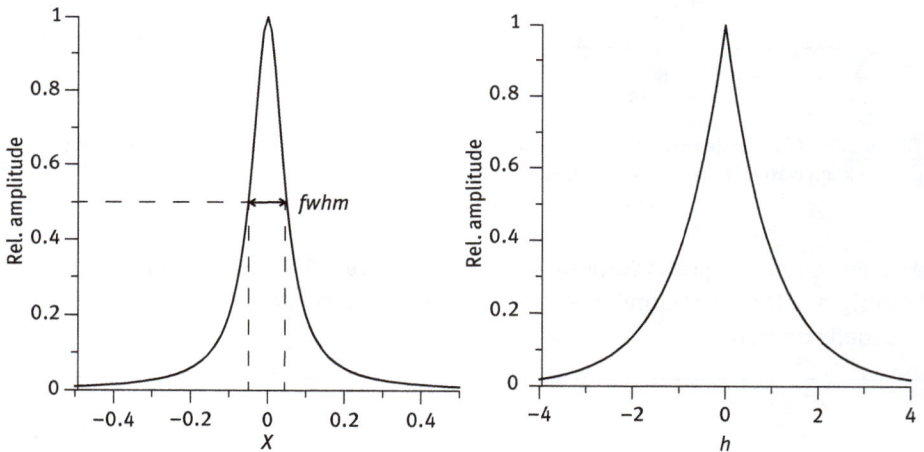

Figure 2.37: The Lorentzian function (left) and the real part of its Fourier transform (right).

The real part of the Fourier transform of a Lorentzian is:

$$LORENTZ(h) = e^{-2\pi\,fwhm\,|h|}. \tag{2.64}$$

The Lorentzian function is commonly used to describe the emission profile from an X-ray tube, as well as crystallite size and strain effects from the sample. Also, in a perfect infinite crystal, Bragg peaks are not δ-functions, but finite Lorentzians with the *fwhm* being the Darwin width.

By analogy with the Gaussian function, convolution of a Lorentzian function with a constant *fwhm* of 0.1° 2θ into all Bragg reflections of the powder pattern can be achieved in TOPAS by:

```
prm fwhm 0.1
user_defined_convolution = (2 Sqrt(Ln(2)/Pi)/fwhm) Exp(-4 Ln(2)(X/fwhm)^2);
```

or

```
prm c 0.1 min 0.00001
lor_fwhm = c;
```

Microstrain Lorentzian broadening can be described with:

```
prm c 0.1 min 0.00001
lor_fwhm = c Tan(Th);
```

and crystallite size with :

```
prm c min 0.00001
lor_fwhm = c /Cos(Th);
```

Again, in order to get physically meaningful values for crystallite size and microstrain, the parameter c must be appropriately interpreted.

2.5.4 The Voigt distribution

The Voigt distribution (Figure 2.38) can be regarded as the convolution of a Gaussian and a Lorentzian:

$$voigt(X) = gauss(X) \circ lorentz(X). \tag{2.65}$$

Convolution of a Voigt function with a constant *fwhm* of 0.1° 2θ for the Gaussian and for the Lorentzian part into all Bragg reflections of the powder pattern can be achieved by:

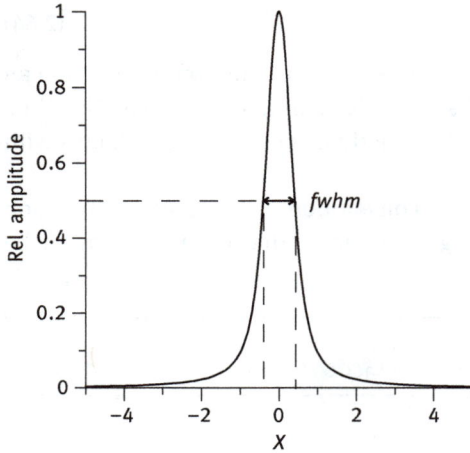

Figure 2.38: Peak profile of the Voigt function with a Lorentzian and a Gaussian *fwhm* of 0.5 each.

```
prm lfwhm   0.1  min 0.00001
prm gfwhm   0.1  min 0.00001
lor_fwhm    = lfwhm;
gauss_fwhm = gfwhm;
```

The assumption that microstrain and crystallite (domain) size generally contain Gaussian as well as Lorentzian contributions leads to the so-called double-Voigt Approach (Balzar, 1999), where the Gaussian and Lorentzian components of two Voigt functions are refined to represent domain size (*csg, csl*) and microstrain (*sg, sl*), respectively. Although this is a simplification, as no physically based models are used, its application is quite common:

```
prm csg 200
prm csl 200
prm sg  0.1
prm sl  0.1
gauss_fwhm = 0.1 Rad Lam / (Cos(Th) csg);
lor_fwhm   = 0.1 Rad Lam / (Cos(Th) csl);
gauss_fwhm = sg Tan(Th);
lor_fwhm   = sl Tan(Th);
```

Physically, more meaningful values for average crystallite size and microstrain can be deduced by calling the following TOPAS code with the refined parameters from above:

```
prm !k =    1
prm !kf =   0.89
prm  lvol   0
prm  lvolf  0
prm  e0     0
LVol_FWHM_CS_G_L(k, lvol, kf, lvolf,,  csg,,  csl)
e0_from_Strain(e0,,  sg,,  sl)
```

where *lvol* is the integral breadth based volume-weighted column height with shape factor *k* (here fixed to 1), *lvolf* is the *fwhm* based volume-weighted column height with shape factor *kf* (defaults to 0.89) and *e0* as the resulting normalized strain.

Voigt (or pseudo-Voigt) functions are also useful for describing the sample contribution to the peak shape in *TOF* experiments. The *fwhm* needs to be converted to *TOF* space:

```
prm lor 0.5 min 0 max 1.0
prm fwhm 0.1
peak_type pv pv_lor lor pv_fwhm = fwhm Constant(t1 0.00001) D_spacing;
```

where *pv_lor* and *pv_fwhm* are the Lorentzian fraction and the *fwhm* of the peak profile, respectively. Note that *t1* is the linear calibration parameter from the *TOF_x_axis_calibration* macro (see Section 2.2.1).[19]

The corresponding predefined macro in TOPAS is:

```
TOF_PV(@, 283.50627`, @, 0.0248895257, t1)
```

Lorentzian and Gaussian size broadening can be similarly convoluted as shown earlier in this chapter:

```
prm  cl 10 min 3 max 10000
prm  cg 10 min 3 max 10000
lor_fwhm   = Constant(t1 .1) D_spacing^2 / cl;
gauss_fwhm = Constant(t1 .1) D_spacing^2 / cg;
```

The same can be achieved by the two predefined macros:

```
TOF_CS_L(@, 10, t1)
TOF_CS_G(@, 10, t1)
```

---

19 The expression *Constant(var)* assigns a fixed value to the quantity *var* at the start of refinement. It is not updated if *var* changes.

2.5.5 The TCHZ pseudo-Voigt function

Another symmetric peak profile function often used for angular dispersive data is the modified Thompson–Cox–Hastings pseudo-Voigt "TCHZ" (Thompson et al., 1987, Young, 1993) that is defined in Chapter 13 (Mathematical basics). The Gaussian $fwhm_G$ and Lorentzian $fwhm_L$ are defined as:

$$fwhm_G = \sqrt{U\tan^2\theta + V\tan\theta + W + \frac{Z}{\cos^2\theta}}$$

$$fwhm_L = X\tan\theta + \frac{Y}{\cos\theta} \tag{2.66}$$

where U, V, W, X, Y and Z are refineable parameters. The overall $fwhm$ can be calculated using eq. (13.89). It should be apparent that U and X are related to microstrain while Z and Y are related to domain size. The shape of the TCHZ pseudo-Voigt function looks practically identical to that of the Voigt function (Figure 2.38). One advantage of the TCHZ pseudo-Voigt function, in contrast to the Voigt function, is the ability to easily report the $fwhm$ of the fitted Bragg reflections. The predefined macro for the TCHZ pseudo-Voigt function in TOPAS is:

```
TCHZ_Peak_Type(pku, 0.0107, pkv, -0.0011, pkw,  0.0005, !pkz, 0, pkx, 0.0418, !pky, 0)
'U, V, W, Z, X, Y
```

In this example, U, V, W and X are refined, while Y and Z are set to zero (no size contribution).

2.5.6 The circles function

A simple approximate function useful for modeling the asymmetry of a Bragg reflection is the so called circles function with the curvature ε_m as an adjustable parameter (Figure 2.39, left):

$$\text{circles}(X) = 1 - \sqrt{\left|\frac{X_m}{X}\right|} \text{ for } 0 \le X \le X_m \tag{2.67}$$

One of the main applications for this function is the phenomenological modeling of the peak asymmetry caused by axial divergence which is mainly due to the increasing curvature of the Debye–Scherrer rings at very low and extremely high angles that are cut by (typically) rectangular receiving slits of finite width (Cheary & Coelho, 1998). In order to model axial divergence with the circles function, a $\tan(2\theta)$-dependence is usually used as given below in

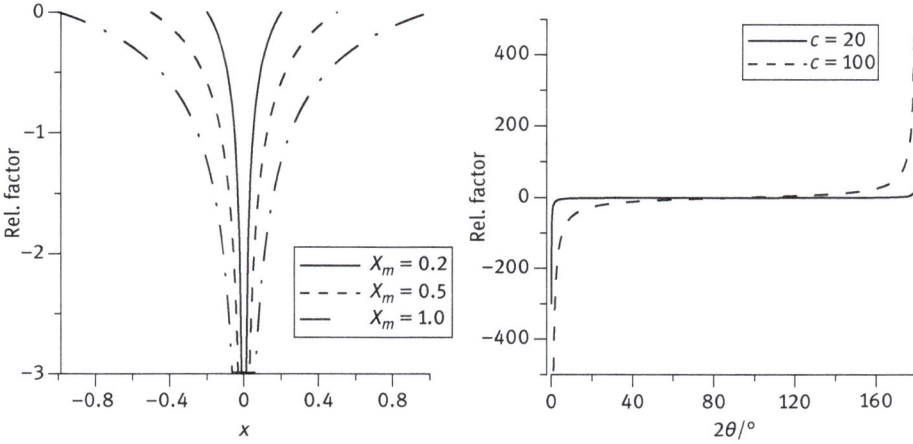

Figure 2.39: Circle function different curvature ε_m (left) and dependence of the parameter ε_m on diffraction angle as typical for axial divergence (right).

TOPAS notation. Please note that the asymmetry is reversed above 90° 2θ (Figure 2.39, right)[20]:

```
prm  c 12 min 0 max 20
circles_conv = -.5 Rad ( c / Rs)^2 / Tan(2 Th);
```

The corresponding predefined TOPAS macro is:

```
Simple_Axial_Model(@, 12)
```

2.5.7 The exponential function

Asymmetry of Bragg reflections can also be described by an exponential asymmetry decay function of the type[21]:

$$exp\_conv\_const = e^{\frac{X}{X_m}\ln 0.001} \quad \text{for } 0 \le X \le X_m \qquad (2.68)$$

---

20 In reality, the asymmetry due to axial divergence has its minimum at around 120° 2θ (Cheary & Coelho, 1998).
21 ln(0.001) is a scaling factor.

that is convoluted into the peak profile. This function can be useful for describing, for example, the highly asymmetric instrumental peak shape of *TOF* data or the effects of transparency on the peak shape in Bragg–Brentano geometry.

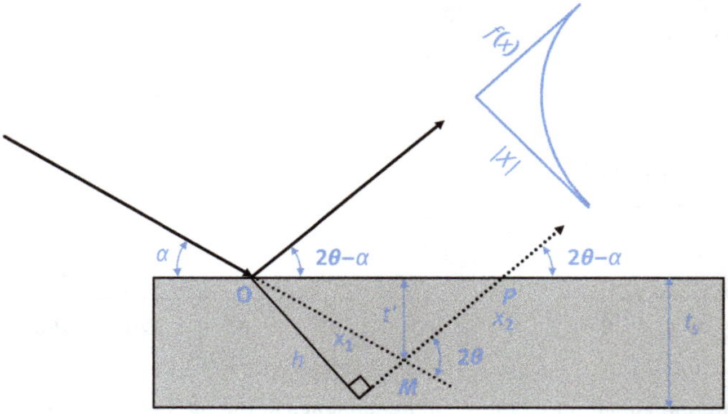

Figure 2.40: Schematic drawing of the aberration due to the transparency effect for (a)symmetric flat plate reflection geometry (in degrees).

In the following, we derive the peak profile aberration due to the transparency effect following the treatment given by Masson et al. (1996) and Rowles and Buckley (2017). Figure 2.40 is a schematic drawing of the aberration due to the transparency effect for (a)symmetric flat plate reflection geometry of an "infinitely thick" sample. The intensity diffracted at point M is reduced by absorption with respect to the intensity diffracted at point O according to eq. (2.37) where the additional pathway x traveled by the X-ray in the sample is:

$$x = x_1 + x_2 = \frac{t'}{\sin\alpha} + \frac{t'}{\sin(2\theta - \alpha)} \tag{2.69}$$

with:

$$t' = x_1 \sin\alpha = \frac{h}{\sin(2\theta)}\sin\alpha \tag{2.70}$$

that leads to:

$$x = \frac{h}{\sin(2\theta)}\left(1 + \frac{\sin\alpha}{\sin(2\theta - \alpha)}\right). \tag{2.71}$$

The angular variable X (in degrees) is here defined by:

$$X = 2\theta' - 2\theta \approx \frac{h}{R_s}\frac{180}{\pi} \tag{2.72}$$

with the angle $2\theta'$ where diffraction is observed at the detector and the specimen–detector distance R_s with $R_s \gg h$. The intensity ratio (eq. (2.37)) thus becomes:

$$\frac{I}{I_0} = \exp\left(-\mu\left[\frac{R_s}{\sin(2\theta)}\left(1+\frac{\sin\alpha}{\sin(2\theta-\alpha)}\right)\frac{\pi}{180}\right]\right). \tag{2.73}$$

After introducing the substitution:

$$-\mu x = \frac{X}{\delta} \tag{2.74}$$

with:

$$\delta = \left[\frac{\mu R_s}{\sin(2\theta)}\left(1+\frac{\sin\alpha}{\sin(2\theta-\alpha)}\right)\right]^{-1}\frac{180}{\pi} \tag{2.75}$$

the normalized absorption profile change induced by the transparency effect is then given by:

$$f(X) = \begin{cases} \frac{1}{\delta}\exp\frac{X}{\delta} & X \le 0 \\ 0 & X > 0 \end{cases} \tag{2.76}$$

Figure 2.41 shows several normalized absorption profiles for different diffraction angles and linear absorption coefficients due to the transparency effect for an infinitely thick specimen in asymmetric flat plate reflection geometry ($\alpha = 10°$). Symmetric flat plate reflection geometry simply leads to a less pronounced asymmetry of the absorption profile.

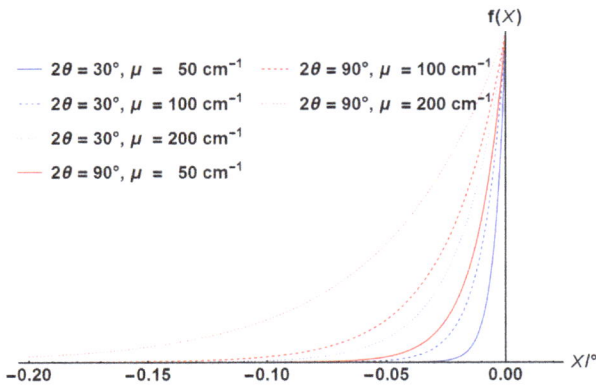

Figure 2.41: Normalized absorption profiles for different diffraction angles and linear absorption coefficients describing the transparency effect for infinitely thick specimen in asymmetric flat plate reflection geometry with an angle between incoming X-rays and specimen normal of 10°.

For thin samples, where the X-ray beam does not get fully absorbed within the sample, the absorption profile will be truncated at X_{min}, which can be calculated according to Rowles and Buckley (2017) by:

$$X_{min} = -\frac{t_S}{R_S}\frac{\sin(2\theta)}{\sin(\alpha)}\frac{180}{\pi}. \tag{2.77}$$

The dependence of X_{min} on the scattering angle for different thicknesses of the sample is shown in Figure 2.42.

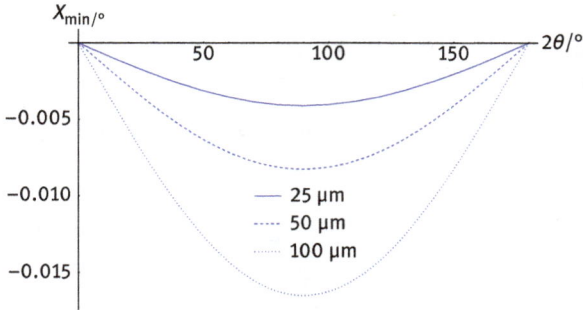

Figure 2.42: Angular dependence of the X_{min} cut-off parameter due to transparency for different thickness of specimen in asymmetric flat plate reflection geometry with an angle between incoming X-rays and specimen normal of 10°.

In TOPAS language, the transparancy effect can be described as follows by the exponential convolution function (*exp_conv_const*):

```
prm c 1 min 3 max 500
exp_conv_const = Ln(0.001) Sin(2 Th) / ( c Rs 0.003490658 );
  ' Note: 0.003490658 = 2 Deg / 10 used for scaling purpose
```

Several predefined macros exist for transparency correction:

```
'scaling factor c, sample thickness d in mm.
 Absorption_With_Sample_Thickness_mm_Shape(@, c, @, d)
 Absorption_With_Sample_Thickness_mm_Intensity(@, c, @, d)
 Absorption_With_Sample_Thickness_mm_Shape_Intensity(@, c, @, d)
```

The significant instrumental asymmetry of *TOF* peaks (Figure 2.43) can also be satisfactorily modeled by convoluting one or more exponential functions into the peak profile. To do this in TOPAS, the usual transformations to TOF space must be performed as:

```
prm a0      774.0  min = Max(Val .3, 1e-6); max = 2 Val + 1;
prm a1         0   min = Max(Val .3, 1e-6); max = 2 Val + 1;
prm wexp     1
prm lr       1            ' can be +1 or -1
exp_conv_const = lr Constant(t1) / ( a0 + a1 / D_spacing^wexp );
```

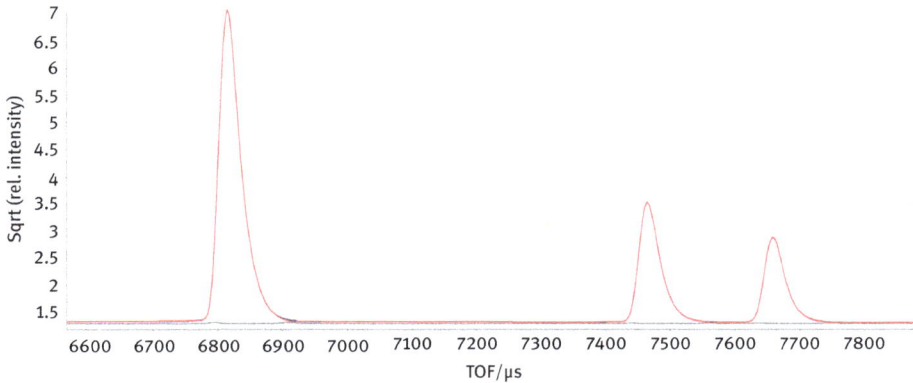

Figure 2.43: Asymmetric peak profile of *TOF* data which was modeled by a pseudo-Voigt sample profile convoluted with two exponential functions one for the left and one for the right side of the reflection.

The predefined macro in TOPAS is:

```
TOF_Exponential(a0, 774.0,,0, 1, t1, +)
```

2.5.8 The 1/X function

Another way to describe asymmetry of a Bragg reflection is using a 1/X decay function of the type:

$$one\_on\_x = \frac{4}{\sqrt{|X\ X_{min}|}} \quad \text{for} \quad x = 0 \text{ to } X_{min} \tag{2.78}$$

where the parameter X_{min} specifies the relative extension of the function on the X axis and can be either positive or negative. It can, for example, be used to describe the effect of divergence slits (either fixed or variable) on the peak profile for angular dispersive data.

In TOPAS, the corresponding function is called *one_on_x_conv*. The angular dependence of the 1/X convolution function for a fixed divergence slit can be defined as (Figure 2.44, right):

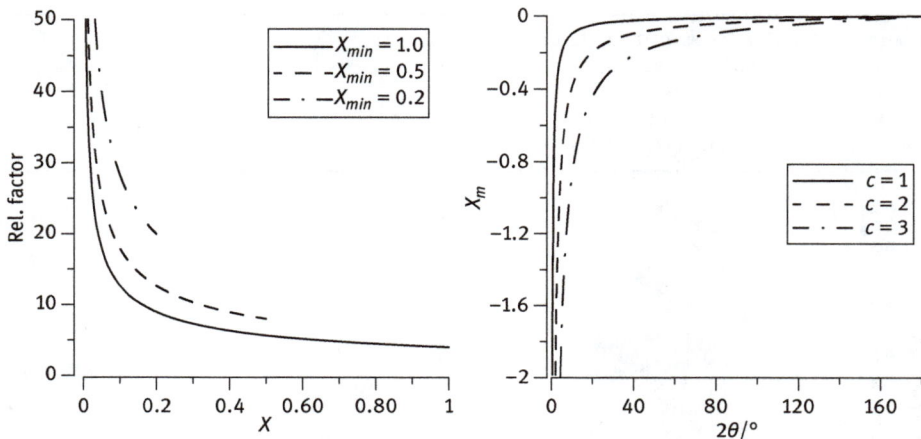

Figure 2.44: $1/X$ correction function (left) and angular dependence of the correction function for modeling the peak asymmetry due to divergence with a fixed divergence slit.

```
prm c 1 min  0.0001 max 6
one_on_x_conv = -c^2 Deg_on_2 / Tan(Th);
```

or alternatively with the predefined macro:

```
Divergence(@, v)
```

The angular dependence for a variable divergence slit is applied with:

```
prm c min 0.0001 max 60
one_on_x_conv = -c^2 Sin(2 Th) Rad/(4 Rs^2);
```

or alternatively with the predefined macro:

```
Variable_Divergence_Shape(@, v)
```

In both cases, a single refinable parameter is used.

2.5.9 The emission profile

The emission profile depends strongly on the X-ray radiation source used. For laboratory X-rays it is usually described as a set of one to five Voigt profiles, representing the distribution of wavelengths, depending on the degree of

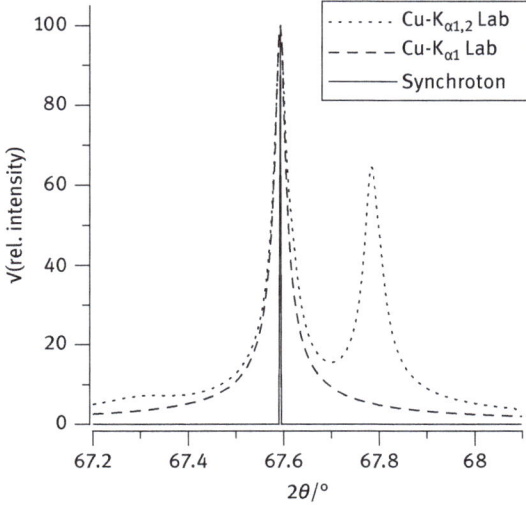

Figure 2.45: Emission profiles for angle dispersive experiments with a wavelength of $\lambda = 1.54059$ Å for a lab source with Ni-filter (dotted), a lab source with primary beam monochromator (dashed) and a synchrotron source (solid) in the angular range between 67.2 and 68.2° 2θ.

monochromatization (Figure 2.45). In addition, tube tails or absorption edges may be present, which need to be modeled separately. A description of the emission profile is the first part of any peak shape description.

In the following text, the parameter names in TOPAS are given in parenthesis. The Voigt lines in the emission profile are defined by wavelength (*lo*), their relative area (*la*), their Gaussian (*lg*) and Lorentzian (*lh*) half widths and the limiting ratio between the maximum and the minimum of a calculated reflection (*ymin_on_ymax*), which determines the x-axis extent over which an emission line is calculated. The area under the emission profile is determined by the sum of the area parameters of the individual lines, which can be scaled to unity. *d*-spacings are calculated from the *lo* value of the emission profile line with the largest *la* value (*Lam*). The following TOPAS script shows the empirical emission profile of Cu-K_α radiation consisting of five Lorentzian lines according to Mendenhall et al. (2017):

```
' official NIST copper kalpha spectrum from JPhysB 2017
lam
    ymin_on_ymax  0.0001
    la sat  0.05026 lo  1.534753  lh  3.6854
    la      3.91    lo  1.5405925 lg  0 lh 0.436
    la      0.474   lo  1.5410769 lg  0 lh 0.558
    la      1.53    lo  1.5443873 lg  0 lh 0.487
    la      0.754   lo  1.5446782 lg  0 lh 0.630
```

For angular dispersive neutron and synchrotron data, a δ-function can be simulated by a Lorentzian profile with a very small half width of, for example, 0.00001 mÅ:

```
Lam
    ymin_on_ymax  0.0001
    la  1 lo  0.2078  lh  0.000001
```

For *TOF* and energy dispersive data, the following macro can be used.

```
TOF_LAM(0.0001)
```

where the argument simply defines *ymin_on_ymax*.

For laboratory instruments, the finite width of the X-ray source is modeled by a simple hat-function. For accurate line profile analysis it may be necessary to modify the simple hat function model to accommodate the so-called tube tails effect (Bergmann et al., 2000) (Figure 2.46). If these "tube tails" are present, the source width aberration function can be better approximated by the combination of a sharp and a broad hat function. The parameters introduced to describe tube tails are the tube filament width [mm], effective width of tube tails in the equatorial plane perpendicular to the X-ray beam in negative and positive z-direction [mm], and the fractional height of the tube tails relative to the main beam:

```
Tube_Tails(, 0.04,,  -1.208450694,,  1.529020892,,  0.00120369033)
```

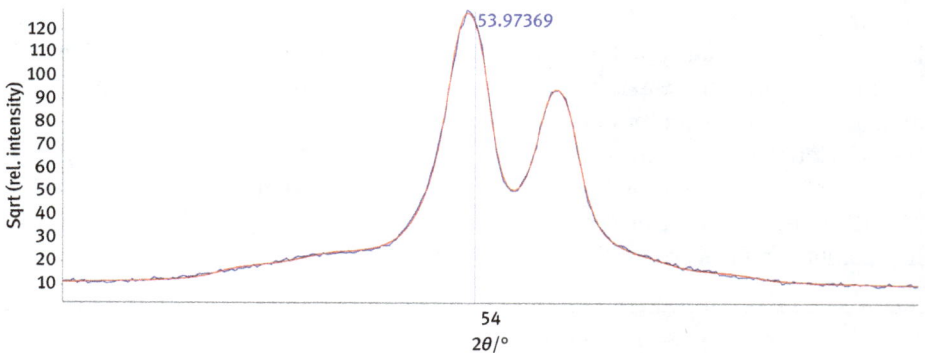

Figure 2.46: TOPAS screen shot showing the fit of a single Bragg reflection of LaB_6 measured with Cu-K_α radiation where the apparent tube tails are fitted using the *Tube_tails* macro.

Metal K_β filters for angle dispersive laboratory data introduce an absorption edge between the K_α and K_β wavelengths, dependent on the wavelength/filter material and

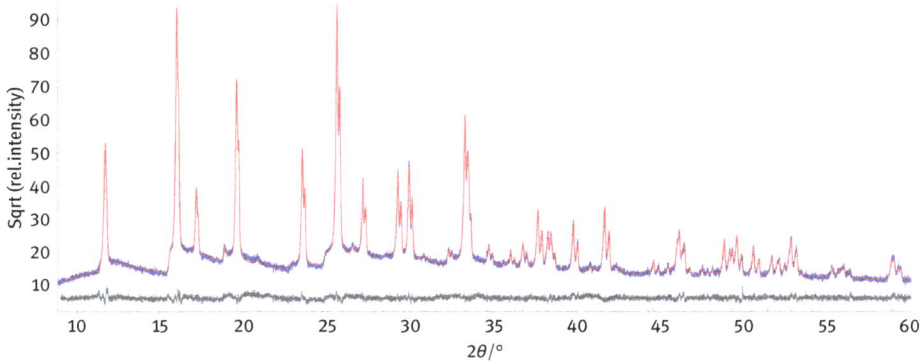

Figure 2.47: Laboratory angle dispersive X-ray powder diffraction pattern of corundum measured with Cu radiation and a Ni-foil K_β filter.

its thickness (Figure 2.47). With position sensitive detectors (PSDs), absorption edges are often visible due to the high signal to noise of the data. To account for this effect, the *Absorption_Edge_Correction* macro can be used in TOPAS. The parameters are: Maximum wavelength (*max_lam*) for extending the source emission profiles to include transmitted K_β and Bremsstrahlung; position of the absorption *edge* of the filter material (edge in Å); remnant Bremsstrahlung curvature (*a_white*, *b_white*), width of the absorption edge (*a_erf*); height of the absorption edge (*cedge_extra*):

```
Absorption_Edge_Correction(
        1.75,       ' max_lam
    @, 1.48668,     ' edge
    @, 0.00121,     ' a_white
    @, 64.51902,    ' b_white
    @, 648.8600,    ' a_erf
    @, 0.00504      ' edge extra
)
```

2.6 The background

The observed background at position i in the powder pattern Bkg_i can be modeled by an analytical or empirical function or it can be manually defined. Sources for the background come from the instrument and the sample, such as disorder, thermal diffuse scattering, incoherent scattering, inelastic scattering and so on.

It is common practice to fit the background with high-order (typically 5–15) orthogonal Chebyshev polynomials of the first kind. The higher the order, the higher are the correlations between background coefficients and between background coefficients and the intensity of overlapping reflections at higher scattering angle. Chebyshev polynomials of first kind are defined by a recursive relation:

$$T_0(x) = 1$$

$$T_1(x) = x$$

$$T_{n+1}(x) = 2xT_n(x) - T_{n-1}(x) \tag{2.79}$$

where the x-axis is normalized between -1 and 1, which is done for an equidistant 2θ-axis according to:

$$x_i = \frac{2(2\theta_i) - (2\theta_{final} + 2\theta_{start})}{2\theta_{final} - 2\theta_{start}} \tag{2.80}$$

The background values are then calculated as:

$$Bkg_i = \sum_{k=0}^{n} c_k T_k(x_i) \tag{2.81}$$

In TOPAS, the predefined keyword *bkg* followed by coefficients of the Chebyshev polynomial can be used (Figure 2.48):

```
bkg @ 0 0 0 0 0 …
```

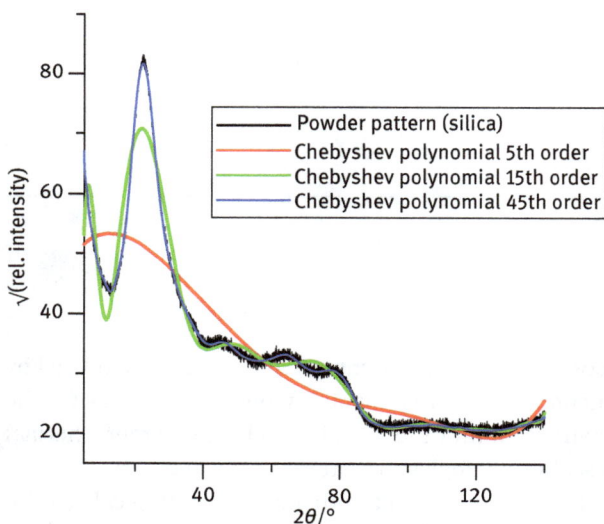

Figure 2.48: Fitting the powder pattern of amorphous silica by the background function in TOPAS using Chebyshev polynomials of different order.

Quite often a steep increase of the background is observed at low scattering angle, particularly if position sensitive detectors with large opening angles are used. Adding a $1/2\theta_i$ term to the Chebyshev polynomial often leads to a reasonable fit:

$$Bkg_i = \frac{c}{2\theta_i} + \sum_{k=0}^{n} c_k T_k(x_i) \tag{2.82}$$

which in TOPAS is done by the predefined macro:

```
One_on_X(@, c)
```

A background can also be defined manually by connecting specified points by straight line segments. In TOPAS, the following list of predefined macros can be used[22]:

```
Bkg_Straight_Line_First(0, x0, y0)
Bkg_Straight_Line(0,  1,  x1,  y1)

...

Bkg_Straight_Line(n-1, n, xn,  yn)
```

The presence of humps in the background due to scattering by amorphous materials (e.g., a glass capillary) can either be modeled by introducing additional artificial reflections that are broadened using a small "crystallite size" or by several (more or less phenomenological) "Debye-like" functions to describe short range order effects (see Section 1.3):

$$Bkg_i = w \frac{\sin(Q_i r)}{Q_i} \tag{2.83}$$

with:

$$Q_i = 2\pi s_i = 4\pi \frac{\sin\theta_i}{\lambda} \tag{2.84}$$

and a refinable weight w and correlation shell radius r. The built-in TOPAS macro "$Bkg\_Diffuse(b, w, bb, r)$" can be used, which can be applied in multiple ways. Another possible background functions is a cosine Fourier series:

$$Bkg_i = \sum_{k=0}^{n} c_k \cos(k2\theta_i) \tag{2.85}$$

which in TOPAS can be realized by[23]:

---

22 Note that the first background point needs to be explicitly defined. $X0$ and Xn refer to the first and last position in the powder pattern, usually $2\theta_{start}$ and $2\theta_{end}$ for angle dispersive diffraction data.
23 The keyword fit_obj fits a user defined function to the observed data (usually a powder pattern).

```
prm bk1 1
prm bk2 1
…
prm bk12 1
fit_obj = bk1 + bk2 Cos(X Deg) +  bk3 Cos(X Deg * 2) + … + bk12 Cos(X Deg * 11);
```

Finally, it is possible to record an experimental background of, for example, an empty capillary and store it as an *xy* ASCII file. This can then be included during the Rietveld fit with the command:

```
user_y back_expt experimental_background.xy 'name then filename
prm back_scale  1.82237'_0.00310
fit_obj = back_scale * back_expt;
Plot_Fit_Obj(back_expt) 'show on screen
```

2.7 The mathematical procedure

In the following, a brief description of the mathematical background of Rietveld analysis is given. More detailed information is given (e.g.) in a book chapter by Robert Von Dreele (2008). The quantity to be minimized (also called the objective function) in Rietveld analysis can be formally written as:

$$S = \sum_{i=1}^{N} w_i \left(y_{obs,i} - y_{calc,i}(\mathbf{p}) \right)^2 \tag{2.86}$$

where $y_{obs,i}$ and $y_{calc,i}$ are the observed and calculated intensities, respectively, at point i in the powder pattern of N data points.

The parameter vector of size P corresponding to the number of independent parameters can be written as:

$$\mathbf{p} = \begin{pmatrix} p_1 \\ \vdots \\ p_P \end{pmatrix}. \tag{2.87}$$

Since $y_{calc,i}(\mathbf{p})$ is a nonlinear function, it must be approximated by a Taylor series which is usually terminated after the first term:

$$y_{calc,i}(\mathbf{p}) \cong y_{calc,i}(\mathbf{p_o}) + \sum_{j=1}^{P} \frac{\partial y_{calc,i}(\mathbf{p_o})}{\partial p_j} \left(p_j - p_{j,0} \right) \tag{2.88}$$

around the initial estimates of the parameters:

$$\mathbf{p_o} = \begin{pmatrix} p_{1,0} \\ \vdots \\ p_{P,0} \end{pmatrix}. \tag{2.89}$$

The vector of the parameter shift is:

$$\Delta\mathbf{p} = \mathbf{p} - \mathbf{p_o}. \tag{2.90}$$

The objective function can then be written as:

$$S = \sum_{i=1}^{N} w_i \left(y_{obs,i} - \left(y_{calc,i}(\mathbf{p_o}) + \sum_{j=1}^{P} \frac{\partial y_{calc,i}(\mathbf{p_o})}{\partial p_j} \Delta p_j \right) \right)^2. \tag{2.91}$$

To find the minimum of the objective function we need the first derivative with respect to the refined parameters, and we introduce subscript k to avoid confusion:

$$\frac{\partial S}{\partial p_k} = -2 \sum_{i=1}^{N} w_i \left(y_{obs,i} - \left(y_{calc,i}(\mathbf{p_o}) + \sum_{j=1}^{P} \frac{\partial y_{calc,i}(\mathbf{p_o})}{\partial p_j} \Delta p_j \right) \right) \frac{\partial y_{calc,i}(\mathbf{p_o})}{\partial p_k}$$

$$= -2 \sum_{i=1}^{N} w_i \left((y_{obs,i} - y_{calc,i}(\mathbf{p_o})) \frac{\partial y_{calc,i}(\mathbf{p_o})}{\partial p_k} - \sum_{j=1}^{P} \frac{\partial y_{calc,i}(\mathbf{p_o})}{\partial p_j} \frac{\partial y_{calc,i}(\mathbf{p_o})}{\partial p_k} \Delta p_j \right). \tag{2.92}$$

At the minimum, the first derivative must be zero:

$$\frac{\partial S}{\partial p_k} = 0, \tag{2.93}$$

from which it follows that:

$$\sum_{i=1}^{N} w_i \sum_{j=1}^{P} \frac{\partial y_{calc,i}(\mathbf{p_o})}{\partial p_j} \frac{\partial y_{calc,i}(\mathbf{p_o})}{\partial p_k} \Delta p_j = \sum_{i=1}^{N} w_i (y_{obs,i} - y_{calc,i}(\mathbf{p_o})) \frac{\partial y_{calc,i}(\mathbf{p_o})}{\partial p_k}. \tag{2.94}$$

Changing the summations on the left side leads to:

$$\sum_{j=1}^{P} \sum_{i=1}^{N} w_i \frac{\partial y_{calc,i}(\mathbf{p_o})}{\partial p_j} \frac{\partial y_{calc,i}(\mathbf{p_o})}{\partial p_k} \Delta p_j = \sum_{i=1}^{N} w_i (y_{obs,i} - y_{calc,i}(\mathbf{p_o})) \frac{\partial y_{calc,i}(\mathbf{p_o})}{\partial p_k}. \tag{2.95}$$

This is equivalent to a linear set of equations in $\Delta\mathbf{p}$:

$$\mathbf{A}\Delta\mathbf{p} = \mathbf{Y} \tag{2.96}$$

with the components of the $P \times P$ matrix $\mathbf{A}$ (each k corresponds to a matrix row and each j corresponds to a column) given by:

$$A_{kj} = A_{jk} = \sum_{i=1}^{N} w_i \frac{\partial y_{calc,i}(\mathbf{p_o})}{\partial p_j} \frac{\partial y_{calc,i}(\mathbf{p_o})}{\partial p_k} \tag{2.97}$$

and the P components of the vector Y by:

$$Y_p = \sum_{i=1}^{N} w_i \left(y_{obs,i} - y_{calc,i}(\mathbf{p_o})\right) \frac{\partial y_{calc,i}(\mathbf{p_o})}{\partial p_k}. \tag{2.98}$$

The equations in $\Delta\mathbf{p}$ of eq. (2.96) are solved for every iteration of refinement. The quantities $\Delta\mathbf{p}$ correspond to the changes in the parameters $\mathbf{p}$ that should minimise eq. (2.86). Unfortunately, due to the Taylor series approximation, the computed shifts $\Delta\mathbf{p}$ don't directly lead to a fully minimized solution but to a hopefully better approximation.[24]

The default algorithm in TOPAS for solving the system of linear equations is the Newton–Raphson nonlinear least squares method with the Marquardt method (1963) included for stability. The Marquardt (1963) method applies a scaling factor to the diagonal elements of the **A** matrix when the solution to the normal equations fails to reduce χ^2:

$$A_{ii,\,new} = A_{ii}(1 + \eta) \tag{2.99}$$

where η is the Marquardt constant. After applying the Marquardt constant the normal equations are solved again and χ^2 recalculated; this scaling process is repeated until χ^2 reduces. The Marquardt constant η is automatically determined each iteration. This determination is based on the actual change in χ^2 and the expected change in χ^2 and the expected change in χ^2. A bound constrained conjugate gradient (BCCG) method (Coelho, 2005) incorporating min/max parameter limits is used for solving the normal equations. Min/max limits are dynamically recalculated during the solution process.

TOPAS allows many options over the algorithms that are used for solving the normal equations. Some of them are listed below:

```
no_normal_equations   ' Prevents the use of the Marquart method
approximate_A          ' Approximate the A matrix without the need for calculating
                       ' the A matrix dot products. Based on the BFGS method (Broyden, 1970;
                       ' Fletcher, 1970; Goldfarb, 1970; Shanno, 1970)
use_LU                 ' LU-decomposition is used instead of BCCG
line_min               ' Steepest decent method
use_extrapolation      ' Parabolic extrapolation of parameters as a function of iteration
A_matrix_memory_allowed_in_Mbytes
A_matrix_elements_tolerance
```

---

24 Least squares can get stuck in a local rather than global minimum if initial parameter approximations are poor. Chapter 7 discusses global optimization protocols for avoiding this.

2.8 Agreement factors

Many different statistical agreement (R-) factors have been proposed for judging the quality of a Rietveld refinement. The most common one is the so-called profile R-factor, which is a measure of the difference between the observed and the calculated profile:

$$R_p = \frac{\sum_{i=1}^{N} |y_{obs,i} - y_{calc,i}(\mathbf{p})|}{\sum_{i=1}^{N} y_{obs,i}} \tag{2.100}$$

This simple sum of all differences relative to the sum of all observed values has several problems. First, it tends to overemphasize the strong reflections and it doesn't take experimental uncertainties into account. Both problems are overcome by applying a weighting scheme, where every data point gets a weight w_i (see below):

$$R_{wp} = \sqrt{\frac{\sum_{i=1}^{N} w_i (y_{obs,i} - y_{calc,i}(\mathbf{p}))^2}{\sum_{i=1}^{N} w_i y_{obs,i}^2}} \tag{2.101}$$

This R-factor is directly related to the Rietveld objective function of eq. (2.86).

The next problem is related to the influence of the background. If the peak to background ratio is low, the profile R-value can be dominated by the well-fitted background points and relatively insensitive to the structural model. To avoid this problem, it is useful to subtract the background from the observed step scan intensities in the denominator:

$$R'_p = \frac{\sum_{i=1}^{N} |y_{obs,i} - y_{calc,i}(\mathbf{p})|}{\sum_{i=1}^{N} |y_{obs,i} - Bkg_i|} \tag{2.102}$$

and

$$R'_{wp} = \sqrt{\frac{\sum_{i=1}^{N} w_i (y_{obs,i} - y_{calc,i}(\mathbf{p}))^2}{\sum_{i=1}^{N} w_i (y_{obs,i} - Bkg_i)^2}} \tag{2.103}$$

Despite these corrections, profile R-values of different refinements can only be compared for identical statistical conditions. The so-called expected R-factor, which is mainly determined by counting statistics, gives a measure of the best possible fit:

$$R_{exp} = \sqrt{\frac{N - P}{\sum_{i=1}^{N} w_i y_{obs,i}^2}} \tag{2.104}$$

and:

$$R'_{exp} = \sqrt{\frac{N - P}{\sum_{i=1}^{N} w_i (y_{obs,i} - Bkg_i)^2}} \tag{2.105}$$

with the number of data points N and the number of parameters P. On an absolute basis, the ratio χ between the weighted profile R-value and the expected R-value (also called goodness of fit, GOF) is a good measure on the quality of the Rietveld refinement:

$$\chi = \frac{R_{wp}}{R_{exp}} = \sqrt{\frac{\sum_{i=1}^{N} w_i (y_{obs,i} - y_{calc,i}(\mathbf{p}))^2}{N - P}} \tag{2.106}$$

A χ between 1 and 1.5 is considered good. For comparison with single crystal data, the Bragg-R-value can be used that is based on integrated reflection intensities rather than step scan intensities:

$$R_{Bragg} = \frac{\sum_{k=1}^{K} |I_{obs,k} - I_{calc,k}|}{\sum_{k=1}^{K} I_{obs,k}} \tag{2.107}$$

$I_{obs,k}$ and $I_{calc,k}$ are the "observed" and calculated intensities of the kth reflection out of K reflections. Rietveld R_{Bragg}-values are often lower than those one would expect in single crystal experiments and should therefore be interpreted with caution. The reason for this is that for overlapping reflections, the intensity is apportioned to individual hkl-reflections according to the ratio of the calculated intensities, averaging out misfits of individual reflection intensities. This leads to a biased or overly optimistic assessment of the Bragg-R-value.

The so called Durbin–Watson statistic (Durbin & Watson, 1971; Hill & Flack, 1987):

$$d = \frac{\sum_{i=1}^{N} (\Delta y_i - \Delta y_{i-1})}{\sum_{i=1}^{N} (\Delta y_i)^2} \tag{2.108}$$

with $\Delta y_i = y_{obs,i} - y_{calc,i}$ measures serial correlations between adjacent data points in the difference curve. For a good refinement in which the difference plot is random, a value of 2.0 is expected. Correlated errors lead to significantly lower values.

All agreement factors can be accessed in TOPAS by keywords at the overall level or for a specific Pawley, LeBail or Rietveld phase. Some TOPAS keywords related to agreement factors are listed below:

```
r_wp        15.6061677 r_exp      12.0417707 r_p        10.0071346
r_wp_dash   17.0540303 r_p_dash   11.1899544 r_exp_dash 13.1589462
gof         1.29600274
r_bragg     1.66505799
weighted_Durbin_Watson 1.46695946
```

2.8.1 Weighting schemes

Depending on the type of observed data (powder diffraction, pair distribution function, single crystal and so on), different weights should be applied to the individual measured intensities. In the ideal situation, the data format (XYE format) contains the associated errors $\sigma(y_{obs,i})$ in the observed intensity at position i in the data set. The default weighting for powder diffraction data in this case is the reciprocal value of the variance of the observed step scan intensities:

$$w_i = \frac{1}{\sigma(y_{obs,i})^2} \qquad (2.109)$$

For Poisson-type counting statistics (as found for scintillation counters), $\sigma(y_{obs,i}) = y_{obs,i}^{1/2}$ such that the weight can be calculated as:

$$w_i = \frac{1}{y_{obs,i}}, \quad y_{obs,i} \geq 1 \qquad (2.110)$$

In general, TOPAS can apply weights as a function of the position X_i, the observed intensity $y_{obs,i}$, the calculated intensity $y_{calc,i}$ and the standard deviation of the observed intensity $\sigma(y_{obs,i})$. Some examples of different weighting schemes in TOPAS are:

```
weighting = If(SigmaYobs < 1, 1, 1/SigmaYobs^2);
weighting = 1 / Max(Yobs, 1);
weighting = If(Yobs <= 1, 1, 1 / Yobs);
weighting = ( Abs(Yobs-Ycalc) / Abs(Yobs+Ycalc) +1) / Sin(X Deg / 2);
```

The weighting is usually calculated at the start of each refinement cycle. In cases where the weight is a function of $y_{calc,i}$, a flag *recal_weighting_on_iter* can be used to recalculate the weighting at the start of refinement iterations.

Figure 2.49 demonstrates the effect of weighting on a refinement, where the weighted difference curve shows that the weak peaks at higher diffraction angle are just as important as the strong ones that are usually found at low angles. The Rietveld plot in Figure 2.49 also shows the cumulative χ^2 function, that is the

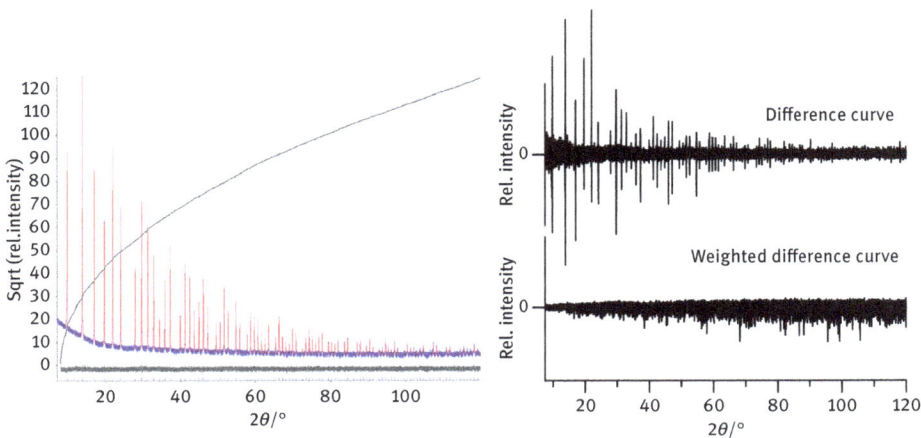

Figure 2.49: Left: Rietveld plot of LaB_6 measured with Mo-$K_{\alpha1}$ radiation. The gray line is the so-called cumulative χ^2. Right: Unweighted (top) and weighted (bottom) difference curve of the Rietveld refinement normalized to the sum of the absolute differences.

weighted sum of the squares of the difference between observed and calculated powder diffraction patterns up to that point in the diffraction pattern. This visualizes the impact of a Bragg peak or a region of the diffraction pattern on the overall fit to the data (David, 2004).

One problem with many Rietveld fits is the presence of unknown crystalline phases that are not included in the refinement. In such a case, a so-called robust refinement scheme can be implemented using an iterative reweighting of the data at each step of the refinement using, for example, the expression (Stone et al., 2009):

$$w_i = \frac{-2\ln[p(M|D, I)]}{(y_{calc,i} - y_{obs,i})^2} \tag{2.111}$$

with $p(M|D, I)$ representing the probability of the model, M, given the data, D, and any other available information I. According to Stone et al., such a function can be replaced by a polynomial and a logarithmic function and a better structural model is obtained for the phase of interest. An appropriate TOPAS macro can be found in their paper and in TOPAS.INC.

2.9 Data collection and refinement strategy

In 1999, a paper called "Rietveld refinement guidelines" was published by McCusker et al. describing good laboratory practice starting from the measurement of a powder pattern to the final Rietveld refinement. These guidelines are still valid and a highly recommended source of information. A more recent practical guide should appear soon (Madsen and Kern, 2019).

Every home-lab Rietveld refinement begins with the measurement of a powder pattern, most commonly in either Debye–Scherrer or Bragg–Brentano geometry. Some general rules apply for both geometries:
- The diffractometer must be careful aligned allowing for good reproducibility of measurements.
- Instrumental aberrations related to misalignment should be avoided.
- The instrumental resolution function (*IRF*) should be accurately measured and regularly monitored.
- The sample should be carefully ground to a grain size of 1–5 μm. Overgrinding destroys crystallinity, leads to peak broadening and should be avoided. Subsequent annealing might remove such grinding-induced broadening.
- At least 5 data points over the *fwhm* should be measured.
- The measured range should include *all peaks* at high *d*-spacing (low diffraction angle) and extend until the peaks die off and become indistinguishable from the background.
- Good counting statistics are essential. Variable counting time, where increasing time is spent at higher diffraction angles, should be considered.

Some recommendations for measurements in Bragg–Brentano geometry:
- Rotating sample holders should be used.
- Preferred orientation should be avoided as far as possible (no pressing of the sample in the sample holder; back or side loading; spray drying etc.).
- Transparency effects on the peak shape can be minimized for very low absorbing materials by using thin slurries of material on single crystal sample holders; the trade-off might be the need for additional corrections of the integrated intensity.
- The contribution of the background from the sample holder should be reduced by using low background single crystal holders (e.g., Si(911) cuts).
- Surface roughness effects can be avoided by either using slurries or by wiping off excessive powder in the cavity of the sample holder with a glass slide.
- The constant volume condition should be ensured at all times thus avoiding overspill effects by using a bigger sample area/smaller slits. The use of variable divergence slits is discouraged. If variable divergence slits are used, proper calculation of the estimated standard deviations and proper scaling must be ensured.

Some recommendations for measurements in Debye–Scherrer geometry:
- The capillary must be carefully aligned, rotating and not wobbling.
- Absorption should be minimized (ideally $\mu R \lesssim 1$) using either an appropriate wavelength, a thin capillary, a sample diluted with a low absorbing amorphous material (e.g., cork or powdered glass), or by mounting on the surface of a thin glass rod.
- The illuminated area of the capillary should be evenly filled with powder.
- The material of the glass capillary should be low absorbing glass (e.g., lithium-borate glass).
- Care must be taken when measuring an empty capillary for background determination, as a capillary filled with sample might have a much stronger absorption leading to a nonlinearly lowered background.

At the infancy of the Rietveld method, a proper least squares refinement strategy was crucial as the radius of convergence for individual parameters was low and refining too many parameters at the same time or in the wrong order led to divergence or program crashes. This situation changed with TOPAS which is a very stable least squares program practically allowing thousands of parameters to be refined simultaneously without being particularly sensitive to the order of parameter turn on. Nevertheless, we can suggest a typical "cautious" recipe for releasing and fixing sets of parameters which is likely to maximize the success rate of complicated Rietveld refinements (Figure 2.50). Some of the key-points are:
- The refinement of the background and the lattice parameters and the peak profile can be separated from the refinement of the crystal structure by starting with a LeBail/Pawley fit, before switching to Rietveld analysis.

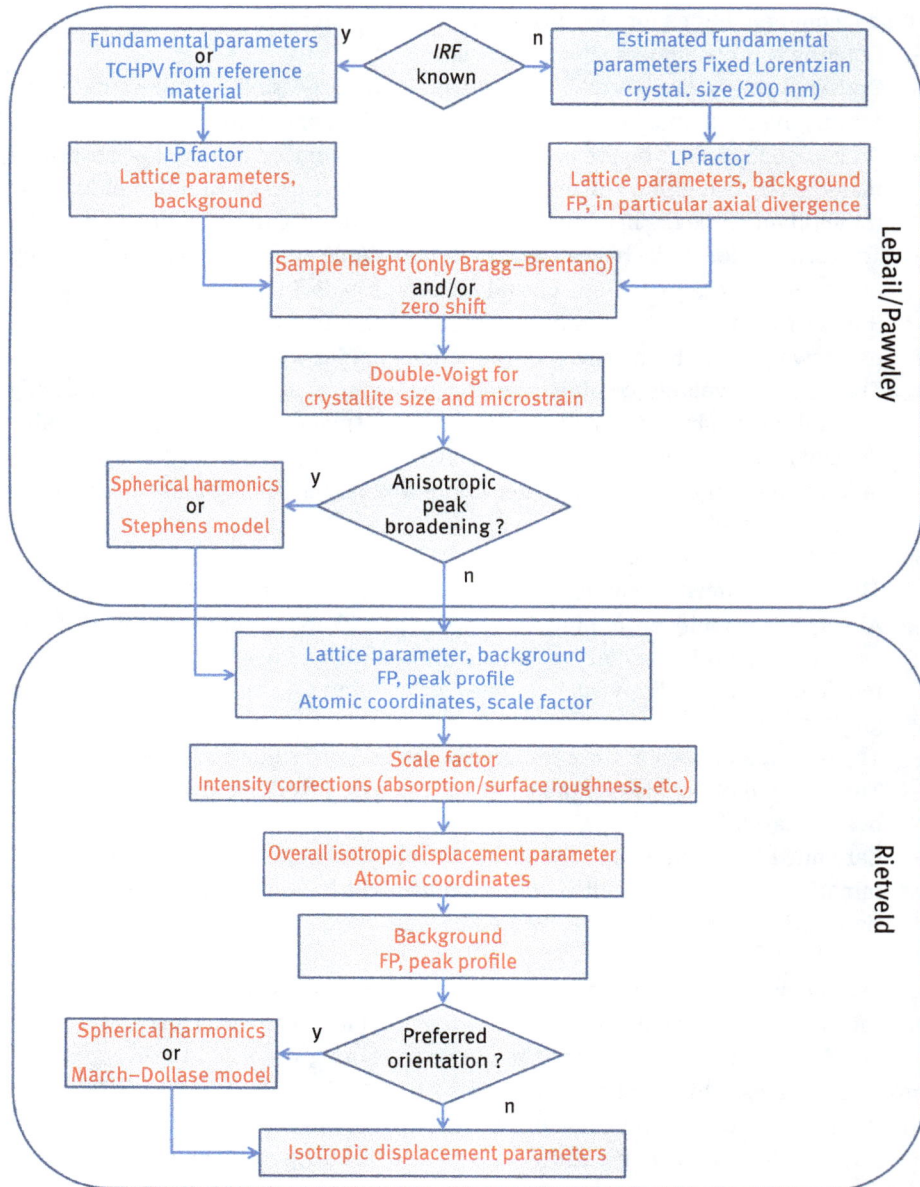

Figure 2.50: Flow chart of a typical Rietveld refinement. Parameters in red are refined, those in blue are fixed (TOPAS notation).

- The LeBail/Pawley fit will give a good indication of the best R_{wp} achievable for a given data set.
- If possible, the *IRF* should be included from the beginning. This has two advantages. Firstly, the *IRF* provides excellent starting parameters for the profile, secondly its use allows the proper separation between instrumental and sample contributions.
- After switching to Rietveld refinement, the background parameters should be refined again to account for possible correlations (in particular at high angles) between the background parameters and peak intensities from a previous Pawley/LeBail refinement.
- At the end of a Rietveld refinement, all refined parameters should be released simultaneously to ensure proper statistics.
- Models and their associated uncertainties should be critically assessed for their reliability and uniqueness.

2.10 Example of a Rietveld refinement

Now it is time to give an example of a full Rietveld refinement. We have selected the laboratory powder pattern of the room temperature phase of the double salt $Mg(H_2O)_6RbBr_3$ for this. Atoms lie on general and special positions and we will use fractional coordinates to describe the crystal structure. Later in the book, this example will be reused to discuss alternative ways of describing a crystal structure, namely rigid bodies (Chapter 6) and symmetry (distortion) modes (Chapter 8).

The room temperature crystal structure of $Mg(H_2O)_6RbBr_3$ has monoclinic $C2/c$ symmetry (Dinnebier et al., 2008). Its structure is characterized by a three-dimensional network of corner-sharing $RbBr_6$ octahedra that contains one $Mg(OH_2)_6$ octahedron in the center of each void (Figure 2.51).

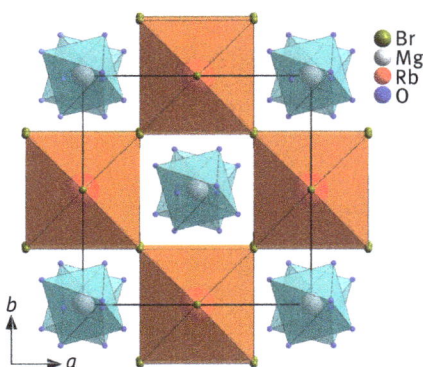

Figure 2.51: Projection of the crystal structure of Mg $(H_2O)_6RbBr_3$ as viewed along the *c*-axis showing green $RbBr_6$ and brown $Mg(OH_2)_6$ octahedra (from Dinnebier et al., 2008).

High-resolution laboratory X-ray powder diffraction data were recorded using a Debye–Scherrer geometry Bruker D8 diffractometer, equipped with a Våntec-1 position-sensitive detector and Cu-K$_{\alpha 1}$ radiation (1.540596 Å) obtained from a primary Ge(111) Johansson monochromator. A small amount of cubic RbBr is present in the powder pattern and is included as a second phase in the Rietveld refinement.

The different parts of the TOPAS INP file[25] are shown below. TOPAS INP files have a treelike structure. At the top level we need to describe the data set using the *xdd* keyword:

```
xdd "RbBrMgBr2_6H2O_295K.raw"          ' Powder pattern in Bruker raw format
'do_errors                             ' flag to calc. standard uncertainties
bkg @ 0 0 0 0 0 0                       ' 6th-order background polynomial
One_on_X(@, 6842.94765)                ' 1/X background term
start_X 10                             ' Start 2θ
```

The next section describes information about the instrument and is discussed in detail in Chapter 4:

```
lam                                    ' Wavelength section
    ymin_on_ymax  0.0001               ' Extent of peak width
    la 1 lo 1.540596 lh  0.401844      ' rel.int., λ , Lorentzian width of emission

Zero_Error(@, 0.00289)                 ' Refined zero shift
LP_Factor( 27.3)                       ' Lorentz-polarisation factor for Ge(111)
Rp 217.5                               ' Primary radius of the diffractometer
Rs 217.5                               ' Secondary radius of the diffractometer
axial_conv                             ' Axial convolution section
    filament_length  8                 ' Length of the filament
    sample_length  8                   ' Length of the sample
    receiving_slit_length  8           ' Length of the receiving slit
    secondary_soller_angle @  2.7      ' Angle of the secondary Soller slit
    axial_n_beta  20                   ' Number of rays in the axial plane used
                                         to describe the axial aberration

Slit_Width( 0.1)                       ' Width of a strip of the PSD
```

The next section describes intensity and positional corrections due to absorption of the capillary sample:

```
prm muR 0.027 min 0.01 max 1.0         ' Refined μeffR for absorption calculations
Cylindrical_I_Correction(muR)          ' Intensity correction
Cylindrical_2Th_Correction(muR)        ' Positional correction
```

---

[25] We refer to the Technical Reference for full TOPAS syntax. Briefly, "@" flags an unnamed parameter to refine; if a parameter has a name (e.g., muR) it is refined, unless its name is prefixed by the "!" symbol, in which case it is fixed. A ' character flags a commentThe jEdit editor (community. dur.ac.uk/john.evans/topas_academic/jedit_main.htm) is one useful tool for editing these files in that it can be easily configured to color-code TOPAS syntax.

The next tree level includes a description of each phase (flagged by *str*) that contributes to the *xdd*. We start with the rubidium bromide impurity phase with fixed coordinates:

```
str                                     ' Tree level for the RbBr impurity phase
  phase_name RbBr                       ' Phase name for the GUI
  Strain_G(!stg, 0.17437)               ' Fixed Gaussian microstrain broadening
  r_bragg  2.478                        ' Bragg-R-factor
  MVW( 661.485, 327.427538, 3.498)      ' Mol. mass of cell contents, cell volume, wt%
  scale @  3.560044e-005                ' Scale factor
  space_group Fm-3m                     ' Hermann Maguin space group symbol
  Cubic(  6.89242)                      ' Fixed cubic lattice parameter
' Atomic site, label, Wyckoff mult., fract. coord., occupancy, displ. param.
  site Rb2 num_posns  4    x=0;   y=0;   z=0;   occ Rb+1  1 beq !B1 2
  site Br3 num_posns  4    x=1/2; y=1/2; z=1/2; occ Br-1  1 beq =B1;
```

It is recommended to write all special positions as equations using integer numbers or fractions to avoid incorrect multiplicities due to round-off errors (e.g., = 1/3;, instead of 0.3333). In order to check the correctness of the multiplicity of the site, the keyword *num_posn 0* should be included at each site. The value of the multiplicity will be automatically updated after the first refinement cycle. An overall fixed (!) displacement factor is defined as parameter "*B1*" and used for all atoms in the list using the equation "*beq = B1;*".

The next section describes the main $Mg(H_2O)_6RbBr_3$ phase. A total of 13 structural coordinates can be refined (H atoms are omitted), the remaining ones are fixed by symmetry:

```
str                                     ' Mg(H2O)6RbBr3 phase
  phase_name Mg(H2O)6RbBr3
  CS_L(@, 800)                          ' Lorentzian size broadening
  CS_G(@, 854)                          ' Gaussian size broadening
  Strain_G(@, 0.123)
  r_bragg  2.910
  MVW( 1781.925, 1311.269, 96.459)
  scale @  8.69364709e-005
  space_group C12/c1
  a  @  9.641313
  b  @  9.865348
  c  @  13.786177
  be @  90.08773
  site Br1 num_posns 4 x=1/2;         y=0;         z=1/2;         occ Br-1 1 beq B2 2
  site Br2 num_posns 8 x @  0.25283 y @  0.74052 z @  0.74863 occ Br-1 1 beq=B2;
  site Mg1 num_posns 4 x=1/2;         y=1/2;       z=1/2;         occ Mg+2 1 beq=B2;
  site Rb1 num_posns 4 x=1/2;         y @ -0.00277 z=3/4;         occ Rb+1 1 beq=B2;
  site O1  num_posns 8 x @  0.40784 y @  0.68248 z @  0.53974 occ O-2 1 beq=B2;
  site O2  num_posns 8 x @  0.31906 y @  0.39673 z @  0.54930 occ O-2 1 beq=B2;
  site O3  num_posns 8 x @  0.41407 y @  0.51740 z @  0.36089 occ O-2 1 beq=B2;
```

The graphical result of the Rietveld refinement is shown in Figure 2.52. The fit gives satisfactory agreement factors (R_{wp} = 5.42 %, R_{wp}' = 8.46 %, GOF = 1.48, R_{Bragg} = 2.91%). Most importantly, the visual fit between calculated and observed patterns is excellent, and all features in the observed pattern are described well. The small misfit due to an unaccounted impurity phase at 20° 2θ shows up as a step in the cumulative χ^2.

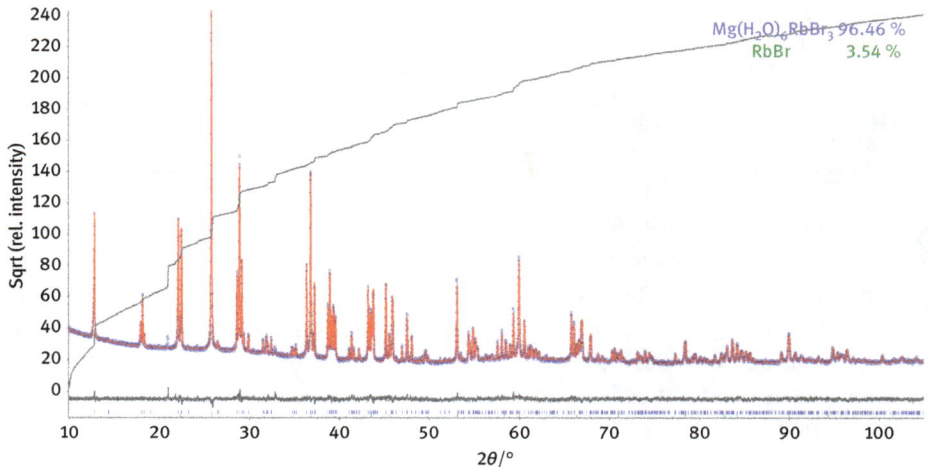

Figure 2.52: Rietveld plot of $Mg(H_2O)_6RbBr_3$ using unconstrained refinement of fractional coordinates. To emphasize weaker features, the square root of the intensity is displayed. A small amount of RbBr is included as second phase. As in most figures in the book, experimental data are shown in blue, the calculated pattern in red and the difference (obs – calc) in gray. Vertical tick marks at the bottom of the plot show positions of allowed *hkl* reflections. The gray line shows the cumulative χ^2.

References

Balzar, D. (1999): *Voigt-function model in diffraction line-broadening analysis* in Defect and microstructure analysis from diffraction, Edited by R.L. Snyder, H.J. Bunge, International Union of Crystallography Monographs on Crystallography, Oxford University Press, New York (US), 785 pages.

Bergmann, J., Kleeberg, R., Haase, A., Breidenstein, B. (2000): *Advanced fundamental parameters model for improved profile analysis.* Mat. Sci. Forum 347–349, 303–308.

Brindley, G.W. (1945): *The effect of grain or particle size on X-ray reflections from mixed powders and alloys, considered in relation to the quantitative determination of crystalline substances by X-ray methods.* Phil. Mag. Ser. 36, 347–369.

Broyden, C.G. (1970): *The convergence of a class of double-rank minimization algorithms.* J. Inst. Maths. Applics. 6, 76–90.

Cheary, R.W., Coelho, A.A. (1992): *A fundamental parameters approach to X-ray line-profile fitting.* J. Appl.Cryst. 25 (2), 109–121.

Cheary, R.W., Coelho, A.A. (1998): *Axial divergence in a conventional X-ray powder diffractometer. I. theoretical foundations*. J. Appl. Cryst. 31, 851–861.

Coelho, A.A. (2005): *A bound constrained conjugate gradient solution method as applied to crystallographic refinement problems*. J. Appl. Cryst. 38, 455–461.

David, W.I.F. (2004): *Powder diffraction: least-squares and beyond*. J Res Natl Inst Stand Technol. 109, 107–123.

Debye, P. (1915): *Zerstreuung von Röntgenstrahlung*. Annalen der Physik, 351, 809–823.

Dinnebier, R.E., Billinge, S.J.L. (eds.) (2008): *Powder diffraction: theory and practice*. RSC publication, Cambridge UK, 574 pages.

Dinnebier, R.E., Liebold-Ribeiro, Y., Jansen, M. (2008): *The low and high temperature crystal structures of [Mg(H$_2$O$_6$)]XBr$_3$ double salts (X = Rb, Cs)*. Z. Anorg. Allg. Chem. 634, 1857–1862.

Durbin, J., Watson, G.S. (1971): *Testing for serial correlation in least square regression. III*. Biometrika 58, 1–19.

Egami, T., Billinge, S. (2012): *Underneath the Bragg peaks - structural analysis of complex materials*, Volume 16, 2nd edition, Pergamon, Oxford (UK), 422 pages.

Fischer, R.X. (1996): *Divergence slit corrections for Bragg–Brentano diffractometers with rectangular sample surface*. Powder Diffraction 11, 17–21.

Fletcher, R. (1970): *A new approach to variable metric algorithms*. Comput. J. 13, 317–322.

Giacovazzo, C., Monaco, H.L., Artioli, G., Viterbo, D., Milanesio, M., Gilli, G., Gilli, P., Zanotti, G., Ferraris, G. (2011): *Fundamentals of crystallography (International Union of Crystallography monographs on crystallography)*. 3rd edition, Oxford University Press, New York (US), 872 pages.

Goldfarb, D. (1970): *A family of variable metric updates derived by variational means*. Math. Comput. 24, 23–26.

Gozzo, F., Cervellino, A., Leoni, L., Scardi, P., Bergamaschi, A., Schmitt, B. (2010): *Instrumental profile of MYTHEN detector in Debye-Scherrer geometry*. Z. Kristallogr 225, 616–624.

Hill, R.J., Flack, H.D. (1987): *The use of the Durbin-Watson d statistic in Rietveld analysis*. J. Appl. Cryst. 20, 356–361.

Hinrichsen, B., Dinnebier, R.E., Jansen,. M (2008): *Two-dimensional diffraction using area detectors*, in Powder diffraction – theory and practice, Edited by R.E. Dinnebier and S.L.J. Billinge. RSC Publishing, Cambridge, 414–438 Pages.

Hull, A.W. (1917): *A new method of X-Ray crystal analysis*. Phys. Rev. 10, 661–697.

Khalifah, P. (2015): *Use of radial symmetry for the calculation of cylindrical absorption coefficients and optimal capillary loadings*. J. Appl. Cryst. 48, 149–158.

Klug, H.P., Alexander, L.E. (1974): *X-ray diffraction procedures for polycrystalline and amorphous materials*, 2nd edition, John Wiley and Sons, New York, 966 pages.

Krüger, H., Fischer, R.X. (2004): *Divergence-slit intensity corrections for Bragg–Brentano diffractometers with circular sample surfaces and known beam intensity distribution*. J. Appl. Cryst. 37, 472–476.

Le Bail, A., Duroy, H., Fourquet, J.L. (1988): *Ab-initio structure determination of LiSbWO$_6$ by X-ray powder diffraction*. Mat. Res. Bull. 23, 447–452.

Levenberg, K. (1944): *A method for the solution of certain problems in least squares*. Quart. Appl. Math. 2, 164–168.

Madsen, I.C., and Scarlett, N.V.Y. (2008): *Quantitative phase analysis*, in Powder diffraction: theory and practice, Edited by. R.E. Dinnebier and S.L.J. Billinge, RSC Publishing, Cambridge UK, 298–331.

Madsen, I.C., Kern, A. (2019): *Guidelines for collecting high quality powder diffraction data on a laboratory instrument*.

Malmros, G., Thomas, J.O. (1977): *Least squares structure refinement based on profile analysis of powder film intensity data measured on an automatic microdensitometer*. J. Appl. Cryst. 10,7–11.

Marquardt, D. (1943): *An Algorithm for least-squares estimation of nonlinear parameters*. J. Appl. Math. 11, 431–441.

Mendenhall, M.H., Henins, A., Hudson, L.T., Szabo, C.I., Windover, D., Cline, J.P. (2017): *High-precision measurement of the X-ray Cu Kα spectrum*. J Phys. B Mol. Opt. Phys. 50, 115004 18pp.

McCusker, L.B., Von Dreele, R.B., Cox, D.E., Louer, D., Scardi, P. (1999): *Rietveld refinement guidelines*. J. Appl. Cryst. 32, 36–50.

Pawley, G.S. (1981): *Unit-cell refinement from powder diffraction scans*. J. Appl. Cryst. 14, 357–361.

Pecharsky, V., Zavalij, P. (2009): *Fundamentals of powder diffraction and structural characterization of materials*, 2nd Edition, Springer US, Softcover, 744 pages.

Pitschke, W., Mattern, N., Hermann, H. (1993): *Incorporation of microabsorption corrections in Rietveld analysis*. Powder Diffraction 8, 223–228.

Press, W.H., Flannery, B.P., Teukolsky, S.A., Vetterling, W.T. (1986): *Numerical recipes*, Cambridge University Press, Cambridge (UK), 819 pages.

Rietveld, H.M. (1967): *Line profiles of neutron powder-diffraction peaks for structure refinement*. Acta Cryst. 22, 151–152.

Rietveld, H.M. (1969): *A profile refinement method for nuclear and magnetic structures*. J. Appl. Cryst. 2, 65–71.

Rowles, M. R., Buckley, C.E. (2017): *Aberration corrections for non-Bragg–Brentano diffraction geometries*. J. Appl. Cryst. 50, 240–251.

Sabine, T.M. (1985): *Extinction in polycrystalline materials*. Aust. J. Phys. 38, 507–518.

Sabine, T.M., Hunter, B.A., Sabine, W.R., Ball, C.J. (1998): *Analytical expressions for the transmission factor and peak shift in absorbing cylindrical specimens*. J. Appl. Cryst. 31, 47–51.

Sabine, T.M., Von Dreele, R.B., Jorgensen, J.E. (1988): *Extinction in time-of-flight neutron powder diffractometry*. Acta Cryst. A 44, 374–379.

Sears, V. (1992): *Neutron scattering lengths and cross section*. Neutron News 3, 26–37.

Shanno, D.F. (1970): *Conditioning of quasi-Newton methods for function minimization*. Math. Comput. 24, 647–656.

Suortti, P. (1972): *Effects of porosity and surface roughness on the X-ray intensity reflected from a powder specimen*. J. Appl. Cryst. 5, 325–331.

Stinton, G.W., Evans, J.S.O. (2007): *Parametric Rietveld refinement*. J. Appl. Cryst. 40, 87–95.

Stone, K.H., Lapidus, S.H., Stephens, P.W. (2009): *Implementation and use of robust refinement on powder diffraction in the presence of impurities*. J. Appl. Cryst. 42, 385–391.

Thompson, P., Cox, D.E., Hastings, J.B. (1987): *Rietveld refinement of Debye-Scherrer synchrotron X-ray data from Al_2O_3*. J. Appl. Cryst. 20, 79–83.

TOPAS version 6 (2017), Bruker-AXS, Karsruhe, Germany.

Von Dreele, R. B. (1999): *Combined Rietveld and stereochemical restraint refinement of a protein crystal structure*. J. Appl. Cryst. 32, 1084–1089.

Von Dreele, R.B. (2008): *Rietveld refinement* in powder diffraction: theory and practice, Edited by R.E. Dinnebier and S.L.J. Billinge, RSC Publishing, Cambridge, 266–281 Pages.

Waasmaier, D., Kirfel, A. (1995): *New analytical scattering-factor functions for free atoms and ions* Acta Cryst. A 51, 416–431.

Wahlberg, N., Bindzus, N., Bjerg, L., Becker, J., Dippel, A.C., Iversen, B.B. (2016): *Synchrotron powder diffraction of silicon: high-quality structure factors and electron density*. Acta Cryst. A 72, 28–35.

Wilson, A.J.C. (ed.) (1995): *International tables for crystallography, Volume C: mathematical, physical and chemical tables*. Corrected reprint of 1st edition. Kluwer Academic Publishers, Dordrecht, The Netherlands.

Young, R.A. (ed.) (1993): *Introduction to the Rietveld method*, The Rietveld Method, IUCr Book Series, Oxford University Press, New York (US), 287 pages.

3 Structure independent fitting

3.1 Introduction

Before starting a Rietveld refinement, it is very helpful to have good starting parameters for peak positions, peak profile (microstructure) and background. If these parameters are known with reasonable accuracy, one can focus on just the structural aspects of the model in the early stages of a Rietveld refinement. The best way to achieve this is to perform a structure-independent whole powder pattern fitting (WPPF) beforehand. The predetermined parameters can then be transferred directly to Rietveld refinement.

Structure-independent WPPF also provides a set of peak intensities, uncertainties and the correlations between them for overlapping reflections. These are often needed for different structure solution methods such as direct methods, simulated annealing (Coelho, 2000; David et al., 2006) or charge flipping (Oszlányi & Süto, 2004). The better the peak shape is described, the more accurately partly overlapping reflections can be separated.

In general, three main WPPF methods are available that differ in the amount of constraints and how the intensities are obtained.

The simplest method is unconstrained *single peak fitting*, where a set of parameters is refined to fit each peak separately. Although this might lead to an excellent fit, the refined parameters can be completely meaningless. Therefore a number of constraints need to be introduced. First, the instrumental resolution function should be used as a basis for all peak shapes. Second, a limited number of overall background parameters should be included in the fitting process. In addition, since shape parameters like asymmetry or strain and crystallite size are usually smooth functions of the scattering angle, it might be useful to constrain them. Single peak fitting is used when no lattice parameters are known. The obtained peak positions are usually of higher quality than typical peak search methods employing derivatives, and can be used, for example, for indexing the powder pattern.

If the lattice parameters are known, they should be included (with a suitable zero shift or sample-displacement function) in the WPPF, thus constraining the peak positions. This drastically reduces the number of parameters needed. The two common approaches for this are the *Pawley* (Pawley, 1981) and the *Le Bail* (Le Bail et al., 1988; Le Bail, 2005) methods; they differ in how integrated reflection intensities are extracted. If the space group is unknown, the refinement is performed in a space group without extinctions, for example, for the primitive orthorhombic lattice this would be extinction group $P-$ ($P222$, $Pmmm$, $Pmm2$, $Pm2m$, $P2mm$). The most probable space groups can then be found either manually by taking all peaks close to zero intensity as extinct, or more systematically by fixing all parameters and

https://doi.org/10.1515/9783110461381-003

refining in all possible space groups one-by-one. The space groups leading to the best R_{wp} agreement factors, having a minimum number of unfitted lines (by visual inspection of the difference curve) and a minimum number of predicted reflections with zero intensity are the most probable ones. It's also possible to apply more statistically sophisticated methods for space group choice (Markvardsen et al., 2001). Programs such as ExtSym will automatically read TOPAS output files to achieve this.

One of the problems in WPPF is the possible correlation between background parameters and peak intensities at high diffraction angle where strong peak overlap occurs. It is crucial to keep the number of background parameters low and to check the oscillations of the background function visually.

Following WPPF, switching to Rietveld refinement or simulated annealing is straightforward. One first fixes all predetermined parameters, adds a crystal structure (or part of it), refines the scale factor and successively introduces structural parameters. Toward the end of a Rietveld refinement it is good practice to free the previously fixed parameters. This is particularly important for the background that may be poorly described in WPPF due to peak overlap at higher diffraction angles.

3.2 Constrained single peak fitting

The simplest method for WPPF is to fit all peaks that can be distinguished by eye individually but with an overall background function and constrained strain and crystallite size parameters (e.g., using the double-Voigt approach).

From a practical viewpoint, the peak positions should be fixed for the first iteration while the intensities are refined freely. The starting value for the intensities should be set to a small number to ensure convergence. After the first iteration, the peak positions can be released for free refinement. Quite frequently, not all overlapping peaks have been identified. Visual inspection of the difference curve sometimes shows unaccounted intensity that belongs to missing peaks. These peaks should be included for the next iteration.

In the following example $K(C_5H_5)$ (KCp) was measured at a synchrotron (Figure 3.1). We will use a simple delta-type function to describe instrumental broadening, as the peak shape is predominantly determined by the sample:

```
xdd "Kcp.raw"
    bkg @ 362.60 -47.51 -57.54 -33.07 64.24 2.02 -9.88 -1.07 1.77 -4.17 0.04
    start_X  6
    finish_X  31.3
    LP_Factor(90)
    convolution_step 2
    Rs 200.5
    Simple_Axial_Model(6.1)
```

```
lam
    ymin_on_ymax  0.001
    la  1 lo  1.14937 lh  1e-006
xo_Is
    xo @ 9.118041043
    peak_type fp
    LVol_FWHM_CS_G_L(1, 135.973, 0.89, 129.658,csg, 146.917,csl, 9918.224)
    e0_from_Strain( 0.0011,,,stl, 0.5116)
    I @ 1.130485033
xo_Is
    xo @ 12.56787108
    peak_type fp
    LVol_FWHM_CS_G_L(1, 135.9734, 0.89, 129.658,(csg), 146.917,(csl), 9918.224)
    e0_from_Strain( 0.0011,,,(stl), 0.5116)
    I @ 0.09691388313_LIMIT_MIN_1e-015
...
xo_Is
    xo @ 30.71553312
    peak_type fp
    LVol_FWHM_CS_G_L(1, 135.973, 0.89, 129.658,(csg), 146.917,(csl), 9918.224)
    e0_from_Strain( 0.0011,,,(stl), 0.5116)
    I @ 4.676600833
```

3.3 The Le Bail method

In the Le Bail WPPF method (Le Bail et al., 1988; Le Bail, 2005), all parameters except for structural parameters (atomic positions, occupancies, displacement factors) are subjected to least squares refinement in a process analogous to regular Rietveld refinement.

The Le Bail WPPF method iterates the Rietveld formula and thus requires only a slight modification of the Rietveld code. Recall that the Rietveld formula (eq. (2.5)) of a single phase for calculating the step-scan-intensity of step i is:

$$y_{calc,i} = S \sum_{\mathbf{s}} \left(|F_{calc,\mathbf{s}}|^2 \cdot \Phi_{\mathbf{s},i} \cdot Corr_{s,i} \right) + Bkg_i. \tag{3.1}$$

Since $|F_{calc,\mathbf{s}}|^2$ cannot be calculated from a crystal structure, all "calculated" peak intensities are initially set to an arbitrary value, for example:

$$|F_{calc,\mathbf{s}}|^2 = 1.0. \tag{3.2}$$

These are then entered in the Rietveld decomposition formula as "calculated" structure factors as if they had been derived from a structural model.

The Rietveld refinement then determines a set of new "calculated" structure factors $|F_{calc,\mathbf{s}}|^2_{new}$ from the decomposition formula according to:

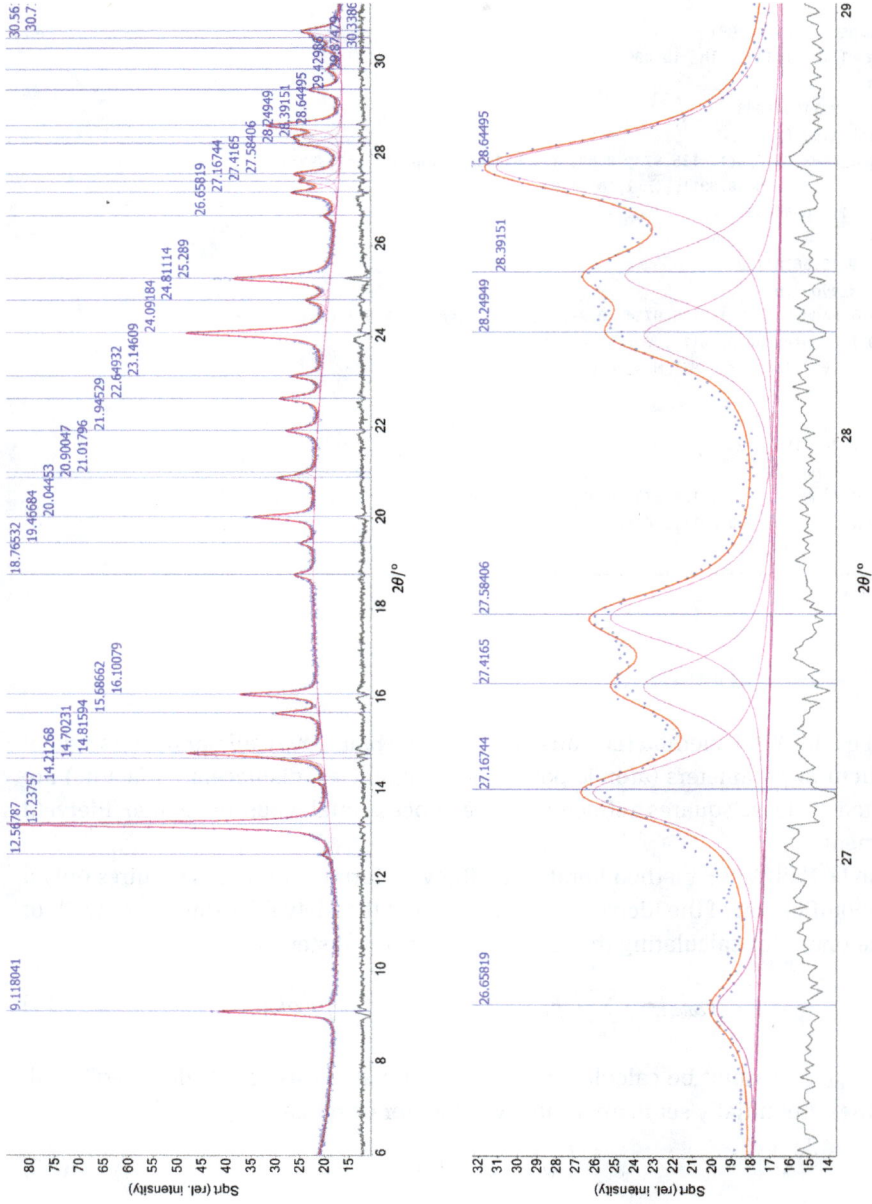

Figure 3.1: Constrained single peak WPPF of KCp. Top: Entire powder pattern; bottom: Zoomed region.

$$|F_{calc,\mathbf{s}}|^2_{new} = \frac{S\sum_i\left(y_{obs,i}|F_{calc,\mathbf{s}}|^2 \cdot \Phi_{\mathbf{s},i} \cdot Corr_{s,i}\right)}{y_{calc,i}} \tag{3.3}$$

This effectively scales $|F_{calc,\mathbf{s}}|^2$ by the ratio between observed and calculated intensities. These are then used as "calculated" structure factors in eq. (3.3) for the next iteration until convergence is reached:

$$|F_{calc,\mathbf{s}}|^2 = |F_{calc,\mathbf{s}}|^2_{new} \tag{3.4}$$

Hence, the intensities of the individual peaks are not treated as least squares parameters and are not directly refined.

If identical peak intensities are used as starting values, the intensities of (almost) fully overlapping reflections tend to be equipartitioned after the refinement converges. Negative intensities are not possible.

Alternatively, if part of the crystal structure is known (e.g., the position of heavy atoms), the intensities from a preliminary Rietveld refinement can be used as starting intensities for a Le Bail WPPF. Intensity ratios of overlapping reflections are then closer to their "true" values often leading to better results if direct methods are subsequently used for structure completion.

The example of a typical Le Bail WPPF of KCp is given below. The small anisotropic peak width due to microstrain is not accounted for (Figure 3.2):

```
xdd "Kcp.raw"
   bkg @ 349.97 -67.75 -59.31 -14.93 64.72 -17.97 -15.28 2.47 4.96 0.119 5.10
   start_X  6
   finish_X  34
   LP_Factor( 90)
   Zero_Error(@, -0.002447297771)
   Rs 200.5
   Simple_Axial_Model(@, 6.022114618)
   lam
      ymin_on_ymax  0.001
      la  1 lo  1.14937 lh  1e-006
   hkl_Is
      phase_name "KCP hkl_Phase"
      lebail  1
      TCHZ_Peak_Type(@, 1, @, -0.3174, @, 0.020,, 0,@, 0.4765,, 0)
      Tetragonal(@ 9.9695,@ 10.4995)
      space_group p-421c
      hkl_m_d_th2 1 0 1 8 7.22960567 9.11856079 I  1.110176902
      hkl_m_d_th2 1 1 0 4 7.04950809 9.3520298 I   0.008123926706
      ...
      hkl_m_d_th2 5 1 0 8 1.95518172 34.186676 I   0.2618402688
```

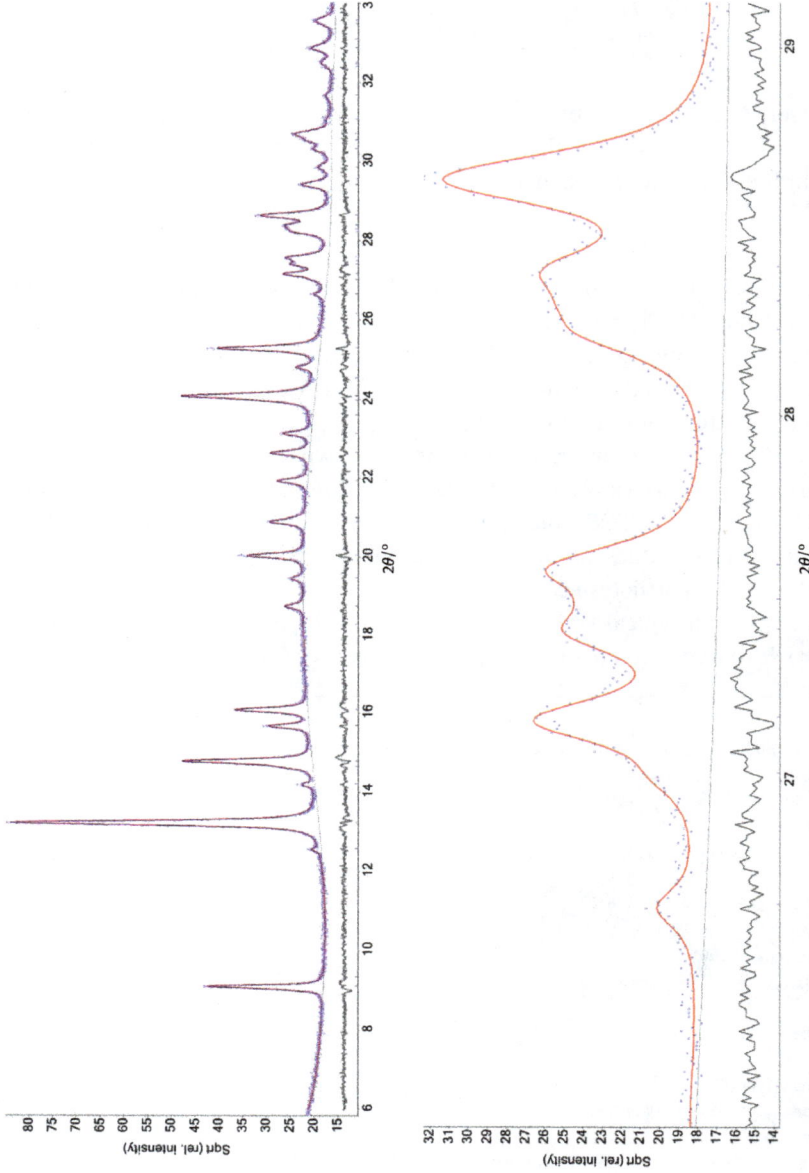

Figure 3.2: Le Bail WPPF of KCp. Top: entire powder pattern; bottom: zoomed region.

3.4 The Pawley method

In the Pawley WPPF method (Pawley, 1981), the unit cell parameters, background parameters, zero point errors, peak shape parameters and *all* reflection intensities are subjected to nonlinear least squares refinement. In principle, the following system of equations must be solved:

$$
\begin{aligned}
y_{calc,0} &= \sum_{\mathbf{s}}(I_{\mathbf{s}} \cdot \Phi_{\mathbf{s},0}) + Bkg_1 \\
y_{calc,1} &= \sum_{\mathbf{s}}(I_{\mathbf{s}} \cdot \Phi_{\mathbf{s},1}) + Bkg_2 \\
&\;\cdots \\
y_{calc,N-1} &= \sum_{\mathbf{s}}(I_{\mathbf{s}} \cdot \Phi_{\mathbf{s},N-1}) + Bkg_n
\end{aligned}
\tag{3.5}
$$

where $y_{calc,i}$ is the calculated step scan intensity at position i in the powder pattern, $I_{\mathbf{s}}$ is the intensity of reflection $\mathbf{s}$, $\Phi_{\mathbf{s},i}$ is the value of the normalized peak shape function of reflection $\mathbf{s}$ at position $i \in [0...N-1]$ in the powder pattern. The reflection positions are determined by the unit cell parameters and the zero point error. The background point Bkg_i is either predetermined or modeled by a polynomial or similar function (see Chapter 2). In terms of a least squares minimization procedure, this requires typically a $(10+n) \times (10+n)$ square matrix, where 10 is the number of background and lattice parameters refined and n is the number of symmetry-independent reflections generated for the 2θ range covered by the data. The correlation between the peak intensities increases with increasing overlap (up to 100% for peaks with identical d-spacing within the limits of resolution of the data). Since there is nothing in the least squares procedure that forces the peaks to have positive intensity, it could be that for two overlapping peaks one peak shows a negative intensity while the other has an intensity higher than the sum of the two peaks. To reduce this problem, Pawley introduced restraints into the least squares procedure such that as the difference between the calculated 2θ values of two adjacent peaks approaches zero, the intensity of the two peaks is set equal with a weight dependent on their separation. Alternatively, refinement can be based on $|F|$ instead of I. This forces the peak intensities to have positive values only.

TOPAS performs a Pawley fit for a *hkl_Is* phase if the keyword *lebail 1* is not defined. TOPAS is quite sophisticated in taking specific care in avoiding unstable (and ultimately singular) least squares matrices. As a result, Le Bail and Pawley WPPF fits in TOPAS lead to identical results in terms of agreement factors. A Pawley fit typically requires fewer least squares cycles for convergence and is slightly faster. With the Pawley method the refined intensities and their uncertainties can be directly used for structure determination by charge flipping (Oszlányi & Süto, 2004). The keyword:

```
Out_for_cf(file)
```

outputs the intensities and their correlations from a Pawley refinement for use in charge flipping.

References

Coelho, A.A. (2000): *Whole-profile structure solution from powder diffraction data using simulated annealing*. J. Appl. Cryst. 33, 899–908.

David, W.I.F., Shankland, K., van de Streek, J., Pidcock, E., Motherwell, W.D.S., Cole, J.C. (2006): *DASH: a program for crystal structure determination from powder diffraction data*. J. Appl. Cryst. 39, 910–915.

Dinnebier, R.E., Billinge, S.J.L. (eds) (2008): *Powder diffraction: theory and practice*, The Royal Society of Chemistry (RCS), 574 pages.

Le Bail, A., Duroy, H., Fourquet, J.L. (1988): *Ab-initio structure determination of LiSbWO$_6$ by X-ray powder diffraction*. Mat. Res. Bull. 23, 447–452.

Le Bail, A., (2005): *Whole powder pattern decomposition methods and applications: a retrospection*. Powder Diffraction 20, 316–326.

Markvardsen, A.J., David, W.I.F., Johnston, J.C., Shankland, K. (2001): *A probabilistic approach to space-group determination from powder diffraction data*. Acta Cryst. A 57, 47–54.

Oszlányi, G., Süto, A. (2004): *Ab initio structure solution by charge flipping*. Acta Cryst. A 60, 134–141.

Pawley, G.S. (1981): *Unit-cell refinement from powder diffraction scans*. J. Appl. Cryst. 14, 357–361.

4 Peak shapes: Instrument ○ microstructure

4.1 Introduction

A diffraction peak measured on a scale x, which is most commonly $x = 2\theta$ (angular dispersive data), can be understood as a convolution of several different contributions. For a diffraction peak located at position x_0 the convolution is best described on the scale $X = x - x_0$. The two most fundamental contributions are the instrumental contribution, $IRF(X)$ (Instrumental Resolution Function) and the sample contribution $MS(X)$ (from MicroStructure). $MS(X)$ is also called *structural line broadening*. The overall, peak profile $\Phi(X)$ of a particular reflection can be described as a convolution of these two contributions:

$$\Phi(X) = (IRF \circ MS)(X). \tag{4.1}$$

As we will show later, both MS and IRF can be regarded as convolutions of several subcontributions. As TOPAS is able to handle convolutions of arbitrary functions, it offers great opportunities to model peak profiles. In particular, it is possible:

- to optimize line-broadening models to improve the quality of Rietveld fits.
- to develop sophisticated models to extract microstructural information from the sample-dependent line broadening.

TOPAS can also handle different scales, like time-of-flight (TOF) or energy (E). Moreover, other scales are important for theoretical considerations (like strain). Generally, the parameters characterizing a peak, like its position x_0 and its width parameters δ (e.g., $fwhm$) will depend on the scale these parameters are referring to. In order to show how widths on different scales are related, let's consider another scale y and let:

$$y = y(x) \tag{4.2}$$

be a bijective (invertible) function of x. If the position of a given reflection hkl is x_0, the position on the y scale is y_0 as calculated by eq. (4.2). If a small shift $x - x_0$ with respect to a position x_0 is considered, on the y scale this shift will be:

$$y - y_0 = \frac{dy}{dx_{x_0}}(x - x_0). \tag{4.3}$$

The shift can correspond to an actual shift of the peak position due to an effect like strain. Alternatively, the (positive) peak width δ can be related to the absolute value of such a shift. The peak width parameter $fwhm$ on the two different scales is then:

$$fwhm_y = \left|\frac{dy}{dx}\right|_{x_0} fwhm_x, \tag{4.4}$$

https://doi.org/10.1515/9783110461381-004

where the subscript gives the scale. In some cases we will also need to model a *hkl*-or other dependence to peak width, and we will flag this using a superscript: $fwhm_{2\theta}^{hkl}$.

In simple cases, where peak broadening is symmetric and isotropic, it is possible to directly use $\Phi(X)$ to describe peak shapes. However, to model more complex effects, such as asymmetry or for a quantitative interpretation of the structural line broadening in terms of, for example, size and microstrain parameters; the IRF^{hkl} and MS^{hkl} terms must be considered separately.

Instrumental line broadening is experimentally assessable using a standard material with no or only negligible structural line broadening. From its diffraction pattern, the IRF is assessed either by fitting and/or by considering the diffraction geometry (see Section 4.2). The IRF will nearly always be isotropic, that is, $IRF^{hkl} = IRF^{x_0}$.[1] The parameters describing the IRF are then fixed when evaluating diffraction data recorded from a different material under the same measurement conditions as those applied for the instrumental standard. The additional broadening MS^{hkl} in eq. (4.1) is then modeled by one or more suitable sample contributions with refined parameters.

Note the following:
(1) The intrinsic properties of an instrumental standard may affect the measured instrumental profile. One property is absorption that can affect the diffracting volume of the specimen and its peak profile (see Chapter 2). Hence, the X-ray absorption coefficients of the instrumental standard and the material to be investigated should be matched if possible.
(2) The diffraction pattern from an instrumental standard only provides direct information about IRF^{x_0} around the positions where diffraction peaks are available. An appropriate functional description of the x_0-dependence of IRF^{x_0} will allow determination of the instrumental resolution around arbitrary x_0 values, but extrapolation to values outside the range of measured peaks can be problematic. Physically realistic descriptions of the IRF will behave better upon extrapolation than purely empirical descriptions.

4.2 Determination of the instrumental resolution function (*IRF*)

All powder diffraction patterns, even those obtained at the highest resolution synchrotron source, have IRF_{x_0} contributions originating from the instrument. It is, therefore important to know this function in detail. This not only helps monitoring changes of

---

[1] This excludes use of narrowly textured or single crystalline materials for determining the instrumental resolution of a powder diffractometer, because certain contributions to the instrumental broadening like divergence are affected by a narrow texture (see Section 4.2).

the diffractometer over time, but also allows explicit determination of the sample-dependent line broadening contributions. The latter is usually dominated by isotropic and/or anisotropic microstrain and domain size. In the following, the most common case of angle-dispersive diffraction data is considered, that is, $x = 2\theta$. The $x = TOF$ scale will be considered in Section 4.4.

The first step in the determination of the *IRF* consists of measuring a high-quality powder pattern of a line profile standard like NIST SRM 660a LaB_6 (current batch is 660c) over the relevant range of diffraction angles under identical experimental conditions to the samples under investigation. The standard is expected to contain only negligible sample contributions, meaning "high crystallinity," no microstrain broadening and a domain size large enough (>500 nm) not to cause any size broadening but small enough to ensure reasonable particle statistics. Misalignment of the diffractometer and the sample should be avoided.

In general, the convolution approach by Klug & Alexander (1974) is used to build up the *IRF* starting from the emission profile. In the so-called *fundamental parameters* (FP) approach (Cheary & Coelho, 1992), the *IRF* is built up essentially from first principles by convoluting the contributions due to relevant instrumental aberrations. Ideally one only uses measurable physical quantities like slit widths, slit lengths, Soller slit opening angles and so on without refinement. In reality, minor refinement might be necessary since the FP process involves some approximations for computational speed and simplicity.

Alternatively, well-established phenomenological line shape functions, like the extended Thompson Cox Hastings (TCHZ) Pseudo-Voigt function (Thompson et al., 1987; Young, 1993) can be used to describe the *IRF* (see Section 2.5.5). The treatment, however, is usually supplemented with models considering divergence-related asymmetry of the diffraction peaks.

In the following, a stepwise recipe is given to determine the *IRF* of different angular dispersive laboratory powder diffractometers using the LaB_6 line profile standard (NIST SRM 660a LaB_6). LaB_6 crystallizes in space group $Pm\bar{3}m$ with a lattice parameter of $a = 4.155$ Å. The lanthanum and boron atoms are located at $(0, 0, 0)$ and $(\sim 0.2, \frac{1}{2}, \frac{1}{2})$ respectively.

4.2.1 Bragg–Brentano geometry (Cu-$K_{\alpha 1,2}$ doublet): FP approach

In this section LaB_6 data from a Bragg–Brenatano diffractometer URD6 (Präzisionsmechanik Freiberg, Germany) are analyzed with the FP approach. The diffractometer has a radius of 250 mm and is equipped with a Cu tube. The primary beam path contained a divergence slit (giving 0.46° divergence) and Soller slits (length 25 mm, plate spacing 0.5 mm). The secondary beam path contained a receiving slit of 0.52 mm and a curved graphite monochromator. The sample was a thin layer of LaB_6 powder on a "zero-background" Si plate with its (510) plane parallel to the surface. Figure 4.1 illustrates

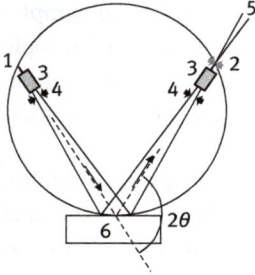

Figure 4.1: Schematic of a Bragg–Brentano diffractometer and the main components which affect the *IRF*. The tube is located near (1) and the detector near (5). The anode material in the tube determines the emission profile seen by the sample and also gives rise to the tube tails. The receiving slit (2) adds an additional contribution. When a position-sensitive detector (PSD) is used at this position, the simplest treatment is to consider a receiving slit with a width corresponding to the PSD channel width. The axial divergence (perpendicular to the plane of the paper) and its contribution to the *IRF* is determined by the length of the tube focus (1), and the acceptance angles of the Soller slits (3). The equatorial divergence (within the paper plane) is determined by the size of the divergence slits (4). Here the primary and secondary diffractometer radii (1)–(6) and (6)–(2) are identical. The large circle is the (para) focusing circle with a 2θ dependent radius. Usually a diffracted beam monochromator is present in front of the detector (5), selecting the $K_{\alpha1,2}$ radiation from the X-rays emitted by the tube so that only the corresponding lines are considered in the emission profile. Alternatively, a primary beam monochromator may be used to select $K_{\alpha1}$, radiation, modifying the emission profile correspondingly. Position (1) would then correspond to the focus of the primary beam monochromator.

the main components of the diffractometer together with the corresponding *IRF* contributions.

Figure 4.2. illustrates the progressive improvement of a Rietveld fit upon adding various contributions to the *IRF*, $g^{2\theta_0} = g_1^{2\theta_0} \circ g_2^{2\theta_0} \circ g_3^{2\theta_0} \ldots$. The scale factor and the background parameters were always adapted to give the correct peak height. The TOPAS INP file is discussed below.

The data (*xdd*) are available in file xy_LaB6_rec-052_109-10_URD6L1_220316.xy in xy-format (two columns: 2θ and number of counts). The keyword *x_calculation_step 0.02* ensures that the calculated data match the experimental step size. The keyword *chi2_-convergence_criteria* sets a stricter refinement convergence limit than the TOPAS default. Six background parameters are set to refine. A Lorentz-polarization factor for a graphite monochromator is defined:

```
xdd xy_LaB6_rec-052_109-10_URD6L1_220316.xy
   r_wp  7.70633377
   chi2_convergence_criteria 0.000001
   x_calculation_step 0.02
   bkg @  180.020452 -14.8395972  33.2174362 -17.2567608  4.27653909 -0.474351033
   LP_Factor( 26.6) ' Graphite diffracted beam monochromator
```

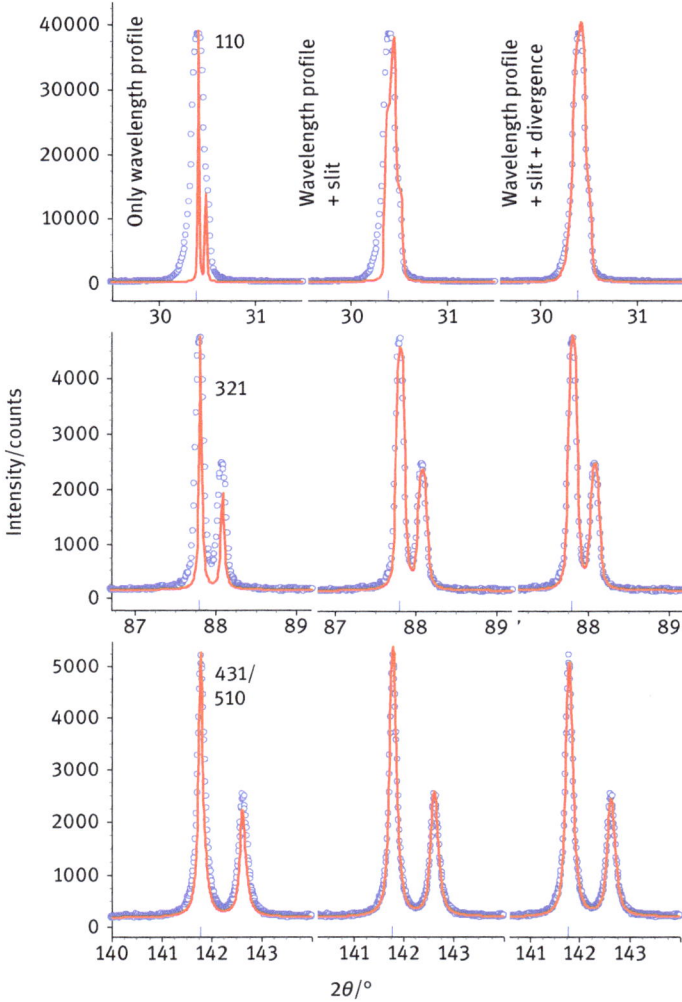

Figure 4.2: (110) (top), (321) (middle row) and (510)/(431) (bottom) reflections of the LaB$_6$ standard measurement considered in Section 4.2.1. Data points are shown as circles and fitted pattern as a continuous line. Left: Rietveld fit with only structural model and wavelength distribution for the Cu-K$_{\alpha1,2}$ doublet as described in the text. Middle column: additional consideration of a convolution due to the detector slit. Right: additional consideration of the contribution due to the divergence slit and the axial divergence. Note that the scale factors were adapted for the left and middle column to yield a calculated line that approximately fits the peak maxima. Lattice parameter and background parameters were taken from an optimum fit considering all contributions.

The next lines provide the first contribution to the *IRF*: the emission profile from the Cu anode in the X-ray tube as modified by the diffracted beam monochromator. It describes the Cu-K$_{\alpha1,2}$ doublet using the contents of the

Cu-Ka4_Holzer.lam[2] file provided in the lam subdirectory of the main TOPAS directory (Hölzer et al., 1997; see Section 2.5.9). Appropriate description of the emission profile is crucial, because a poor description of its shape cannot be compensated by other line broadening contributions:

```
Lam                    ' Equivalent command: CuKa4(0.0001)
   ymin_on_ymax  0.0001
   la  0.579 lo  1.5405909 lh  0.4374157
   la  0.080 lo  1.5410639 lh  0.643214
   la  0.236 lo  1.5443990 lh  0.5128764
   la  0.105 lo  1.5446855 lh  0.6872322
```

One might wonder why two major and two minor profile functions are used to describe the effect of the $K_{\alpha 1}$ + $K_{\alpha 2}$ doublet. The two major contributions describe the $K_{\alpha 1}$ and $K_{\alpha 2}$ lines at their literature positions. The two minor contributions are slightly shifted to describe the experimentally observed asymmetry of both lines.

The next lines define the primary and secondary radius (typically equal for ordinary Bragg–Brentano diffractometers) in mm. These default to 173 mm if not defined. The macro *Specimen_Displacement* corrects 2θ shifts caused by the specimen being displaced from its ideal position (see Section 2.1.1). The value is in mm and depends on primary radius *Rp*. Peak shape effects due to defocusing caused by the specimen displacement aren't considered by this macro:

```
Rp 250                      ' Primary radius (mm)
Rs 250                      ' Secondary radius (mm)
Specimen_Displacement(@,0)  ' Displacement (mm)
```

The peaks of the LaB_6 phase can be generated in alternative ways. Using the keyword *hkl_Is* either a Le Bail (Le Bail et al., 1988) or a Pawley (1981) (*lebail 0* or omission of the keyword) refinement can be performed:

```
hkl_Is
   lebail  1        ' Omit this line for a Pawley refinement
   phase_name "LaB6 line profile standard"
   space_group Pm-3m
   Cubic(@ 4.156)
```

---

[2] An increased number of lines allows a more detailed description of the wave-length distribution. Depending on the masking by convolution with other *IRF* contributions, a two-line classical description (e.g., CuKa2.lam) may suffice.

In this way each peak is fitted with an optimum intensity (see Chapter 3). After successful refinement, a list of *hkl* peaks is added automatically inside {} parentheses below the generated command *load hkl_m_d_th2 I*. This list is reused in subsequent refinements if it is left in the INP file. If the evaluated diffraction range or the lattice parameters are changed significantly, the list of reflections and the keyword *load hkl_m_d_th2 I* should be deleted to force the generation of a new list. Note that the reflection intensities are treated as truly refined parameters for a Pawley refinement.

Alternatively, using the keyword *str*, one can perform a Rietveld refinement (Rietveld, 1969):

```
str
   phase_name "LaB6 line profile standard"
   space_group Pm-3m
   scale @  0.00220081863
   Cubic(@  4.156579)
   site La num_posns  1 x 0      y 0      z 0     occ La  1 beq bb 0.3
   site B   num_posns  6 x @  0.2 y =1/2;  z =1/2; occ B   1 beq = bb;
```

Note that the atomic displacement parameters of the La and B atoms have been coupled together, and are thus forced to assume equal values.

In this example fits of almost identical quality can be achieved with the Le Bail, Pawley or Rietveld methods. Generally, it is expected that the Le Bail or Pawley approach will give better fits than the Rietveld approach, because they can compensate for inappropriately modeled atomic structure or texture. Problems only occur in cases of severe peak overlap, where improper partitioning between background and peak intensity can be expected, and can lead to inappropriate peak profile parameters. This is unlikely to be an issue for the simple materials typically used for measurements for the *IRF*.

Figure 4.2 (left) shows the Rietveld fit for three peaks using just the wavelength distribution. We see that the width of the wavelength distribution increases on the 2θ scale with increasing diffraction angle. In fact, the width increases with $\tan\theta_0$, with θ_0 as the (half) diffraction angle of the peak.

The next contribution we consider is the receiving slit size of 0.52 mm. This can be included with the macro:

```
Slit_Width( 0.52)
```

prior to the *hkl_Is* or *str* line. This macro convolutes each peak (at this stage only the wavelength distribution) with a box (*hat* in TOPAS) function of $2\theta_0$-independent width of 180°/π × (slit width)/(secondary diffractometer radius). Figure 4.2 (middle) shows the effect of this additional convolution. Due to the $2\theta_0$ independence of this contribution and the increase of the width of the wave length contribution with

$2\theta_0$ (Figure 4.2, left), the relative importance of the slit width decreases with increasing $2\theta_0$.

The effects due to equatorial and axial divergence can be considered by:

```
Divergence( 0.46)
axial_conv
    filament_length  10
    sample_length  15
    receiving_slit_length  12
    primary_soller_angle @ 2.251032597        ' Only parameter refined
    axial_n_beta  30
```

The macro *Divergence* applies an appropriate convolution for equatorial divergence of 0.46° as defined by the primary divergence slit. The lines following *axial_conv* indicate the filament length, sample length (actually the width of the sample perpendicular to the beam direction), the length of the receiving slit (all in mm) and the acceptance angle of the Soller slit (in °) in the primary beam. The number after keyword *axial_n_beta* controls the numerical accuracy with which the convolutions belonging to the *axial_conv* keyword are performed. Figure 4.2 (right) shows the improvement of the profile description achieved with these commands, especially with respect to description of the asymmetry of reflections at low $2\theta_0$.

Note that in many cases the divergence-related asymmetry effects can instead be described in an empirical way by refining the single argument of the macro *Simple_Axial_Model*:

```
Simple_Axial_Model(@, 14.81471)
```

Two further improvements are possible, the effect of these is demonstrated in Figure 4.3. First, the effects of so-called tube tails are refined using the numbers in the macro *Tube_Tails* (Bergmann et al., 2000; see Section 2.5.9). This considers diffraction by X-rays not coming from the actual focus of the tube:

```
Tube_Tails(@, 0.04493,@, -0.98542,@, 1.24953,@, 0.00152)
```

Moreover, a tiny Cu-K_β contribution to the diffraction line can be discerned. Its intensity depends on the characteristics of the graphite monochromator and can be refined together with its width and exact wavelength:

```
la @ 0.00198654126 lo @ 1.392030917 lh @ 0.4525269504
```

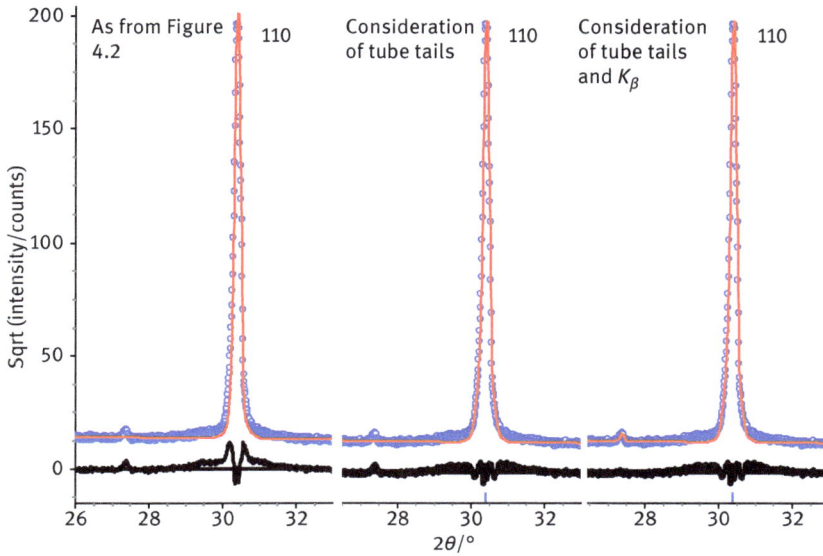

Figure 4.3: 110 peak of the LaB$_6$ standard measurement considered in Section 4.2.1. Note the square root (Sqrt) intensity scale in contrast to Figure 4.2. Left: As evaluated in Figure 4.2. Middle: Additional consideration of tube tails. Right: Additional consideration of a Cu-K$_\beta$ contribution to the profile. Note that the final fit still misses some features in the tails of the peak.

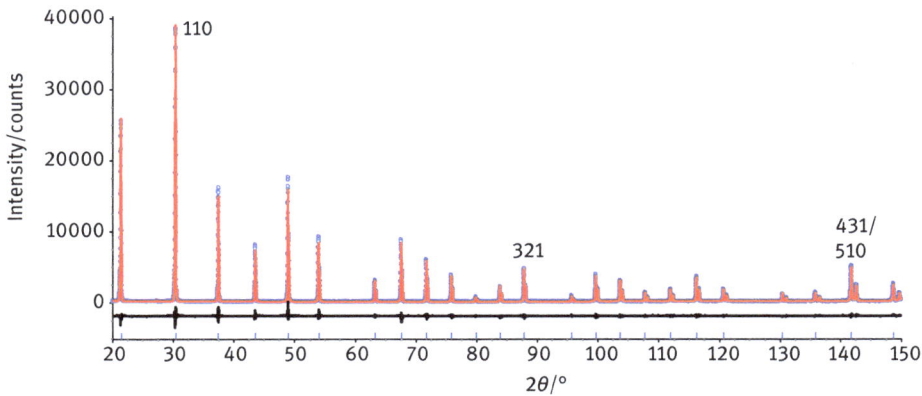

Figure 4.4: Final fit to the LaB$_6$ standard measurement considered in Section 4.2.1. The reflections shown in Figures 4.2 and 4.3 are labeled.

This line is added as an additional line along with four other *la* lines of the original wavelength distribution. The final fit to the whole diffraction pattern is shown in Figure 4.4.

Note that in this FP approach, only the acceptance angle of the primary beam Soller slit, the parameters pertaining to the tiny Cu-K$_\beta$ line and parameters of the tube tails have been refined. All other parameters were based on the actual geometry of the instrument. This approach works well for a well-adjusted Bragg–Brentano diffractometer. Unnecessary refinement of well-established instrument parameters can lead to severe correlations during least squares refinement and to unphysical refined values. This should be avoided as unphysical values can sometimes cause:

– improper extrapolation of the *IRF* outside the $2\theta_0$ range available from the LaB$_6$ measurements. Such extrapolation is sometimes unavoidable in case of low $2\theta_0$ values, that is, at lower angles than the LaB$_6$-100-reflection.
– long computation times and other artefacts.

Once one has arrived at a proper description of the *IRF*, it can be used for Rietveld (or Le Bail or Pawley) fitting of diffraction data from substances that show line broadening with respect to the *IRF*. This requires that their diffraction data have been measured under conditions leading to the same *IRF*. If this is the case, the keywords and macros determining the *IRF* should be copied with fixed parameters into the INP file to be used for evaluating the broadened data. This can conveniently be done by replacing the description of the LaB$_6$ phase by descriptions of one or several new phases in the INP file.

4.2.2 Bragg–Brentano geometry (Cu-K$_{\alpha1,2}$ doublet): extended TCHZ approach

As a simple alternative to the treatment in Section 4.2.1, one can describe the *IRF* on the basis of a pseudo-Voigt function (Thompson et al., 1987) along with a simplified description of the beam-divergence-induced peak asymmetry:

```
' xdd level
Simple_Axial_Model(@, 12.02815)
' str or hkl_Is level
TCHZ_Peak_Type(!pku,0,!pkv,  0,pkw, 0.00624,!pkz, 0,!pkx, 0.02280,pky, 0.01927)
```

The TCHZ macro has to be entered at the *str* or *hkl_Is* level, where the LaB$_6$ phase is described. The macro *TCHZ_Peak_Type* has 12 entries separated by commas storing the six numerical values of parameters named *pku, pkv, pkw, pkz, pkx* and *pky* (traditionally named *U, V, W, Z, X* and *Y*, see Section 2.5.5). The first four values (*pku, pkv, pkw, pkz*) describe the $2\theta_0$ dependence of the Gaussian contribution. The effect of *pkz* can be described in terms of a combination of *pku* and *pkw*. As a result, *pkz* should not be refined at the same time as *pku* and *pkw* without additional constraints (for details see TOPAS technical reference manual). *pkx* and *pky* describe the $2\theta_0$ dependence of the Lorentzian contribution. The $2\theta_0$-dependent shape of the pseudo Voigt function is then determined as described by eq. (2.66).

The wavelength definition used in the FP approach (Section 4.2.1) should not be used in combination with the TCHZ function (or only if the wavelength distribution dominates the *IRF*). Instead a simplified emission spectrum should be used:

```
lam                      ' Equivalent to the line CuKa2(0.0001) plus a Kbeta component
   ymin_on_ymax  0.0001
   la  2 lo  1.5405909 lh  1
   la  1 lo  1.544399 lh  1
   la @  0.00602 lo  1.392030917 lh  1
```

This approach is better as TOPAS interprets the *lam* commands differently when FP is not used. In this case the *U–Y* (*pku–pky*) parameters are used to describe the width of the strongest profile contribution and the *lh* values are factors describing the relative width of each contribution. This works as non-equal widths of $K_{\alpha1/2}$ and K_β components are usually masked by other broadening effects. Use of an emission profile as in the FP approach (Section 4.2.1) can give unrealistically different widths for the $K\alpha_1$ and $K\alpha_2$ components leading to a poor profile description. This becomes irrelevant if only a single wavelength component is present (e.g., when using a Johansson monochromator in the laboratory or monochromatic synchrotron radiation).

4.2.3 Bragg–Brentano geometry with Cu-K$_{\alpha1}$ radiation

In our third example of the determination of an *IRF*, an extended TCHZ approach has been employed on LaB$_6$ diffraction data recorded on a Bruker diffractometer equipped with a Johansson monochromator in the primary beam and a 3.5° position sensitive silicon strip detector. The complete TOPAS INP file reads:

```
xdd LaB6-Standard-020712-2.raw
   r_wp  3.38593058
   chi2_convergence_criteria 0.000001
   bkg @ 1714.9 -182.8 67.0 -151.7 76.9 -121.5 57.7 -80.4 51.7 -46. 9 32.4 -16.7
   start_X  18
   finish_X  118
   LP_Factor( 27.3)
   Specimen_Displacement(@, -0.04561)
   Simple_Axial_Model(@, 7.73601)
   lam
```

```
  ymin_on_ymax  1e-003
  la  1 lo  1.540596 lh  1
hkl_Is
  phase_name "LaB6"
  TCHZ_Peak_Type(@, 0.00233,@, -0.00403,@, 0.00354,, 0,@, 0.05742,, 0)
  r_bragg  0.180027153
  space_group "Pm-3m"
```

In the *lam* section, the *la* value (set to 1) acts as an overall scale factor for intensities. As a *TCHZ_Peak_Type* is used together with a single wavelength, the value of the line width *lh* and/or *lg* has no effect on the peak shape and width. The macro *Simple_Axial_Model* describes some minor divergence-related peak asymmetry. The final Pawley fit is shown in Figure 4.5. Note that a Rietveld fit to the same data leads to a somewhat improper fit of the angle-dependent integrated intensities, likely caused by nonnegligible effects of absorption (see Section 2.3.3).

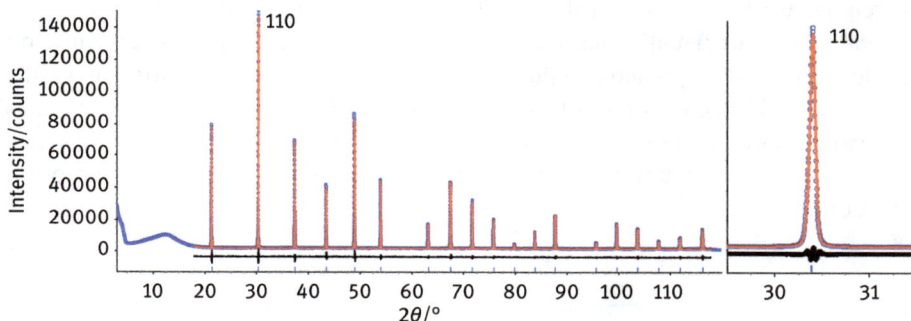

Figure 4.5: Final fit to the LaB_6 standard measurement considered in Section 4.2.3 (Cu-$K_{\alpha 1}$ radiation). Right: zoomed 110 reflection.

Use of the FP approach on data from a diffractometer with primary beam monochromator and/or position-sensitive detectors may require more individual treatment. The wavelength distribution of the $K_{\alpha 1}$ component may be narrower and less Lorentzian shaped than is observed without a monochromator (Cline et al., 2015). This can be considered by an additional Gaussian component of the wave-length line with a width given by the parameter *lg* (not to be used with the TCHZ approach):

```
lam
ymin_on_ymax  1e-003
la  1 lo  1.540596 lh  0.3 lg 0.2
```

To a first approximation, the effect of the equatorial channel width of a position-sensitive detector may be considered as a convolution with a box (*hat*) function with 2θ-independent width[3]:

```
Rp 217.5
Rs 217.5
prm c 0.1 min 0.000001 max 1
hat = Rad c / Rs;
```

---

where c is the width of a channel in mm. This effect corresponds to that of a receiving slit in classical Bragg–Brentano geometry (see Section 4.2.1, Cline et al., 2015).

Note that misalignment of diffractometer components, especially of primary beam monochromators, can lead to line-profile effects that are difficult to model. This again illustrates the importance of a proper assessment of the instrumental profile.

4.3 Treatment of sample-dependent line broadening

Like in Section 4.1 we first focus on the effects on the $x = 2\theta$ scale, whereas the $x = TOF$ scale will be considered within Section 4.4. Line broadening originating from the sample can arise from multiple origins of differing complexity (see Figure 4.6). As compared to the *IRF* (which can be regarded as a minimum observable line broadening in view of eq. (4.1)), the effects can be small or large; that is, the measured reflections can be slightly or considerably broader than a peak described by the *IRF*. Parameters describing sample line broadening are often strongly correlated, and it is, therefore, important to use a simple and appropriate model that is matched to the data quality available.

We will restrict ourselves to size and microstrain as causes for structural line broadening. These effects lead to modifications in the diffraction patterns that can be described by broadening of the individual peaks, so that one can rewrite *MS* in eq. (4.1) as:

$$MS = (Size \circ Strain), \tag{4.5}$$

where the contributions from size and microstrain are assumed to be independent. The situation is somewhat more complex in the case of irregular stacking of layers, which is discussed separately in Chapter 10. Moreover, although microstrain broadening can be significantly asymmetric (see Figure 4.6), only symmetric microstrain broadening will be considered here. Both contributions can be either isotropic or

---

3 The macro *Rad* is a predefined constant equal to 57.2957795130823, that is, *180/Pi*, where the constant *Pi* is also predefined.

anisotropic. Isotropy of a line broadening contribution means that shape and width of the contribution varies solely as a function of the diffraction angle $2\theta_0$ of the hkl peak. Anisotropy means an additional dependence of width and/or shape on hkl. Isotropic and anisotropic line widths are illustrated as a function of $2\theta_0$ in Figure 4.6. Figure 4.7 anticipates details from the next sections by highlighting the different $2\theta_0$ dependences of size and microstrain broadening due to the dependence of their widths (in angle-dispersive diffraction) on $1/\cos\theta_0$ and $\tan\theta_0$ respectively. Roughly

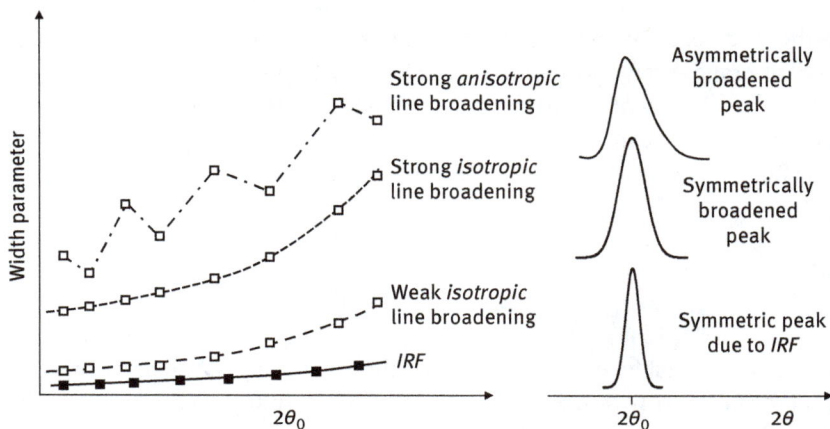

Figure 4.6: Left: Typical evolution of reflection widths with increasing diffraction angle for the instrumental resolution (*IRF*), and for materials showing *weakly* or *strongly* isotropically broadened peaks, as well as anisotropic line broadening, where the total widths cannot be described as a continuous function in 2θ. Right: symmetric and asymmetric line broadening relative to the *IRF*.

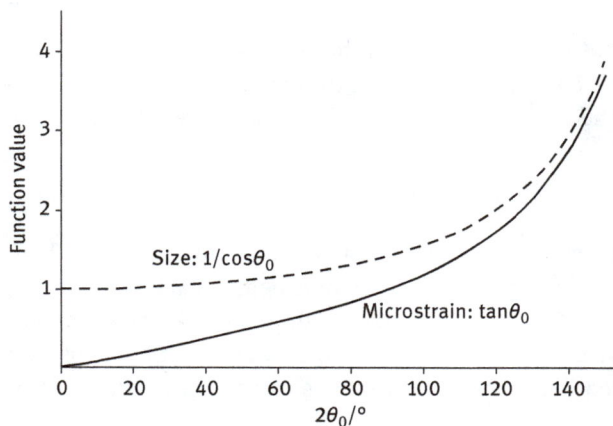

Figure 4.7: 2θ dependence of the functions $1/\cos\theta_0$ and $\tan\theta_0$ illustrating the relative importance of size and microstrain broadening at low and high angles.

speaking, the size broadening is important at all angles, whereas microstrain broadening becomes much more important at high diffraction angles.

The commands/macros described in the next sections are applied at the *str* or *hkl_Is* phase level, whereas for multiphase mixtures the *IRF* is phase independent and described at the global level. Nevertheless, specific macros like the *TCHZ_Peak_Type* for the description of the *IRF* need to be defined (identically) for each phase or via a construct such as *for strs*.

It should also be noted that from Version 5 on the so-called Whole Powder Pattern Modeling (WPPM; Scardi & Leoni, 2002) approach can be applied within TOPAS. This in particular opens the possibility to describe different *MS* contributions on the *x* scale (2θ or also *s*) in terms of the Fourier transforms. This is useful because there are many situations (e.g., explicit consideration of size distributions) where analytical expressions for the line broadening contribution only exist on the Fourier scale. Macros for the WPPM approach are available on http://topas.dur.ac.uk/topas wiki/doku.php?id=wppm_macros&s[]=wppm. The WPPM approach will not be dealt with in the present chapter.

4.3.1 Size broadening

As discussed in Section 1.2, size broadening is caused by a finite size of the crystallites (or more precisely, of the coherently diffracting domains) contributing to the diffraction pattern. Typically these crystallites are not all identical but vary in size and shape. This is frequently considered in terms of a crystallite size distribution, which can be characterized by different parameters, the most important of these are average size values.

On the 1/*d* scale the *fwhm* of the size broadening for a reflection *hkl* is proportional to the reciprocal of the crystallite sizes. For convenience, we introduce D_{hkl} as *some* average of the crystallite-size distribution perpendicular to the *hkl* lattice planes:

$$fwhm_{1/d}^{hkl} = \frac{K_{fwhm}}{D_{hkl}}, \tag{4.6}$$

where K_{fwhm} is a shape constant of the order of 1 (and set to 1 from hereon).

Transforming this expression to the 2θ scale via eq. (4.4), gives the famous Scherrer equation derived in Section 1.2:

$$fwhm_{2\theta}^{hkl} = \frac{\lambda}{D_{hkl}\cos\theta_0}, \tag{4.7}$$

where $fwhm_{2\theta}^{hkl}$ is the *fwhm* (in radians) of the size broadening contribution to the overall peak shape of the reflection *hkl* at the position 2θ_0. Although use of the Scherrer equation is frequently criticized, it is basically correct, and the criticism should more concentrate on the way in which the analysis is done.[4] Practical application of eq. (4.7) requires some description of the *hkl*-dependence of D_{hkl} and of the *hkl*-dependence of the shape of the peaks. The peak shape depends on the shape of the crystals and on the crystallite size distribution and is taken as *hkl*-independent in most Rietveld applications. For approaches to consider more complex shapes due to realistic size distributions, which also take into account specific crystallite shapes, see, for example, David et al. (2010).

If the size is isotropic, the same size parameter value D_0 applies for each reflection *hkl*:

$$D_{hkl} = D_0. \qquad (4.8)$$

Adopting a Gaussian peak shape, size broadening can be realized by the following commands (see Section 2.5):

```
prm D0 50
gauss_fwhm = 0.1 Rad Lam / (Cos(Th) D0);
```

with a starting value of *D0* = 50 (in nm). This is equivalent to the built-in macro:

```
CS_G(@, 50)
```

For a Lorentzian shape, the equivalent commands would read:

```
prm D0 50
lor_fwhm = 0.1 Rad Lam / (Cos(Th) D0);
```

or:

```
CS_L(@, 50)
```

In the scripts above, the command *gauss_fwhm* (*lor_fwhm*) convolutes each peak of the corresponding phase with a Gaussian (Lorentzian) with a *fwhm* as given in the argument. The reserved parameter name *Lam* is the wavelength (see Section 4.2.1) and the factor 0.1 converts from Å into the usual unit for *D0* of nm.

In reality there is no crystallite shape and size distribution, which can lead to an exact Gaussian size broadening. Typically, size broadening can be reasonably well

---

4 The most frequent flaws are not considering the *IRF* upon determination of $fwhm_{2\theta}^{hkl}$, not realizing that the *fwhm* in eq. (4.7) is given in radians, the use of a single reflection and an inadequate interpretation of the determined value D_{hkl}.

approximated by an intermediate Gaussian–Lorentzian function, although broad size distributions can also lead to super-Lorentzian profiles (which we won't consider further). This description for size broadening can be introduced by simply refining the two parameters *DOG* and *DOL* simultaneously:

```
prm D0G 100
prm D0L 100
gauss_fwhm = 0.1 Rad Lam / (Cos(Th) D0L);
lor_fwhm   = 0.1 Rad Lam / (Cos(Th) D0G);
```

The same is achieved by a combination of the predefined macros *CS_G* and *CS_L*:

```
CS_G(@, 100)
CS_L(@, 100)
```

or *CS_G(DOG, 100)* and *CS_L(DOL, 100)* if the refined parameters need to be accessible for further calculations.

Getting a good fit after refining different values for *DOG* and *DOL* raises the question of how to interpret the two values. It is clear that there is not a separate Gaussian or Lorentzian distribution of domain sizes. Instead the ensemble exhibits overall size broadening given by the shape and the total width of the function resulting from the convolution of the Gaussian with the Lorentzian (i.e., a Voigt function, which is, usually approximated in TOPAS by a pseudo-Voigt).[5] A well-defined average measure for the domain size, and thus a physically meaningful parameter, is the *volume average column height* (length), called D_{vol} here. That value corresponds to the *integral breadth* (see eq. (13.97) and following) of the overall size broadening on the $1/d$ scale, which is reflection-order independent:

$$\beta_{1/d} = \frac{1}{D_{vol}}. \qquad (4.9)$$

Taking into account eq. (4.4) and the relations between the width parameters integral breadth and *fwhm*, the value of D_{vol} can be calculated and reported by[6]:

```
CS_G(D0G, 50)
CS_L(D0L, 50)
prm Dvol = 1/Voigt_Integral_Breadth_GL (1/D0G, 1/D0L);:22.44
```

---

5 If a more accurate approximation to the Voigt function than the pseudo-Voigt is required, it can be achieved with the TOPAS keyword *more_accurate_Voigt*.

6 Omitting "*:22.44*" would mean that the value is calculated by TOPAS but not reported. Report is achieved by adding the ":" and a dummy number, which is replaced by the final value after refinement (and the standard deviation, if *do_errors* is active).

hkl-dependent size broadening due to anisotropically sized domains can be modeled by having a *hkl*-dependent size value D_{hkl} instead of D_0. D_{hkl} depends only on the direction, but not on the reflection order n. Hence $D_{hkl} = D_{h'k'l'}$ with $h' = nh$, $k' = nk$, $l' = nl$. One way of ensuring such properties for D_{hkl} is to multiply an average size parameter DO with a spherical harmonics function using the *spherical_harmonics_hkl* keyword (see Section 13.19)[7]:

```
prm D0  50
spherical_harmonics_hkl Ahkl
sh_order  4
prm Dhkl = Max(D0 Ahkl, 1);
```

TOPAS extends the *sh_order 4* keyword by a symmetry adapted series of parameters that are the coefficients of a spherical harmonics expansion describing the *hkl*-dependence of the factor *Ahkl*. In the case of $m\bar{3}m$ Laue symmetry this reads:

```
prm D0 50
spherical_harmonics_hkl Ahkl
sh_order  4 load sh_Cij_prm {
    k00   !Ahkl_c00  1.00000
    k41    Ahkl_c41  0.00000
}
prm Dhkl = Max(D0 Ahkl, 1);
```

Since *Ahkl* can be negative in some directions in space (i.e., for certain *hkl*), a positive lower limit is imposed through the expression *Max(DO Ahkl, 1)*. The values of *Dhkl* then can be used for size broadening with:

```
prm !zeta 1 min 0 max 1
prm DhklG = Dhkl/(1-zeta);
prm DhklL = Dhkl/zeta;
CS_G(DhklG)
CS_L(DhklL)
```

By using (1-*zeta*) and *zeta* as weighting factors it is possible to associate an *hkl*-independent shape with *Dhkl*, which is Gaussian for *zeta* = 0 and Lorentzian for *zeta* = 1. Note that by dividing the *Dhkl* by *zeta* or *(1-zeta)* the *fwhm* values of the Gaussian and Lorentzian broadening executed by the *CS_G* and *CS_L* macros become proportional to the weighing factors divided by *Dhkl* (see above and eq. (4.7)). The parameter *zeta* resembles the mixing parameter η (or *lor* as

---

[7] Laue symmetry is automatically considered in the spherical harmonics expansions although, in general, the crystallite size does not need to be symmetry invariant.

frequently used in TOPAS) of a pseudo-Voigt (see Section 13.17.4) but is not identical to it, except for the fact that the extreme values of *zeta* = 0 and 1 lead to a Gaussian and Lorentzian respectively.[8] A similar approach will be used in Section 4.3.2 for microstrain broadening.

hkl-dependent values for the *volume average column height* (length) can be calculated as follows:

```
prm Dvolhkl = Dhkl/Voigt_Integral_Breadth_GL ((1-zeta), zeta);
```

Since *Dvolhkl* is different in different directions the easiest way to visualize its value is to plot it as a surface, where the direction-dependent distance from the center of the plot indicates the direction-dependent value of *Dvolhkl*:

```
normals_plot = Dvolhkl;
  normals_plot_min_d 0.3
```

To generate such a plot (see, e.g., Figure 4.8), TOPAS performs a direction-dependent interpolation between *Dvolhkl* values calculated for all reflections *hkl* considered in the refinement. The command *normal_plot_min_d 0.3* restricts the number of reflections by considering only *hkl* with a minimum *d* spacing of (in this case) 0.3 Å. Although the *normals_plot* keyword provides easy access to plots, one should be aware that the *Dvolhkl* values cannot be read directly from the plot. It is, however, possible to extract values with a construct such as:

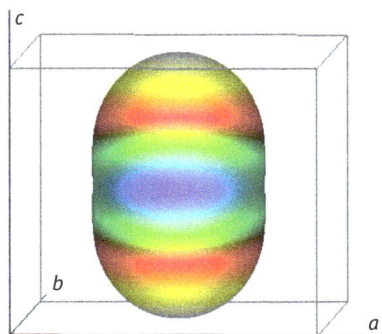

Figure 4.8: Example of a surface depicting the direction dependence of the coherently diffracting domain size (e.g., of the *volume average column height*) drawn using the *normals_plot* keyword.

---

8 *zeta* corresponds to ζ in Section 13.17.5 since, due to the 1/*size* dependence of the *fwhm*, it acts as weighting parameter for the Gaussian and Lorentzian convolutions conducted by the macros *CS_G* and *CS_L*.

```
phase_out Dvolhkl_values.out load out_record out_fmt out_eqn   {
   "%3.0f" = H;  " %3.0f" = K;  " %3.0f" = L;  " %11.4f\n" = Dvolhkl;}
   }
```

This will write the *Dvolhkl* values for different *hkl* into a file called Dvolhkl_values.out.

Generally, direction-dependent domain size values should be interpreted with care and shapes like in Figure 4.8 should not be equated with true crystallite shapes. Use of microscopic methods is generally recommended to supplement line-broadening analysis data. For interpretation of such data in terms of actual crystallite shapes, see more advanced literature (e.g., Scardi & Leoni, 2001).

4.3.2 Microstrain broadening

Microstrain broadening is caused by local variations of *d*-spacings in the crystallites of a diffracting specimen. A specific deviation from some reference *d*-spacing d_0^{hkl} (e.g., the average) is typically referred to as *strain* ε^{hkl} (see Figure 4.9):

$$\varepsilon^{hkl} = \frac{d^{hkl} - d_0^{hkl}}{d_0^{hkl}}.$$ (4.10)

With respect to the reflection position $2\theta_0^{hkl}$ expected for the *d* spacing d_0^{hkl}, a peak due to a spacing d^{hkl} associated with a strain ε^{hkl} will shift by (with angles in radians!):

$$2\theta^{hkl} - 2\theta_0^{hkl} = -2\varepsilon^{hkl}\tan\theta_0^{hkl}$$ (4.11)

Here, eq. (4.3) has been used to describe the effect of a change in *d* spacing on the 2θ scale.

The term microstrain means a coherent or incoherent superposition of the diffraction from the regions in the specimen with different strain, thus, it is microstrain broadening that appears as *Strain* in eq. (4.5). The strain underlying microstrain broadening can be of elastic nature, that is, due to microstress/stress of second kind as well as from the strain fields of dislocations or of nonelastic nature due to local composition or concentration variations (see Leineweber, 2011).

In any real sample the strain will exhibit some kind of probability density function on the ε^{hkl} scale. This can be mapped onto the 2θ scale (in radians) via eq. (4.4) (Leineweber & Mittemeijer, 2010):

$$fwhm_{2\theta}^{hkl} = 2\,fwhm_{\varepsilon}^{hkl}\tan\theta_0$$ (4.12)

The quantity $fwhm_{\varepsilon}^{hkl}$ is an unusual measure for microstrain (in contrast to the square root of the variance of the microstrain $\sigma_\varepsilon = \text{var}\left(\varepsilon^{hkl}\right)^{1/2}$, or $\tilde{e}^{hkl}$ corresponding to half of the integral breadth of the microstrain distribution [Delhez et al., 1988]), but $fwhm_{\varepsilon}^{hkl}$ can easily be related with the *fwhm* of the associated line broadening without taking the actual shape of the peak into account.

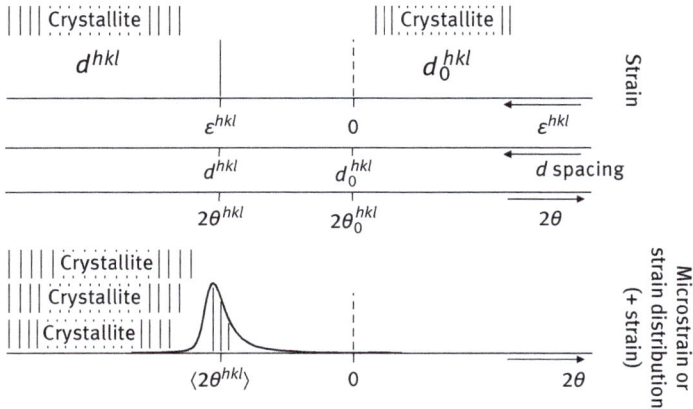

Figure 4.9: Illustration of the effects of strain and microstrain on diffraction pattern for a given reflection *hkl*. Top: Effect of *strain*, that is, a constant change of the *d*-spacing throughout (all crystallites of) the sample in the sense of eq. (4.10), shown on the strain (ε), *d* spacing (*d*) and diffraction angle (2θ) scales with the corresponding reflection indicated as a vertical line. On the 2θ scale a shift with respect to the reference value $2\theta_0^{hkl}$ occurs in accordance with eq. (4.11). Bottom: Microstrain corresponds to inhomogeneous strain (e.g., different in different crystallites or varying within the crystallites) leading to *microstrain broadening*. As an example three sets of lattice planes with slightly different strain contributing to the overall profile (three vertical lines) are shown, whereby here an average strain is superposed by microstrain. In the current chapter the average *d*-spacing is taken as identical with the reference *d*-spacing. Moreover, only symmetric microstrain broadening is considered.

Microstrain broadening can be modeled in TOPAS in various ways. Similar to isotropic crystallite size broadening, isotropic microstrain broadening with a Gaussian line shape can be introduced by:

```
prm E0 0.001
gauss_fwhm = 360 / Pi Tan(Th) E0;
```

which is alternatively achieved by the macro *Strain_G*:

```
prm E0 0.00100
prm arg_E0 = 360/Pi E0;:0.11459
Strain_G( arg_E0 )
```

The equivalent for a Lorentzian line shape is:

```
prm E0 0.001
lor_fwhm = 360 / Pi Tan(Th) E0;
```

or:

```
prm E0 0.00100
prm arg_E0 = 360/Pi E0;;0.11459
Strain_L( arg_E0 )
```

For all cases the parameter *E0* corresponds to $fwhm_\varepsilon^{hkl}$. Note that arguments of the macros *Strain_G/L* have been defined in TOPAS such that they lead to unrealistic values of strain measures (in contrast to *CS_G/L*). That has been compensated here by the factors *360/Pi*.

Microstrain broadening of intermediate Gaussian–Lorentzian shape can be modeled with:

```
prm E0G   0.00100
prm E0L   0.00100
prm arg_E0G = 360/Pi E0G;;0.11459
prm arg_E0L = 360/Pi E0L;;0.11459
Strain_G( arg_E0G )
Strain_L( arg_E0L )
```

From the values *E0G* and *E0L* one can calculate the microstrain $fwhm_\varepsilon^{hkl}$, which is the *fwhm* of a (approximate) Voigt function describing the distribution of ε through:

```
prm E0 = Voigt_FWHM_GL(E0G, E0L);;0.00164
```

Anisotropic microstrain broadening can be modeled in different, more or less equivalent fashions. The most prominent approach has been introduced by Rodriguez-Carvajal et al. (1991) and is frequently used in the forms described by Popa (1998) or Stephens (1999). The approaches have a sound statistical basis (at least concerning the direction dependence of the variance of the microstrain), that is, they are related with the statistics of the microstrain distribution. Invariance of the line width with respect to the crystal's Laue class is typically adopted. Different macros are therefore available in TOPAS.INC for the different Laue classes: *Stephens_triclinic* ($\bar{1}$), *Stephens_monoclinic* ($2/m$), *Stephens_orthorhombic* (mmm), *Stephens_tetragonal_low* ($4/m$), Stephens_ tetragonal_high ($4/mmm$), Stephens_trigonal_low ($\bar{3}$), *Stephens_trigonal_high* ($\bar{3}m1$), *Stephens_trigonal_high_2* ($\bar{3}1m$), *Stephens_hexagonal* ($6/m$ and $6/mmm$), *Stephens_cubic* ($m\bar{3}$ and $m\bar{3}m$). Note that for all trigonal Laue classes the macros assume the hexagonal setting of the unit cell. There are no ready-made macros for rhombohedral unit cells, but symmetry restrictions have been reported by Leineweber (2006). The user should note that these different macros employ different numbers of arguments due to the different number of independent parameters to be refined.

In the following this approach is illustrated using X-ray data recorded from tetragonal Pb_3O_4. This polymorph typically exhibits pronounced microstrain broadening of the $h00$ reflections, whereas $hh0$ reflection are relatively narrow (for some background, see Leineweber & Dinnebier [2010]). The data were recorded with the same instrumental settings as the LaB_6 data used for evaluation of the *IRF* in Section 4.2.3 (including Cu-$K_{\alpha 1}$ radiation). After replacement of the diffraction data (*xdd* keyword), fixing the parameters of the *TCHZ_Peak_Type* and the *Simple_Axial_Model* macros, introducing an appropriate model for the crystal structure and introducing the macro *Stephens_tetragonal_high*, the INP file looks like (the impurity lines are not shown in the listing):

```
xdd Pb3O4.raw
   r_wp  8.96620702
   chi2_convergence_criteria 0.000001
   bkg @  655.26 -349.0 313.1 -109.4 109.0 -27. 8 35.8 20.7 -24.6 24.8
   start_X  18
   finish_X  118
   LP_Factor( 27.3)
   Specimen_Displacement(@,  0.00334)
   Rp 217.5
   Rs 217.5
   Simple_Axial_Model( 7.73601)
   lam
       ymin_on_ymax   1e-003
       la  1 lo  1.540596 lh  1
   str
       phase_name "Pb3O4"
       TCHZ_Peak_Type(, 0.00233,, -0.00403,, 0.00354,, 0,, 0.05742,, 0)
       Stephens_tetragonal_high(zeta, 1.0, s400, 336.5,s004, 10.4,s220, -670.0,s202, 25.5)
       r_bragg  4.546
       space_group P42/mbc
       scale @  0.000285947836
       Tetragonal(@  8.813227,@  6.564940)
       site Pb   num_posns 4 x      0        y   0.5       z  0.25 occ Pb 1 beq @   0.9
       site Pb1 num_posns 8 x      0.14     y @ 0.16381 z  0       occ Pb 1 beq @   1.2
       site O1   num_posns 8 x o1xx 0.68135 y =o1xx-0.5; z  0.25 occ O  1 beq bbo 3.0
       site O2   num_posns 8 x @    0.08847 y @ 0.62008 z  0       occ O  1 beq bbo 3.0
```

For simplicity a common isotropic displacement parameter for the two O sites (*bbo*) is used. The symmetry restriction for the fractional coordinates is ensured by equating the values for the fractional coordinates x and y. After refinement of all parameters one gets the fit shown in Figure 4.10. The only line broadening with respect to the *IRF* is achieved by the macro *Stephens_tetragonal_high* (with the usual "@" substituted by user-set parameter names).

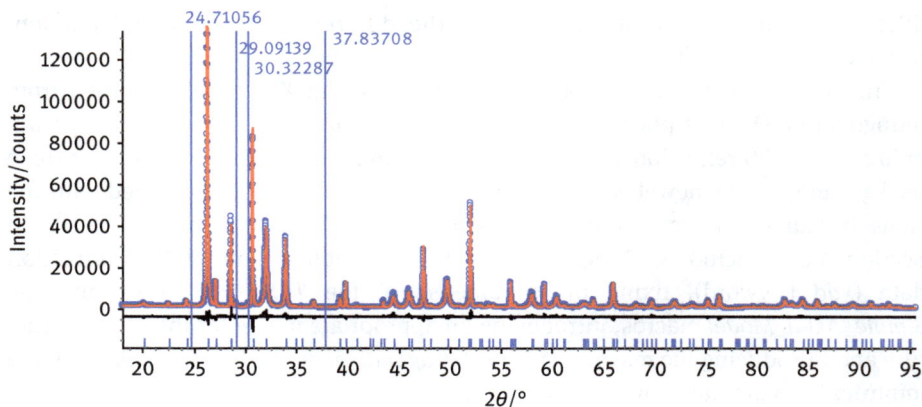

Figure 4.10: Final fit to powder diffraction data of Pb_3O_4 recorded with Cu-K$_{\alpha 1}$ using the macro *Stephens_tetragonal_high* to describe anisotropic microstrain broadening. Details of the effects of that macro are illustrated in Figure 4.11. Vertical lines indicate reflections of minor unknown impurities treated as a peak phase.

In Figure 4.11 we can see an improving fit on introducing increasingly sophisticated line broadening models: Using only the *IRF* gives too narrow fitted reflections. Isotropic microstrain (e.g., use of the *Strain_L* macro) broadening gives a poor fit since the line broadening is pronouncedly anisotropic. Only use of the macro *Stephens_tetragonal_high* to describe the anisotropy of the microstrain broadening leads to reasonable agreement.

The macro *Stephens_tetragonal_high* applies the following lines:

```
prm zeta   1.00000 min 0 max 1
prm s400   336.53970 min 0
prm s004   10.39557 min 0
prm s220   -670.02572
prm s202   25.49056
prm mhkl = s400 (H^4 + K^4) + s004 L^4 + s220 H^2 K^2 + s202 (H^2 L^2 + K^2 L^2);
prm pp = D_spacing^2 * Sqrt(Max(mhkl,0)) Tan(Th) 0.0018/Pi;
gauss_fwhm =  pp (1-zeta);
lor_fwhm   =  pp zeta;
```

We can see that the macro convolutes each Bragg reflection with a Voigt-like function. The width parameter E of each *hkl* reflection (on the 2θ scale in °) in the sense of eq. (13.101) is given by the parameter pp, while the shape is defined by the parameter *zeta* (variable ζ in eq. (13.101)). The shape of the microstrain broadening is independent of *hkl*, but the width pp is direction (*hkl*) dependent.

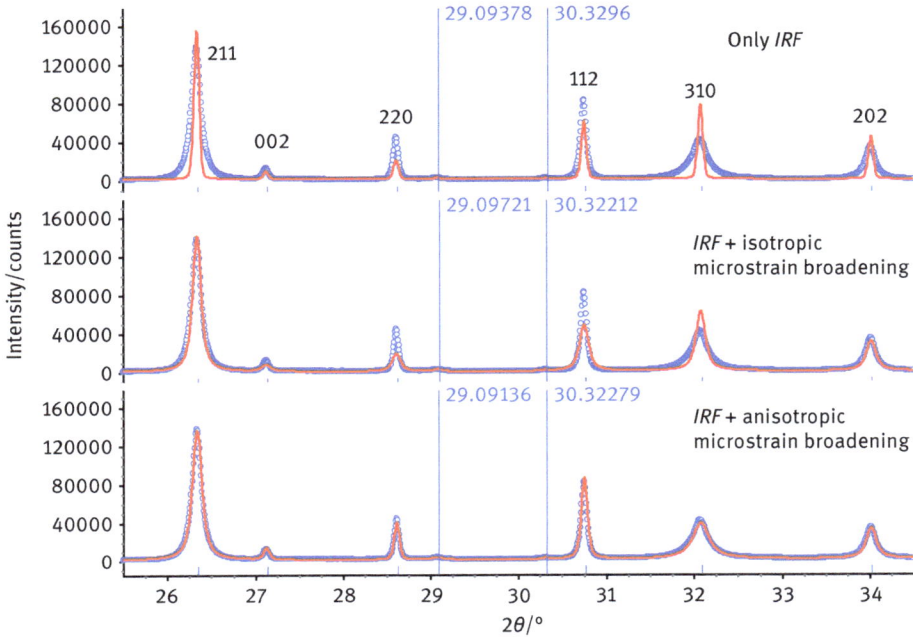

Figure 4.11: Part of the Pb_3O_4 data from Figure 4.10 with *hkl* indices for the different reflections, showing the improvement of fit with increasing complexity of the line broadening model: top: only *IRF* (scale factor adapted). Middle: isotropic microstrain broadening. Bottom: anisotropic microstrain broadening as in Figure 4.10. Vertical lines indicate reflections of minor unknown impurities treated as a peak phase.

The shape corresponds, in the present case, to a Lorentzian limit (*zeta* = 1). The direction-dependence of the microstrain broadening is quantified by the values of *s400*, *s004*, *s220* and *s202* (called *sHKL* parameters) as coefficients of a symmetry invariant (here with respect to 4/*mmm* symmetry) fourth-order polynomial in the Laue indices *hkl*. Following the original ideas of the theory (e.g., Stephens, 1999), this fourth-order polynomial (parameter *mhkl*) should be ≥ 0 for each *H*, *K* and *L* used to calculate the value of *pp*. Except for high symmetries, it is cumbersome to develop *general* limits for the *sHKL* parameters to ensure this. We can avoid negative values pragmatically by calculating *Sqrt(Max(mhkl,0))* instead of *Sqrt(mhkl)*. In fact, the present experimental data imply (nearly) vanishing line broadening for *hh*0 reflections. This can be seen as *s220* ≈ −2 × *s400*. More negative values of *s220* would result in *mhkl* < 0 for *H* = *K*. Note that underlying physical models of microstrain distributions/lattice parameter distributions require *mhkl* ≥ 0 (Leineweber, 2011).

The direction dependence of the microstrain broadening is best analyzed by the direction dependence of some suitable microstrain measure like $fwhm_\varepsilon^{hkl}$. That value can be calculated as (compare eq. (4.12)):

$$fwhm_\varepsilon^{hkl} = \frac{\left(d^{hkl}\right)^2 \sqrt{mhkl}}{2 \times 10000} \times corr \tag{4.13}$$

where *corr* is the ratio $fwhm/E$ in accordance with eq. (13.101), which is 1 for the present case with $\zeta = zeta = 1$. The factor 10000 in the denominator of eq. (4.13), takes into account the factor 0.0018 instead of 180 (degrees) used in the calculation of *pp*. This factor is used to obtain easy-to-handle values of *sHKL*, which could otherwise be very small and incorrectly reported in TOPAS OUT files.[9] Figure 4.12 shows the direction dependence of $fwhm^{hkl}$ resulting from the refined values of *sHKL*, which is plotted by:

```
prm !corr = ((1-zeta)^5 + 2.69269 (1-zeta)^4 zeta +
    2.42843 (1-zeta)^3 zeta^2 +4.47163 (1-zeta)^2 zeta^3 +
    0.07842 (1-zeta)zeta^4 + zeta^5)^.2;:1.00000
normals_plot = D_spacing^2 mhkl^.5/20000 corr;
    normals_plot_min_d 0.3
```

These lines include the factor $corr = fwhm/E$ according to eq. (13.101). Note that in the present case the parameter *zeta* refines to its upper limit of $zeta = 1$. Hence the values of $corr = 1$.

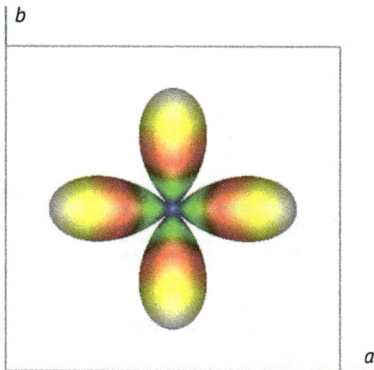

Figure 4.12: Surface representing the direction-dependence of the extent of microstrain of Pb_3O_4 as considered in Section 4.3.2. The maximum extent of $fwhm_\varepsilon$ occurs along the $\langle 100 \rangle$ directions and amounts $fwhm_\varepsilon = 0.007$.

---

9 TOPAS handles small floating point numbers correctly internally but converts them to fixed-point number with a limited number of digits in the OUT file.

As also discussed for an anisotropic size-broadening model in Section 4.3.1, this easily-implemented *sHKL* model has some practical shortcomings. For example, isotropy of the microstrain broadening is not easily recognizable from the *sHKL* parameters (Leineweber, 2006; Leineweber, 2011). This has led to introduction of an alternative Cartesian parametrization of the microstrain broadening. In this model one refines the coefficients of a fourth-order polynomial in the components x_1, x_2 and x_3 of a unit vector parallel to the diffraction vector described using a standard Cartesian coordinate system. Note that the definition of the unit-vector components *x1, x2* and *x3* is more complex for lower symmetries. The basis vectors of the Cartesian of a tetragonal crystal are usually chosen along the [100], [010] and [001] directions. A core piece of the treatment is the polynomial:

$$E_\varepsilon = Z^E_{1111}\left(x_1^4 + x_2^4\right) + Z^E_{3333}\ x_3^4 + 6Z^E_{1122}x_1^2x_2^2 + 6Z^E_{1133}\left(x_1^2x_3^2 + x_2^2x_3^2\right) \tag{4.14}$$

defining the width parameter E on the strain ε scale. The coefficients Z^E_{ijpq} are the refinable parameters (the number of parameters, again, depends on the Laue class, Leineweber, 2006). The polynomial is used in the following way to describe the microstrain broadening of Pb_3O_4:

```
prm zeta   1.00000 min 0 max 1
prm z_e1111_scsc   50.49880 min 0
prm z_e1122_scsc   -16.75325
prm z_e3333_scsc   0.48388 min 0
prm z_e1133_scsc   0.65699
prm !sc = 10^-3;
prm x1 = D_spacing H/Lpa;
prm x2 = D_spacing K/Lpb;
prm x3 = D_spacing L/Lpc;
prm ee_e = sc^2 (z_e1111_scsc (x1^4 + x2^4) + z_e3333_scsc x3^4
   + 6 z_e1122_scsc x1^2 x2^2 + 6 z_e1133_scsc (x1^2 x3^2 + x2^2 x3^2);
prm ee_2Th = (180/Pi)^2 (-2 Tan(Th))^2 ee_e;
```

In this series of commands, the Z^E_{ijpq} parameters (*z_e1111_scsc* and others) are scaled by the parameter sc^2 to keep their numerical values to a reasonable magnitude. Parameters, which retain their "true" values lack the "_scsc" label, for example, the (*H, K* and *L* dependent) squared width parameter E on the 2θ scale (in °), which is used for convolution of each peak.

Using the additional lines:

```
prm !corr = ((1-zeta)^5 + 2.69269 (1-zeta)^4 zeta +
2.42843 (1-zeta)^3 zeta^2 +4.47163 (1-zeta)^2 zeta^3 +
0.07842 (1-zeta)zeta^4 + zeta^5)^.2;:1.00000
prm !z_fwhm1111_scsc = z_e1111_scsc corr^2;:50.49880
```

```
prm !z_fwhm3333_scsc = z_e3333_scsc corr^2;:0.48388
prm !z_fwhm1122_scsc = z_e1122_scsc corr^2;:-16.75325
prm !z_fwhm1133_scsc = z_e1133_scsc corr^2;:0.65699
```

one can calculate new (scaled) parameters Z_{ijpq}^{fwhm} pertaining to:

$$(fwhm_\varepsilon)^2 = Z_{1111}^{fwhm}\left(x_1^4 + x_2^4\right) + Z_{3333}^{fwhm}x_3^4 + 6Z_{1122}^{fwhm}x_1^2x_2^2 + 6Z_{1133}^{fwhm}\left(x_1^2x_3^2 + x_2^2x_3^2\right) \qquad (4.15)$$

Using the Z_{ijpq}^E parameters, isotropy of the microstrain broadening is ensured by $Z_{1111}^E\left(= Z_{2222}^E\right) = Z_{3333}^E = 3Z_{1122}^E = 3Z_{1133}^E\left(= 3Z_{2233}^E\right)$. The Z_{ijpq}^E parameters (or Z_{ijpq}^{fwhm} or parameters defining the square root of the variance of the microstrain, Lineweber, 2011) are dimensionless and are ideally suited to quantitatively relate observed microstrain broadening with anisotropic tensor properties. It is possible, to construct one-to-one relations between the *sHKL* and *zE_ijpq_scsc* parameters, meaning that the treatments are equivalent and should lead to exactly the same quality of fit.

The *spherical_harmonics_hkl* keyword can also be used to describe anisotropic microstrain broadening. The most convenient implementation is to fix the spherical harmonic to an average value of 1 and scale it by an overall *average* microstrain.[10] Note that from here on the fitting parameter *zeta* was fixed to 1:

```
prm !zeta  1 min 0 max 1
prm average  13.26144
spherical_harmonics_hkl ahkl
sh_order  4
prm !sc = 10^-3;
prm ee_e_scsc = average ahkl;
prm ee_2Th = sc^2 (180/Pi)^2 (-2 Tan(Th))^2 ee_e_scsc;
gauss_fwhm = (1-zeta) Max(ee_2Th,0)^.5;
lor_fwhm = zeta Max(ee_2Th,0)^.5;
```

After the first cycle, the third line automatically expands to the correct format for the Laue symmetry. Refinement of the spherical harmonic coefficients and the parameter *average* leads to:

```
prm !zeta  1 min 0 max 1
prm average  13.26144
spherical_harmonics_hkl ahkl
sh_order  4 load sh_Cij_prm {
y00   !ahkl_c00  1.00000
y20   ahkl_c20  -1.36314
```

---

10 Alternatively one can leave out the parameter *average* and remove the (set by default) "!" in front of the parameter name *ahkl_c20-ahkl_c00*, which then takes the role of *average*; the other parameters become *average* × *ahkl_c20-ahkl_c44p*.

```
y40     ahkl_c40    0.39057
y44p    ahkl_c44p   1.82325
}
```

Note that the products such as *average* × *ahkl_c20-ahkl_c44p* can be directly related to the parameters of the two other models described above (Leineweber, 2011), since the number of parameters for a given symmetry is the same in each of them. This 1:1 relation wouldn't exist, if, for example, the E_ε (the square root of *ee_e*) had been expanded in a similar way.

4.4 Neutron diffraction data with focus on the *TOF* method

When evaluating neutron diffraction data use of the *neutron_data* keyword is essential. This activates use of neutron scattering lengths. It is also important to ensure that the correct isotope (e.g., ^{11}B) is specified in the *site* list. For constant wavelength neutron diffraction data, determination of the *IRF* can proceed using the TCHZ approach of Section 4.2.3 using a suitable standard material.[11] Constant wavelength neutron *IRF*s are usually relatively simple. The situation is more complex for *TOF* neutron diffraction data and the following should be kept in mind:

- Data are usually collected in several detector banks, each containing reflections over a limited *d*-spacing range and each having a different *IRF*. TOPAS can analyze these patterns simultaneously so that common (e.g., structural) parameters can be evaluated simultaneously from all banks.
- The profile shape is very dependent on the specific *TOF* instrument. It is therefore advisable to consult the instrument scientist for guidance.
- Data are traditionally provided on the *TOF* scale. Appropriate calibration constants are needed to convert *TOF* to *d*-spacing.
- The complex instrument detector geometry and neutron wavelength distribution means it is essential to retain experimental standard uncertainties through the calculations. The use of three column .xye files is recommended.

In the following INP file the *IRF* of a detector bank of HRPD, ISIS, UK is evaluated using data from a CeO_2 standard:

```
xdd hrp51690_1.xye
neutron_data
start_X 25000
finish_X 125000
x_calculation_step = Yobs_dx_at(Xo);
TOF_LAM(0.001)
```

---

11 Not $La^{10}B_6$, which has an enormous neutron absorption cross section!

```
TOF_x_axis_calibration(t0,-11.31260,!t1, 48281.14,t2,-7.19610)
scale_pks = D_spacing^4;
bkg @  9.37575017  7.38370967  0.924665853  0.114324576  0.285302449  0.0835731213
TOF_Exponential(a1, 193.96941,a2, 38.72409, 4, t1, +)
hkl_Is
   phase_name CeO2
   TOF_PV(b1, 56.95525, lor_b1, 0.269485101, t1)
   Cubic(aa  5.413263)
   space_group "Fm-3m"
```

The *x_calculation_step* command used ensures that the non-constant stepwidth in *x* (*TOF*) is properly treated during analysis.[12] The *TOF_LAM(0.001)* line determines the intensity range for which the peak profile will be calculated (like *ymin_on_ymax*). The macro *TOF_x_axis_calibration* defines the coefficients t_0, t_1 and t_2 in eq. (2.3) relating the *d* spacing with the *x* (*TOF*) coordinate in the xye file, which is repeated here for convenience:

$$TOF = t_0 + t_1 d + t_2 d^2 \tag{4.16}$$

where the numerical values of t_0–t_2 must correspond to *d* in Å. t_1 relates to the instrument flight path, t_0 to a (usually electronic) zero-time offset and t_2 is an empirical quadratic correction factor. The variable names *zero, difc* and *difa* are often used for t_0, t_1 and t_2, respectively. In the INP file above the data were used to fine-tune t_0 and t_2 (which should be done with care). The *scale_peaks* line gives the Lorentz factor for *TOF* neutron diffraction.

The profile function is described using the macro *TOF_Exponential*, which applies to all phases, and the macro *TOF_PV*, which applies to each phase since it describes the sample contribution. *TOF_PV* defines a pseudo-Voigt of *d*-spacing independent shape *lor_b1* = η and *fwhm* which is proportional to the refined parameter *b1* and the *d*-spacing. *t1* is used as a scaling parameter for the refined parameter *b1*. *TOF_Exponential* convolutes each peak with a (single) right-hand ("+") exponential of *d*-spacing dependent width, which is described by the two fitting parameters *a1* and *a2*. The functional form of the *d*-spacing dependence is determined by the exponent "4." Again, the constant *t1* is used by the macro to scale the fitting parameters. More advanced macros exist to describe the details of the peak shapes, some of them dedicated for specific instruments and based on a FP approach. See, for example, the TOPAS wiki (http://topas.dur.ac.uk/topaswiki/doku.php?id=time_of_flight_tof_isis_instrument_standard_files).

Once the standard material has been used to determine the *IRF*, its parameters are held fixed when evaluating data measured on a real sample under the same conditions. To impose size or microstrain broadening, one has to convolute the line

---

[12] Typically such data are collected or provided in steps with constant $\Delta d/d$ or $\Delta TOF/TOF$.

broadening on the *TOF* scale instead of on the 2θ scale as considered in Section 4.3. For size broadening, eqs. (4.7) and (4.4) lead to:

$$fwhm_{TOF}^{hkl} = t_1 \left(d^{hkl}\right)^2 \frac{1}{D_{hkl}}. \tag{4.17}$$

This commonly-accepted form neglects the (usually small) t_2 term in eq. (4.16). Analogously, for microstrain one can arrive at:

$$fwhm_{TOF}^{hkl} = t_1 d^{hkl} fwhm_\varepsilon^{hkl}. \tag{4.18}$$

For isotropic size broadening analogues of the macros *TOF_CS_L* and *TOF_CS_G* can be used. For example:

```
TOF_CS_L(D0,50)
```

applies the following:

```
prm D0 50
lor_fwhm = t1 .1 D_spacing^2 /D0;
```

The factor 0.1 again ensures *DO* values in nm rather than in Å.

For intermediate Gaussian–Lorentzian shape, the same recipes as applied for constant-wavelength X-ray diffraction data in Section 4.3.1 can be used. Similar treatments for anisotropic size broadening are possible.

Currently, no ready-made macros are available in TOPAS.INC for microstrain broadening. An analogue of what is achieved by the *Strain_G* macro (see Section 4.3.2) is achieved by the following commands:

```
prm E0 0.01
gauss_fwhm = t1 D_spacing E0;
```

where *EO* takes the role of the *fwhm* of the microstrain. Like in Section 4.3.2, the treatment of microstrain can be extended to intermediate Gaussian–Lorentzian shapes and to anisotropic microstrain broadening.

We'll finish the chapter with an example of describing anisotropic strain broadening in the orthorhombic cementite (Fe_3C) making use of the CeO_2 *IRF* determined above (Leineweber, 2016). The measure of the microstrain broadening used was the traditionally employed squareroot of the variance of the microstrain $var(\varepsilon^{hkl})$. This is possible due to the pure Gaussian shape of the microstrain broadening (having a finite variance). The expression for the direction-dependent microstrain for the orthorhombic symmetry of the cementite is:

$$\operatorname{var}(\varepsilon^{hkl}) = Z_{1111}x_1^4 + Z_{2222}x_2^4 + Z_{3333}x_3^4 + 6Z_{1122}x_1^2x_2^2 + 6Z_{1133}x_1^2x_3^2 + 6Z_{2233}x_2^2x_3^2 \qquad (4.19)$$

The INP file including the *IRF* and t_0-t_2 as determined from CeO_2 (and therefore fixed) is:

```
xdd hrp51694_b1_TOF.xye
    neutron_data
    x_calculation_step = Yobs_dx_at(Xo);
    scale_pks = D_spacing^4;
    bkg @  2.75  0.71 -0.19 -0.05 -0.04 -0.11 -0.09 -0.01 -0.012 -0.04
    start_X 25000
    finish_X 125000
    TOF_LAM(0.001)
    TOF_x_axis_calibration(!t0,-11.31260,!t1, 48281.14,!t2,-7.19610)
    TOF_Exponential(!a1, 193.96941,!a2, 38.72409, 4, t1, +)
    hkl_Is
        TOF_PV(!b1, 56.95525, !lor_b1, 0.269485101, t1)
        phase_name Fe3C
        a @  5.091282
        b @  6.729827
        c @  4.525684
        prm z1111_scsc  0.37466 min 0
        prm z2222_scsc  0.52398 min 0
        prm z3333_scsc  0.40968 min 0
        prm z1122_scsc  0.04499
        prm z1133_scsc  0.06015
        prm z2233_scsc  0.71732

        prm x1 = D_spacing H/Lpa;
        prm x2 = D_spacing K/Lpb;
        prm x3 = D_spacing L/Lpc;
        prm !sc =  10^(-3);
        prm vareps = sc^2 (z1111_scsc x1^4 +z2222_scsc x2^4 + z3333_scsc x3^4 + 6 z1122_scsc
x1^2 x2^2 + 6 z1133_scsc x1^2 x3^2 + 6 z2233_scsc x2^2 x3^2);
        prm vartof = (t1)^2 D_spacing^2 vareps;
        gauss_fwhm = 2(2 Ln(2) Max(vartof,0))^.5;
        prm sze  548.03587 max 5000
        lor_fwhm = D_spacing^2 t1 0.1/sze;
        space_group "Pnma"
```

The factor $2(2\,Ln(2))^{.5}$ ensures interconversion between the *fwhm* and the square-root of the variance. The final fits to the CeO_2 and Fe_3C data are shown in Figure 4.13. Using the *normals_plot* command, the direction dependence of $\sigma_\varepsilon = \operatorname{var}(\varepsilon^{hkl})^{1/2}$ determined from the Z_{ijpq} parameters is shown in Figure 4.14. This reflects thermal microstress leading to anisotropic thermal microstrain via the elastic anisotropy of Fe_3C (Leineweber, 2016).

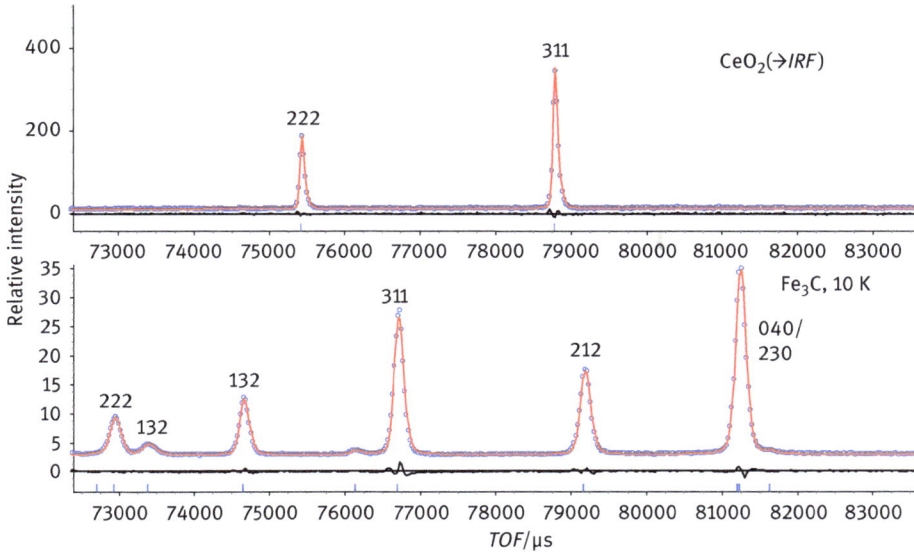

Figure 4.13: Zoomed regions of the Pawley fits to *TOF* neutron powder-diffraction data from CeO_2 (top: used to determine the *IRF*) and Fe_3C (bottom: exhibiting microstrain broadening).

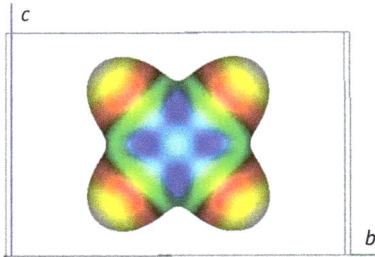

Figure 4.14: Direction-dependence of the microstrain as quantified by $\text{var}(\varepsilon^{hkl})^{1/2}$ in Fe_3C at 10 K due to thermal microstrain. The maximum extent of the microstrain amounts to 0.011.

References

Bergmann, J., Kleeberg, R., Haase, A., Breidenstein, B. (2000): *Advanced fundamental parameters model for improved profile analysis*, in Proceedings of the 5th European Conference on Residual Stresses, Delft-Noordwijkerhout, The Netherlands, September 29–30, 1999, A.J. Böttger, R. Delhez, and E.J. Mittemeijer, Eds. Trans Tech Publications, 347–349, pp. 303–308.

Cheary, R.W., Coelho, A.A (1992): *A fundamental parameters approach to X-ray line-profile fitting*. J. Appl.Cryst. 25, 109–121.

Cline, J.P., Mendenhall, M.H., Black, D., Windover, D., Henins, A. (2015): *The optics and alignment of the divergent beam laboratory X-ray powder diffractometer and its calibration using NIST standard reference materials*. J. Res. NIST 120, 173–222.

David, W.I.F., Leoni, M., Scardi, P. (2010): *Domain size analysis in the Rietveld method*. Mater. Sci. Forum 651, 187–200.

Delhez, R., de Keijser, T., Mittemeijer, E.J., Langford, J.I. (1988): *Size and strain parameters from peak profiles: sense and nonsense*. Aust. J. Phys. 41, 213–227.

Hölzer, G., Fritsch, M., Deutsch, M., Härtwig, J., Förster, E. (1997): $K_{\alpha 1,2}$ and $K_{\alpha 1,2}$ X-ray emission lines of the 3d transition metals. Phys. Rev. A 56, 4554–4568.

Klug, H.P, Alexander, L.E. (1974): *X-ray diffraction procedures for polycrystalline and amorphous materials*, John Wiley and Sons, New York, USA, 992 pages.

Le Bail, A., Duroy, H., Fourquet, J.L. (1988): *Ab-initio structure determination of $LiSbWO_6$ by X-ray powder diffraction*,. Mat Res. Bull., 23, 447–452.

Leineweber, A. (2006): *Anisotropic diffraction-line broadening due to microstrain distribution: parametrization opportunities*. J. Appl. Cryst. 39, 509–518.

Leineweber, A. (2011): *Understanding anisotropic microstrain broadening in Rietveld refinement*. Z. Kristallogr. 226, 905–923.

Leineweber, A. (2016): *Thermal expansion anisotropy as source for microstrain broadening of poly-crystalline cementite, Fe_3C*. J. Appl. Cryst. 49, 1632–1644.

Leineweber, A., Dinnebier, R.E. (2010): *Anisotropic microstrain broadening of minium, Pb_3O_4, in a high-pressure cell: interpretation of line-width parameters in terms of stress variations*. J. Appl. Cryst. 43, 17–26.

Leineweber, A., Mittemeijer, E.J. (2010): *Notes on the order-of-reflection dependence of microstrain broadening*. J. Appl. Cryst. 43, 981–989.

Pawley, G. S. (1981): *Unit-cell refinement from powder diffraction scans*. J. Appl. Cryst. 14, 357–361.

Popa, N.C. (1998): *The (hkl) dependence of diffraction-line broadening caused by strain and size for all laue groups in Rietveld refinement*. J. Appl. Phys. 31, 176–180.

Rodriguez-Carvajal, J., Fernandez-Diaz, M.T., Martinez, J.L. (1991): *Neutron diffraction study on structural and magnetic properties of La_2NiO_4*. J. Phys.: Condens. Matter 3 (1991) 3215–3234.

Rietveld, H.M. (1969): *A profile refinement method for nuclear and magnetic structures*. J. Appl. Cryst. 2, 65–71.

Scardi, P., Leoni, M. (2001): *Diffraction line profiles from polydisperse crystalline systems*. Acta Cryst. A 57, 604–613.

Scardi, P., Leoni, M. (2002): *Whole powder pattern modelling*. Acta Cryst. A 58, 190–200.

Stephens, P.W. (1999): *Phenomenological model of anisotropic peak broadening in powder diffraction*. J. Appl. Cryst. 32, 281–289.

Thompson, P., Cox, D.E., Hastings, J.B. (1987): *Rietveld refinement of Debye-Scherrer synchrotron X-ray data from Al_2O_3*, J. Appl. Cryst. 20, 79–83.

Young, R.A. (1993): *Introduction to the Rietveld method*. In: The Rietveld method. In: edited by R.A. Young, IUCr Book Series, Oxford University Press, UK, 1–39.

5 Quantitative phase analysis

5.1 Introduction

Quantitative phase analysis (QPA) is a method to determine the absolute weight fractions of all crystalline phases present in a multiphase sample. Powder diffraction as a *phase*-sensitive method is highly suited for this task and various procedures using the intensities of single peaks, group of peaks or the entire powder pattern have been developed over time. A significant breakthrough in QPA came when the suitability of the Rietveld method as a precise phase-sensitive analytical technique for QPA was realized. This is the focus of this chapter.

If all phases in the mixture are crystalline with known crystal structure, no external or internal standard material is required, making Rietveld QPA an easy-to-use standard-less method. Complications arise if an amorphous phase or a phase with an unknown or only partially known crystal structure is present. In this case, an external (or better) internal, well-defined standard can be used to renormalize the weight fractions on an absolute basis. Once the amount of the amorphous or unknown crystalline phase is quantified, this phase can be stored as a new standard (so-called PONKCS) phase for further use. External and PONKCS standards depend on the experimental conditions, which should be kept constant for a series of analyses. QPA becomes more and more unreliable the more unknown phases are present. In all cases, the intensity-affecting factors like absorption, overspill or preferred orientation should be kept to a minimum.

In the following sections, the theory of QPA and its application either standard-less or using internal, external or PONKCS standards is described. The integrated intensity I of a reflection $\mathbf{s} = (hkl)$ for phase α in a multi-phase mixture measured on a flat plate sample (Bragg–Brentano geometry) with "infinite thickness" is:

$$I_{\mathbf{s},\alpha} = \left[\frac{I_0\lambda^3}{32\pi R}\frac{e^4}{m_e^2 c^4}\right]$$

$$\cdot \left[\frac{M_{\mathbf{s}}}{2V_\alpha^2}|F_{\mathbf{s},\alpha}|^2\left(\frac{1+\cos^2 2\theta\cos^2 2\theta_m}{\sin^2\theta\cos\theta}\right)\exp\left(-2B_\alpha\left(\frac{\sin\theta}{\lambda}\right)^2\right)\right]\cdot\left[\frac{W_\alpha}{\rho_\alpha\mu_m^*}\right] \qquad (5.1)$$

with instrument (blue), sample (green) and phase (red) dependent parameters: incident beam intensity I_0, the distance R between specimen and detector, the X-ray wavelength λ, the classical electron radius $e^2/m_e c^2$, the multiplicity $M_{\mathbf{s}}$ of reflection $\mathbf{s}$ of phase α, the weight fraction W_α of phase α, the mass absorption coefficient μ_m^* of the entire sample, the volume V_α of the unit cell of phase α in Å^3, the structure factor $F_{\mathbf{s},\alpha}$ of reflection $\mathbf{s}$ of phase α, the diffraction angle 2θ of reflection $\mathbf{s}$ of phase α, the diffraction angle $2\theta_m$ of the monochromator, the density ρ_α of phase α, and the overall atomic displacement parameter B_α of phase α. Similar relations exist for other geometries.

https://doi.org/10.1515/9783110461381-005

The relation can be greatly simplified by introducing two constants, one for the experimental setup C_{instr}, and a second one for the intensity of peak **s** for phase α $C_{\mathbf{s},\alpha}$:

$$I_{\mathbf{s},\alpha} = C_{instr} \cdot C_{\mathbf{s},\alpha} \cdot \frac{1}{V_\alpha^2} \cdot \left[\frac{W_\alpha}{\rho_\alpha \mu_m^*}\right]. \tag{5.2}$$

In Rietveld analysis, all reflection intensities of a phase α are proportional to the corresponding scale factor S:

$$I_{\mathbf{s},\alpha} \propto S_\alpha. \tag{5.3}$$

This allows the introduction of a single scaling factor K that depends exclusively on the instrumental condition and not on the sample/phases:

$$S_\alpha = K \cdot \frac{1}{V_\alpha^2} \cdot \left[\frac{W_\alpha}{\rho_\alpha \mu_m^*}\right]. \tag{5.4}$$

The phase density (g cm^{-3}) can be calculated as:

$$\rho_\alpha = 1.6604 \cdot \frac{Z_\alpha M_\alpha}{V_\alpha} \tag{5.5}$$

with the molar mass M (g mol^{-1}) of one formula unit and Z the number of formula units within a unit cell of phase α. The scale factor of phase α can be thus be rewritten as:

$$S_\alpha = K \cdot \frac{1}{V_\alpha} \cdot \left[\frac{W_\alpha}{1.6604 \cdot Z_\alpha M_\alpha \mu_m^*}\right] \tag{5.6}$$

5.2 External standard method

The external standard method was described by O'Connor and Raven (1988). From eq. (5.6), the weight fraction of phase α can be calculated as:

$$W_\alpha = \frac{S_\alpha (ZMV)_\alpha \mu_m^*}{K'} \tag{5.7}$$

$(ZMV)_\alpha$ is an abbreviation of the product of $Z_\alpha M_\alpha V_\alpha$. K' is an experimental constant that is specific for the experimental setup used and depends only on instrumental and data collection conditions. K' can be determined from a single measurement of an external standard, for which μ_m^* must be known:

$$K' = \frac{S_\alpha (ZMV)_\alpha \mu_m^*}{W_\alpha} \tag{5.8}$$

With the knowledge of K' and μ_m^*, the correct weight fractions for all phases of a mixture for which the product $(ZMV)_\alpha$ and the S_α scale factor exists can be calculated.

The mass attenuation coefficient (MAC) μ of a sample (see Section 2.3.3) is usually calculated from standard data in units of length^{-1}. The tabulated values of the MAC are usually those divided by the density of the material (Hubbell & Seltzer, 2004):

$$\mu^* = \frac{\mu}{\rho}. \tag{5.9}$$

It should be noted that the sample must fulfill the "infinite thickness" criterion (Section 2.3.3) that is on the order of 0.1 mm for quartz and Cu-K$_\alpha$ radiation. The (average) mass absorption coefficient μ_m^* of the entire sample (MAC) is defined as the sum of linear attenuation coefficients of the elements multiplied by their weight fractions. This sum runs over all phases i in the sample:

$$\mu_m^* = \sum_i w_i \mu_i^*. \tag{5.10}$$

As a practical example, the MAC of a pure quartz (SiO$_2$) sample for Cu-K$_\alpha$ radiation is:

$$\mu_{Quartz}^* = \frac{28.086}{28.086 + 2 \cdot 15.999} \, 60.6 \, \frac{cm^2}{g}$$

$$+ \frac{2 \cdot 15.999}{28.086 + 2 \cdot 15.999} \, 11.5 \, \frac{cm^2}{g} = 34.45 \, \frac{cm^2}{g}. \tag{5.11}$$

The same calculation for a pure alumina (Al$_2$O$_3$) sample is:

$$\mu_{Alumina}^* = \frac{2 \cdot 26.982}{2 \cdot 26.982 + 3 \cdot 15.999} \, 48.6 \, \frac{cm^2}{g}$$

$$+ \frac{3 \cdot 15.999}{2 \cdot 26.982 + 3 \cdot 15.999} \, 11.5 \, \frac{cm^2}{g} = 31.14 \, \frac{cm^2}{g}. \tag{5.12}$$

The MAC of a mixture of 50% alumina and 50% quartz would then be:

$$\mu_m^* = 0.5 \cdot 34.45 \, \frac{cm^2}{g} + 0.5 \cdot 31.14 \, \frac{cm^2}{g} = 32.79 \, \frac{cm^2}{g}. \tag{5.13}$$

5.3 Rietveld method

Under the assumption that all phases in a mixture are crystalline, the following normalization relation can be used:

$$\sum_i W_i = 1. \tag{5.14}$$

This relationship allows elimination of the instrument constant and the mass absorption coefficient of the sample through:

$$W_\alpha = \frac{S_\alpha (ZMV)_\alpha}{\sum_i S_i (ZMV)_i}.$$

(5.15)

Quantitative Rietveld refinement can therefore be performed without knowledge of K' and μ_m^* (Madsen & Scarlett, 2008). The drawback of this standard-less method is that the weight fractions of the phases are relative and not absolute numbers. They are therefore insensitive to unknown or amorphous content. To overcome this problem, an external or internal standard must be used.

5.4 Internal standard method

The amount of amorphous or unknown content can be determined by adding an internal standard *std*:

$$W_{\alpha(meas)} = W_{std(meas)} \frac{S_{\alpha(meas)} (ZMV)_{\alpha(meas)}}{S_{std(meas)} (ZMV)_{std(meas)}}.$$

(5.16)

The absolute weight fractions of the known materials can then be can calculated by:

$$W_{\alpha(abs)} = W_{\alpha(meas)} \frac{W_{std(known)}}{W_{std(meas)}}.$$

(5.17)

The weight fraction of the unknown or amorphous material follows directly from:

$$W_{unknown(abs)} = 1.0 - \sum_k W_{k(abs)}.$$

(5.18)

where k runs over all phases of the mixture except the unknown or amorphous material.

5.5 PONKCS method

In cases where Bragg reflections from an unknown phase can clearly be identified, a partial or no known crystal structure (PONKCS) phase can be created from a pure sample or from a sample where the amount of the unknown phase is known (e.g., by mixing a known amount of a standard with the sample). First, a set of intensities (group of single peaks, Pawley or Le Bail) needs to be determined with an overall fixed scale factor (=1.0). Second, the Lorenz-polarization correction must be applied, which is done automatically in TOPAS using the appropriate keywords and macros (see Chapter 2):

$$I'_{meas} = \frac{I_{meas}}{LP}.$$
(5.19)

With the help of an internal standard *std*, an artificial value of *ZM* for the unknown phase α can be calculated:

$$(ZM)_\alpha = \frac{W_\alpha}{W_{std}} \frac{S_{std}}{S_\alpha} \frac{(ZMV)_{std}}{V_\alpha}.$$
(5.20)

The volume is either known in the case of Pawley or Le Bail fits, or set to $V_\alpha = 1$ for group of peaks. Note that the obtained $(ZM)_\alpha$ value has no physical meaning and is only valid for the chosen experimental configuration. The "correct" $(ZM)_\alpha$ value requires knowledge of the density of the unknown material:

$$(ZM)_{\alpha(true)} = \frac{\rho_\alpha V_\alpha}{1.6604}.$$
(5.21)

Peak intensities can then be scaled by:

$$\frac{(ZM)_{\alpha(true)}}{(ZM)_\alpha}.$$
(5.22)

5.6 Some correction factors

QPA is affected by many sources of error such as instrument configuration, particle statistics, counting error, preferred orientation and microabsorption. The application of correction factors like the Brindley-correction for spherical particles (Brindley, 1945) can be necessary for mixtures with strongly different mass absorption coefficients for the different phases. However, the assumption that all phases in a sample consist of spherical particles of identical diameter is unrealistic, and should be used with extreme caution. The Brindley model for correction of microabsorption effects (Brindley, 1945) in TOPAS notation is:

```
prm !R 0.002  min 0         '  radius of the particle in [cm]
prm !PD 0.6   min 0 max 1   '  packing density.
Apply_Brindley_Spherical_R_PD( R, PD)
```

The methods described above for quantitative phase analysis can also be applied to powder diffraction data measured in Debye–Scherrer geometry, though there are several important differences. On the positive side, preferred orientation and graininess is typically less of a problem than in reflection geometry. Measurement of an empty capillary will generally lead to a higher background than a filled capillary due to the much lower absorption at low diffraction angles. This leads to severe quantification errors in amorphous content. In order to determine the background function reliably, it is therefore necessary to measure a highly crystalline sample of similar absorption and packing density as the sample under investigation or to use high energy radiation where absorption is not significant.

5.7 Elemental composition

TOPAS has various built-in tools for working out the overall composition in a multiphase fit. For example, the following keywords exist:

```
Get_Element_Weight(Element) ' Returns the weight% a specific element in the sample
Get(sum_smvs_minus_this)    ' Returns the sum of SMVs minus the phase where it is defined.
element_weight_percent $ELEMENT $NAME #  ' An xdd dependent keyword that returns the
  weight percent of an element within the corresponding str's of the xdd
```

If the overall amount of an element is known from other sources (e.g., XRF, EDAX, ICPMS analysis), it is possible to set up a constraint to force the refinement to reflect this information using the keywords above (see the technical manual of TOPAS for more information).

A more intuitive way to include a known weight% of an element is through a restraint where the difference between known and refined weight% is minimized. In the following example "zr" is the name given to the weight% of element Zr^{+4} in the sample. A restraint is used to minimize the difference between the refined and the known value (65 %). The refinement obeys the restraint according to the value set for the keyword "$penalties\_weighting\_K1$". A high value for $penalties\_weighting\_K1$ should be avoided as the restraints then behave more like constraints that might contradict the information from the powder pattern. Restraints are covered in more detail in Chapter 6.

```
penalties_weighting_K1 .1
xdd
    element_weight_percent Zr+4 zr 0 ' After refinement   65.0275252
    restraint = (zr - 65); : 0        ' After refinement    0.0275251
```

5.8 Practical application

In the remainder of the chapter we discuss several exercises on QPA in detail. The mixtures considered consist of crystalline alumina, crystalline quartz and amorphous silica flour. The data are taken from the work of Madsen, Scarlett & Kern (2011). XRPD data were measured using a Philips X'Pert diffractometer in Bragg–Brentano geometry (173 mm radius, Cu-K_α radiation from secondary graphite monochromator $2\theta_m = 26.6°$, fixed 1° divergence slit, 0.3 mm receiving slit, 1° antiscatter slit, primary Soller slits with 2.5° opening). All data were recorded from 10°–140° 2θ in 0.02° 2θ steps for 3 seconds per step.

5.8.1 Determination of the *IRF* and background

As discussed in Chapter 4, it is a good practice to determine the instrumental resolution function (*IRF*) from a line profile standard before a Rietveld refinement. Here, Y_2O_3 was used as the line profile standard. The background was described by a Chebyshev polynomial of eighth order and the line profile was fitted by a TCHZ-Pseudo-Voigt function refining GW, LX and LY parameters only (Figure 5.1).

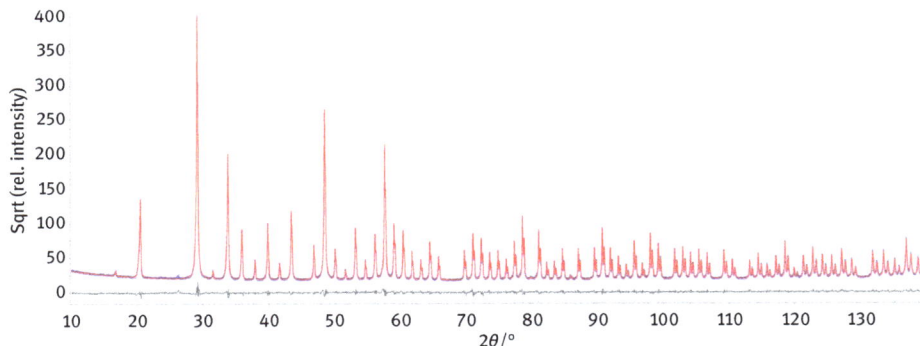

Figure 5.1: Rietveld plot of the line profile standard Y_2O_3 for determination of the *IRF*.

The background and the TCHZ pseudo-Voigt peak profile parameters need to be fixed and copied to further refinements. The TOPAS script file for the Rietveld refinement of the line profile standard Y_2O_3 is given below:

```
xdd Y1080223.xy
   CuKa2(0.00001)
   x_calculation_step 0.02
```

```
bkg @ 428.40 -119.34 171.44 -86.61 75.03 -40.29 26.92 -8.74 1.36
LP_Factor( 26.6)
Specimen_Displacement(@, -0.00658)
Rp 173
Rs 173
axial_conv
    filament_length         12
    sample_length        @ 18.247
    receiving_slit_length   12
    primary_soller_angle @  2.629
    axial_n_beta  30
str
    phase_name "Y2O3 standard"
    TCHZ_Peak_Type(, 0,, 0,@, 0.005555,, 0,@, 0.034624,@, 0.01553)
    space_group Ia-3
    scale @ 0.0001726791627
    Cubic(@ 10.60344)
    site Y1 num_posns 24 x @ -0.03233 y   0       z   0.25    occ Y+3 1 beq @ 0.338
    site Y2 num_posns 8  x    0.25    y   0.25    z   0.25    occ Y+3 1 beq @ 0.382
    site O1 num_posns 48 x @  0.39102 y @ 0.15207 z @ 0.38113 occ O-2 1 beq @ 0.453
```

5.8.2 Standard-less Rietveld QPA

A diffraction pattern from a mixture of 50 weight% quartz and 50 weight% alumina is subjected to unconstrained Rietveld refinement giving the fit shown in Figure 5.2.

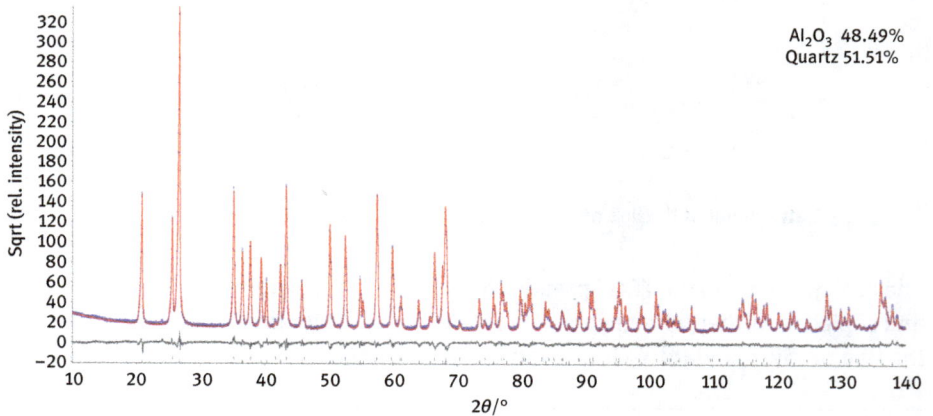

Figure 5.2: Rietveld plot of a 50/50 mixture of alumina and quartz.

The corresponding TOPAS INP file is shown below. In TOPAS, the weight fractions are automatically reported in the *MVW* macro. The refined weight fractions of 48.5% and 51.5% are close to the expected values:

```
xdd NS_I1.xy
   CuKa2(0.00001)
   x_calculation_step 0.02
   bkg @ 252.95 -159.07 155.55 -76.97 63.85 -46.05 25.43 -14.66 9.64
   start_X  10
   LP_Factor( 26.6)
   Specimen_Displacement(@, -0.0430481624)
   Rp 173
   Rs 173
   axial_conv
      filament_length  12
      sample_length  18.24674385
      receiving_slit_length  12
      primary_soller_angle  2.628582881
      axial_n_beta  30
   Absorption(@, 36.00689955)
   str
      e0_from_Strain( 8.402217427e-005,,,@, 0.03853369506)
      TCHZ_Peak_Type(, 0,, 0,, 0.005555009,, 0,, 0.03462355,, 0.01552971)
      phase_MAC 31.59020452
      phase_name "Al2O3"
      MVW( 611.767656, 254.7642651, 48.48921677)
      space_group R-3c
      scale @ 0.007759970864
      Phase_LAC_1_on_cm( 125.9649831)
      Phase_Density_g_on_cm3( 3.987469693)
      Trigonal(@ 4.758757821,@ 12.9903446)
      site Al1 num_posns 12 x  0       y  0 z  0.35214 occ Al+3  1 beq  0.34
      site O1  num_posns 18 x  0.30656 y  0 z  0.25     occ O-2  1 beq  0.33
   str
      e0_from_Strain( 5.149340352e-005,,,@, 0.02361556489)
      TCHZ_Peak_Type(, 0,, 0,, 0.005555009,, 0,, 0.03462355,, 0.01552971)
      phase_MAC 35.81264014
      phase_name "Quartz"
      MVW( 180.2529, 112.973826, 51.51078323)
      space_group P3221
      scale @ 0.06309253838
      Phase_LAC_1_on_cm( 94.88338204)
      Phase_Density_g_on_cm3( 2.649438346)
      Trigonal(@ 4.913002985,@ 5.404470483)
      site Si1 num_posns 3 x 0.47074  y =0;    z =2/3;    occ Si+4  1 beq  0.8822
      site O1  num_posns 6 x 0.41648 y 0.267569 z 0.791080 occ O-2  1 beq  1.5434
```

From the parameters in red in the INP file, the weight fractions can also be manually calculated using eq. (5.16):

$$W_{Quartz} =$$

$$\frac{(0.06309 \cdot 180.253 \cdot 112.974)}{(0.06309 \cdot 180.253 \cdot 112.974) + (0.00776 \cdot 611.768 \cdot 254.764)} = 0.515 \qquad (5.23)$$

and

$$W_{Alumina} =$$

$$\frac{0.00776 \cdot 611.768 \cdot 254.764}{0.06309 \cdot 180.253 \cdot 112.974 + 0.00776 \cdot 611.768 \cdot 254.764} = 0.485. \qquad (5.24)$$

5.8.3 External standard method for determining unknown or amorphous content

To apply the external standard method, the diffractometer constant K' must be determined. A measurement of, for example, alumina (Figure 5.3) can be used.

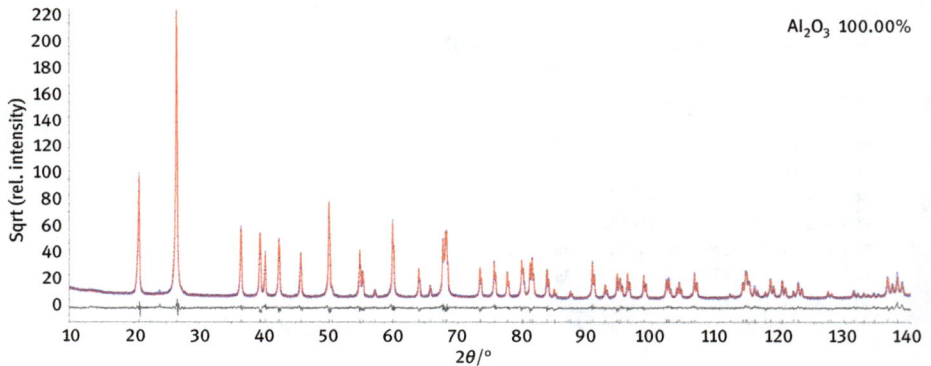

Figure 5.3: Rietveld plot of the diffraction data of alumina.

The TOPAS script file for the refinement is:

```
xdd Al2031Cu-2.xy
  CuKa2(0.00001)
  x_calculation_step 0.02
  bkg @ 252.95 -159.08 155.55 -76.97 63.85 -46.05 25.43 -14.66 9.64
  start_X 10
  LP_Factor( 26.6)
  Specimen_Displacement(@, -0.06813183736)
  Rp 173
  Rs 173
  axial_conv
    filament_length 12
    sample_length 18.24674385
```

```
      receiving_slit_length  12
      primary_soller_angle  2.628582881
      axial_n_beta  30
   Absorption(@, 28.13477304)
   str
      phase_name "Al2O3"
      LVol_FWHM_CS_G_L( 1, 857.5025984, 0.89, 1190.952787,, 9981.763646,@, 1364.793828)
      e0_from_Strain( 5.695107198e-005,,,@, 0.02611852478)
      TCHZ_Peak_Type(, 0,, 0,, 0.005555009,, 0,, 0.03462355,, 0.01552971)
      phase_MAC 31.59020452
      MVW( 611.767656, 254.7217567, 100)
      space_group R-3c
      scale @ 0.01650053178
      Phase_LAC_1_on_cm( 125.9860043)
      Phase_Density_g_on_cm3( 3.988135128)
      Trigonal(@ 4.758530635,@ 12.98941733)
      site Al1 num_posns 12 x    0      y  0 z @ 0.35214 occ Al+3  1 beq @ 0.34
      site O1  num_posns 18 x @ 0.30656 y  0 z    0.25    occ O-2  1 beq @ 0.33
```

From the parameters in red in the INP file, K' can be calculated using eq. (5.8):

$$K' = \frac{0.0165 \cdot \left(611.768 \cdot 254.722 \text{ Å}^3\right) \cdot 31.59 \text{ cm}^2/\text{g}}{1} = 81224. \tag{5.25}$$

Note that in TOPAS, a conversion factor of 1.66/100 is added leading to a K' factor of 1348. The corresponding code in TOPAS to retrieve this value is:

```
K_Factor_WP(1348)
```

Alternatively, a Rietveld refinement of quartz (Figure 5.4) can be used to retrieve K'. The corresponding TOPAS script is:

```
xdd QuartzCu.xy
   CuKa2(0.00001)
   x_calculation_step 0.02
   bkg @ 310.80 -150.21 157.92 -57.89 86.41 -28.46 32.75 0.084 17.95
   start_X  10
   LP_Factor( 26.6)
   Specimen_Displacement(@, -0.008354646507)
   Rp 173
   Rs 173
   axial_conv
      filament_length  12
      sample_length  18.24674385
      receiving_slit_length  12
      primary_soller_angle  2.628582881
      axial_n_beta  30
   Absorption(@, 51.37923198)
   str
```

```
e0_from_Strain( 3.350261639e-005,,,@, 0.01536474883)
TCHZ_Peak_Type(, 0,, 0,, 0.005555009,, 0,, 0.03462355,, 0.01552971)
phase_MAC 35.81264014
phase_name "Quartz"
MVW( 180.2529, 112.9561208, 100)
space_group P3221
scale @ 0.1168561698
Phase_LAC_1_on_cm( 94.89825444)
Phase_Density_g_on_cm3( 2.649853629)
Trigonal(@ 4.912656195,@ 5.404386419)
site Si1 num_posns 3 x @ 0.4707 y =0; :  0 z =2/3;    occ Si+4  1 beq @ 0.8822
site O1  num_posns 6 x @ 0.4165 y @ 0.2676 z @ 0.7911 occ O-2  1 beq @ 1.5434
```

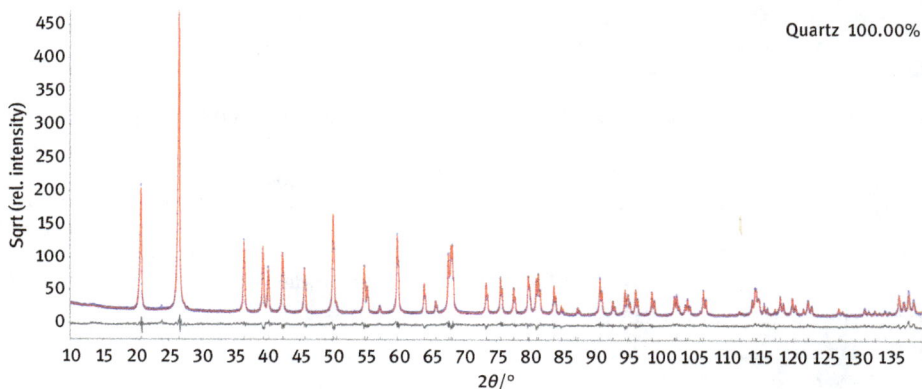

Figure 5.4: Rietveld plot of the diffraction data of quartz.

From the parameters in red in the INP file, K' can be calculated using eq. (5.8):

$$K' = \frac{0.11686 \cdot \left(180.253 \cdot 112.956 \text{ Å}^3\right) \cdot 35.81 \text{ cm}^2/\text{g}}{1} = 85204. \quad (5.26)$$

Note that the values of K' for alumina and quartz are not fully equal as they theoretically should be.

We can use these K' values to analyze materials with amorphous content. Here we will analyze a mixture of alumina and 50 weight% of silica flour. Several options exist for fitting the slowly oscillating background from the amorphous phase, a constrained group of single peaks, a Pawley (Le Bail) phase, or a Rietveld phase (see Chapters 2 and 3). The Rietveld-like approach uses the crystal structure of a closely related compound (e.g., cristobalite) while applying a small crystallite size on the order of 2–3 nm. This will only work if the positions and relative intensities of the amorphous humps and the strongest Bragg reflections approximately match.

In the present case, a Pawley (or Le Bail) fit without known lattice parameters is used (Figure 5.5). In order to distribute positions of peaks equally over the entire 2θ range, an quasi one-dimensional orthorhombic unit cell in *Pmmm* with one large *a* and two very small *b* and *c* (e.g., 0.1 Å) lattice parameters is used. Thereby, only equidistant (*h*00) reflections are predicted in the powder pattern. An overall small crystallite size (e.g., 1.5 nm) is set to ensure broad reflections. In order to have the first peak position located at the beginning of the pattern, the corresponding *d*-value (or multiples thereof) is taken as the lattice parameter *a* (here ≈ 4 × 8.8 Å). The scale factor is fixed at unity. The TOPAS INP file is given below:

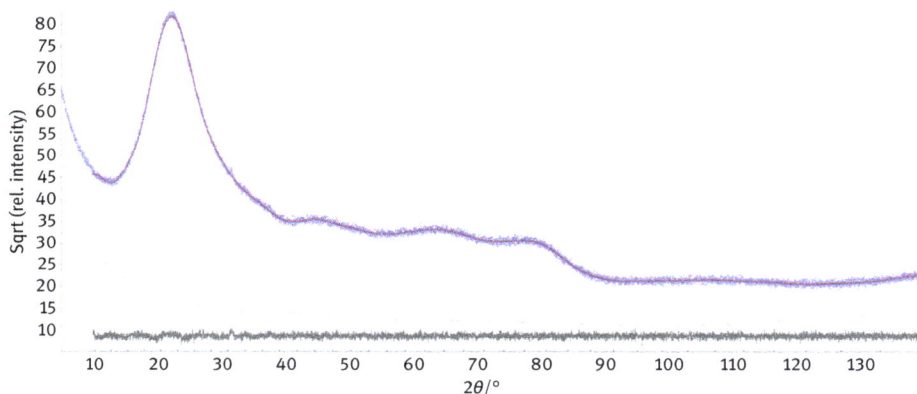

Figure 5.5: Fitting of a powder pattern of silica flour using a Pawley fit of a pseudo one-dimensional orthorhombic phase. The background is denoted by a thin gray line.

```
xdd SilicaCu-2.xy
    CuKa2(0.00001)
    x_calculation_step 0.02
    bkg @ 252.954 -159.07 155.55 -76.97 63.85 -46.05 25.43 -14.66 9.64
    start_X  10
    LP_Factor( 26.6)
    Specimen_Displacement(, 0.08476106561)
    Rp 173
    Rs 173
    axial_conv
        filament_length  12
        sample_length  18.24674385
        receiving_slit_length  12
        primary_soller_angle  2.628582881
        axial_n_beta  30
```

```
Absorption(, 37.51979846)
hkl_Is
    phase_name "amorphous"
    LVol_FWHM_CS_G_L( 1, 1.755642682, 0.89, 1.663253122,@, 1.868823733,,)
    phase_MAC 31.59020452
    scale  1
    MVW( 0, 0.344, 0)
    a  34.4    b  0.1   c  0.1
    space_group Pmmm
    hkl_m_d_th2 4 0 0 2 8.60000038 10.2776861 I @ 24.82109266
    ...
    hkl_m_d_th2 42 0 0 2 0.81904763 140.264679 I @ 789.2849601
```

Alternatively, a constrained group of 15 single peaks with identical *fwhm* (defined by an overall Lorentzian crystallite size parameter) and an overall scale factor fixed to unity can be used (Figure 5.6).

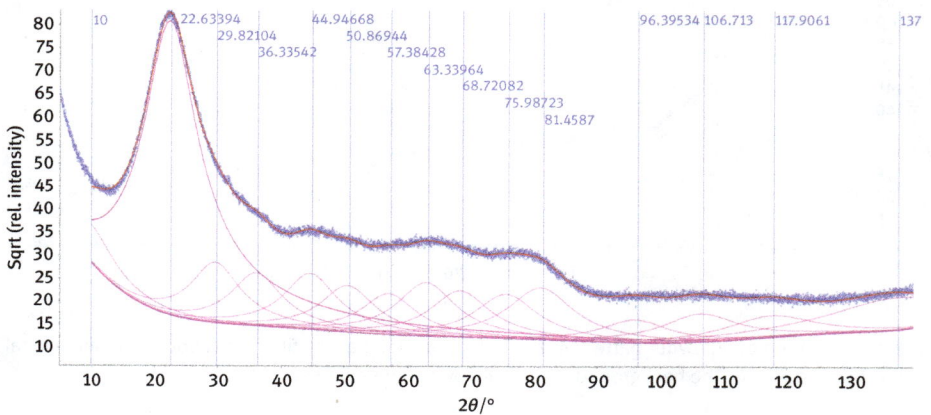

Figure 5.6: Fitting a powder pattern of silica flour using a constrained group of single peaks.

The following TOPAS script shows only the peak phase that substitutes the Pawley phase of the script above:

```
...
pk_grp phase_name "Peaks Phase:0"
scale  1.0

xo_Is
  xo @ 10
  peak_type fp
  LVol_FWHM_CS_G_L( 1, 0.7044361144, 0.89, 0.9848078382,,,CS1, 1.106525661)
  I @ 28.65612154
...
  xo_Is
```

```
xo @ 137.5886737
peak_type fp
LVol_FWHM_CS_G_L( 1, 0.7044361144, 0.89, 0.9848078382,,,(CS1), 1.106525661)
I @ 1898.616265
```

If we return to the data from the powder containing 50% alumina and 50% silica flour, a regular Rietveld refinement leads to 100 weight% of alumina (Figure 5.7). In order to get the correct amount of alumina it is necessary to know the diffractometer constant K' and the mass absorption coefficient μ_m^* of the entire sample. In our case, according to eq. (5.7) this amounts to:

$$W_{Alumina} = \frac{0.007772 \cdot 611.77 \cdot 254.72 \cdot 32.79}{81224} = 0.49 \tag{5.27}$$

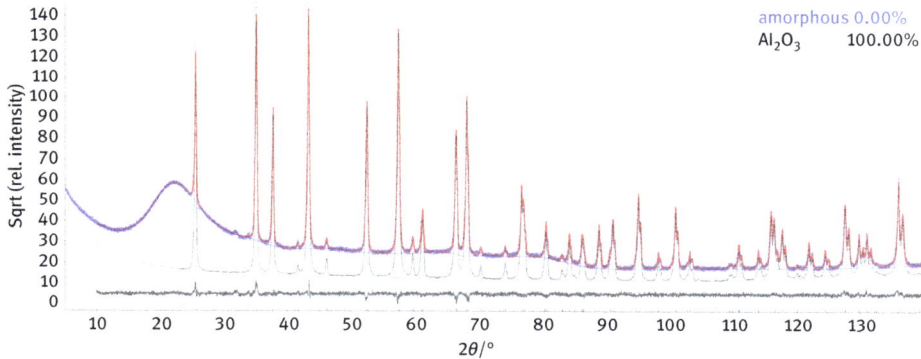

Figure 5.7: Rietveld plot of the 50/50 mixture of alumina and amorphous silica flour.

A variety of predefined keywords and macros to perform this task exist in TOPAS. The amount of silica flour can then be easily deduced by subtracting the amount of alumina from one as 51 weight%. The TOPAS script file for the Rietveld refinement of the 50/50 mixture of alumina and silica flour is:

```
xdd NS_A1.xy
   CuKa2(0.00001)
   x_calculation_step 0.02
   bkg  252.95 -159.07 155.55 -76.97 63.85 -46.05 25.43 -14.66 9.64
   start_X  10
   LP_Factor( 26.6)
   Specimen_Displacement(@, -0.03934)
   Rp 173
   Rs 173
   axial_conv
      filament_length  12
```

```
    sample_length  18.247
    receiving_slit_length  12
    primary_soller_angle  2.6286
    axial_n_beta  30
 Absorption(@, 12.59461479)
 str
    phase_name "Al2O3"
    e0_from_Strain( 5.483701953e-005,,,@, 0.025148992)
    TCHZ_Peak_Type(, 0,, 0,, 0.005555009,, 0,, 0.03462355,, 0.01552971)
    phase_MAC 31.59020452
    MVW( 611.767656, 254.7418766, 100)
    space_group R-3c
    scale @ 0.007772
    Phase_LAC_1_on_cm( 125.9760538)
    Phase_Density_g_on_cm3( 3.987820139)
    Trigonal(@ 4.758633079,@ 12.98988403)
    site Al1 num_posns 12 x  0       y  0 z  0.35214 occ Al+3  1 beq  0.34
    site O1  num_posns 18 x  0.30656 y  0 z  0.25    occ O-2  1 beq  0.33
 hkl_Is
    phase_name "amorphous"
    LVol_FWHM_CS_G_L( 1, 1.646417734, 0.89, 1.559776067,, 1.752557378,,)
    phase_MAC 31.59020452
    scale  @ 1.0
    MVW( 0, 0.344, 0)
    a  34.4    b  0.1    c  0.1
    space_group Pmmm
    hkl_m_d_th2 4 0 0 2 8.60000038 10.2776861 I  14.17381819
    ...
    hkl_m_d_th2 42 0 0 2 0.81904763 140.264679 I  408.5852987
```

5.8.4 Internal standard for determining unknown or amorphous content

Nowadays amorphous content is most commonly determined using internal standard methods. The unknown substance will be available in either pure form (ideal case) or in a mixture (real life) and is spiked by an internal standard of similar absorption. One might use alumina or quartz in our example, a less absorbing material like LiF for an organic or TiO_2 for a first-row transition metal oxide. Here we will consider a mixture of crystalline and amorphous SiO_2 spiked by 50 weight% of alumina, which gave the Rietveld fit shown in Figure 5.8. The refined Al_2O_3 weight% was 61.5% and SiO_2 38.5%. We can calculate corrected weight fractions as:

$$W_{alumina} = 0.615 \frac{0.5}{0.615} = 0.5 \text{ (spiked phase)}$$

$$W_{quartz} = 0.385 \frac{0.5}{0.615} = 0.31 \tag{5.28}$$

and consequently the amorphous content is:

$$W_{unknown} = 1.0 - (0.5 + 0.39) = 0.19. \tag{5.29}$$

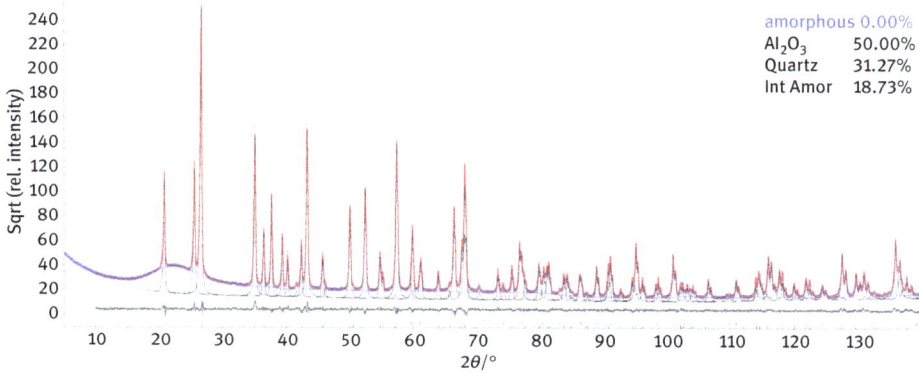

Figure 5.8: Rietveld plot of the mixture of 50 weight% alumina, 30 weight% quartz and 20 weight% silica flour.

The TOPAS script file for the Rietveld refinement of the mixture is given below:

```
xdd NS_C1.xy
   CuKa2(0.00001)
   x_calculation_step 0.02
   bkg 252.95 -159.07 155.55 -76.97 63.85 -46.05 25.43 -14.66 9.64
   start_X 10
   LP_Factor( 26.6 )
   Specimen_Displacement(@, -0.04810104121)
   Rp 173
   Rs 173
   axial_conv
      filament_length  12
      sample_length  18.24674385
      receiving_slit_length  12
      primary_soller_angle  2.628582881
      axial_n_beta  30
   Absorption(@, 19.58052177)
   str
      phase_name "Al2O3"
      e0_from_Strain( 4.88542844e-005,,,@, 0.0224052295)
      TCHZ_Peak_Type(, 0,, 0,, 0.005555009,, 0,, 0.03462355,, 0.01552971)
      phase_MAC 31.59020452
      MVW( 611.767656, 254.7217502, 61.52078011)
      space_group R-3c
      scale @ 0.007745045593
      Phase_LAC_1_on_cm( 125.9860076)
```

```
      Phase_Density_g_on_cm3( 3.98813523)
      Trigonal( 4.7585306, 12.98941719)
      site Al1 num_posns 12 x  0        y  0 z  0.35214 occ Al+3  1 beq  0.34
      site O1  num_posns 18 x  0.30656 y  0 z  0.25     occ  O-2  1 beq  0.33
  str
      phase_name "Quartz"
      e0_from_Strain( 9.401987686e-006,,,@, 0.004311877544)
      TCHZ_Peak_Type(, 0,, 0,, 0.005555009,, 0,, 0.03462355,, 0.01552971)
      phase_MAC 35.81264014
      MVW( 180.2529, 112.9561208, 38.47921989)
      space_group P3221
      scale @ 0.03707566215
      Phase_LAC_1_on_cm( 94.89825445)
      Phase_Density_g_on_cm3( 2.649853629)
      Trigonal( 4.912656195, 5.404386419)
      site Si1 num_posns 3 x 0.470737 y =0;      z =2/3;    occ Si+4  1 beq  0.88
      site O1  num_posns 6 x 0.41648  y 0.26757 z  0.79108 occ  O-2  1 beq  1.54
  hkl_Is
      phase_name "amorphous"
      LVol_FWHM_CS_G_L( 1, 1.755642682, 0.89, 1.663253122,, 1.868823733,,)
      phase_MAC 31.59020452
      scale  1
      MVW( 0, 0.344, 0)
      a  34.4       b  0.1    c  0.1
      space_group Pmmm
      hkl_m_d_th2 4 0 0 2 8.60000038 10.2776861 I @ 5.439444087
      ...
      hkl_m_d_th2 42 0 0 2 0.81904763 140.264679 I @ 320.1672535
```

The correct weight% of the spiked phase can be added to the corresponding structure phase in the INP file:

```
str
  phase_name "Al2O3"
  spiked_phase_measured_weight_percent  50
```

By adding the keyword *corrected_weight_percent* to all Peak, Rietveld, Pawley and Le Bail phases and rerunning the refinement, the corrected weight% will be automatically displayed:

```
str
  phase_name "Quartz"
  corrected_weight_percent  31.27375797
```

We can also add *"weight_percent_amorphous 0"* at the *xdd* level to directly report the amorphous content.

5.8.5 Application of the PONKCS method

For a successful application of the PONKCS method, the unknown phase must be present in a pure form or in a significant amount in a mixture of known phases. In the following, we reuse the amorphous SiO_2 and 50 weight% alumina sample from the previous section where the weight fraction of the amorphous phase was determined to be 19%.

In the PONKCS approach, the unknown amorphous phase is fitted with an orthorhombic pseudo Pawley phase with the scale factor fixed to unity, all peak intensities are fixed and the phase is stored as the so-called PONKCS phase. A pseudo *ZM* value can then be calculated:

$$(ZM)_{amorphous} = \frac{0.1873}{0.5}\frac{0.007745}{1}\frac{611.77 \cdot 254.72}{0.344} = 1314.26 \; \frac{g}{mol}. \tag{5.30}$$

This value can now be reused for a different sample. Here we analyze a mixture where 5 weight% of the amorphous SiO_2 was mixed with 45 weight% quartz and 50 weight% of alumina (Figure 5.9). The previously-stored PONKCS phase is included in the INP file with the scale factor as the only refined parameter and all intensities fixed. Using the *corrected_weight_percent* keyword immediately reveals the correct weight% of the PONKCS phase.

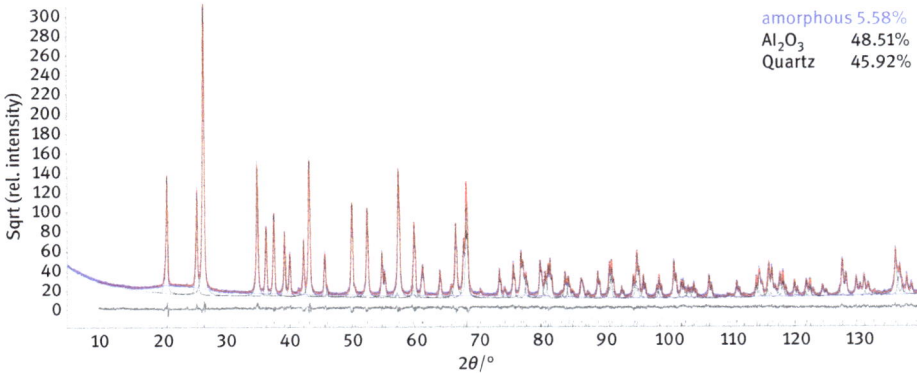

amorphous 5.58%
Al_2O_3 48.51%
Quartz 45.92%

Figure 5.9: Rietveld plot of the mixture of 50 weight% alumina, 45 weight% quartz and 5 weight% silica flour.

References

Brindley, G.W. (1945): *The effect of grain or particle size on X-ray reflections from mixed powders and alloys, considered in relation to the quantitative determination of crystalline substances by X-ray methods.* Phil. Mag. 36, 347–369.

Madsen, I.C., Scarlett, N.V.Y. (2008): Chapter 11 in *Powder diffraction: theory and practice*, Royal society of chemistry, Dinnebier, R. E. and Billinge, S. (eds.)

Madsen, I.C., Scarlett, N.V.Y., Kern, A. (2011): *Description and survey of methodologies for the determination of amorphous content via X-ray powder diffraction*. Z. Kristallogr. 226, 944–955.

Hubbell, J.H., Seltzer, S.M. (2004): *Tables of X-ray mass attenuation coefficients and mass energy-absorption coefficients* (version 1.4). Available: http://physics.nist.gov/xaamdi, National Institute of Standards and Technology, Gaithersburg, MD (Originally published 1995 as NISTIR 5632).

O'Connor, B.H., Raven, M.D. (1988): *Application of the Rietveld refinement procedure in assaying powdered mixtures*. Powder Diffraction 3, 2–6.

6 Restraints, constraints and rigid bodies

Powder diffraction analysis is frequently limited by the low information content in the experimental data. In particular, the ratio between the number of experimentally observed intensities (or groups of intensities) and refined structural parameters is usually much lower than achievable in single crystal diffraction. Introduction of non-diffraction based information (e.g., from chemical or structural knowledge) can, therefore, be extremely valuable. It can stabilize a refinement by either reducing the number of parameters or by keeping parameters close to reasonable values. Meaningless atomic positions due to flat minima of the minimization function can be avoided, and the chance of refining to the correct global minimum increases. Ways of including external information can include:

– Imposing linear or non-linear functional dependencies between parameters (constraints).
– Grouping of atoms which move as a rigid or semi-rigid entity (rigid bodies – a specific form of constraint).
– Guiding parameters to an expected value or expected range using a restraint or penalty function that increases with increasing departure from the expected value (restraints, sometimes called soft constraints).

Constraints and restraints are not limited to structural parameters but can be applied to any of the parameters used during Rietveld refinement.

Constraints are incorporated into the least-squares model in a mathematically precise way such that they must always be exactly obeyed. They can, therefore, be thought of as "hard" information. Restraints, however, are treated in a similar way to experimental observations and merely guide the refinement. They therefore act as "softer" information and the degree to which they're obeyed is balanced against the model's need to fit the experimental data. TOPAS does this by adding extra terms to the objective function that is expressed as the sum of contributions from observations (χ_0^2), penalties (χ_P^2) and restraints (χ_R^2):

$$\chi^2 = \chi_0^2 + \chi_P^2 + \chi_R^2 \tag{6.1}$$

with

$$\chi_0^2 = K \sum_{i=1}^{N} \left(w_i (y_{calc,i} - y_{obs,i})^2 \right), \quad K = \frac{1}{\sum_{i=1}^{N} w_i y_{obs,i}^2},$$

$$\chi_P^2 = K K_1 K_P \sum_{p=1}^{N_P} P_p, \quad \chi_R^2 = K K_1 K_R \sum_{r=1}^{N_R} R_r^2, \tag{6.2}$$

where $y_{calc,i}$ and $y_{obs,i}$ are the calculated and observed data, respectively, at point i of N data points; w_i is the weighting given to data point i; P_p are penalty functions; N_P is

https://doi.org/10.1515/9783110461381-006

the number of penalty functions; R_r are restraints; N_R is the number of restraints; and K_1, K_P and K_R are scaling terms applied to the penalty functions and restraints, respectively, and discussed in Section 6.2. The distinction between penalties and restraints is discussed below, though the terms are often used interchangeably. We'll use (soft-) restraints to describe both types.

The way in which penalties can help guide a refinement, restricting it to "sensible" solutions, can be best understood through simple examples. One is the "anti-bump" penalty which can be used to prevent atoms approaching more closely than is chemically plausible. This can be expressed by using a penalty that feeds into the χ_P^2 term of the objective function of the form:

$$P_p = AB_i = \begin{cases} \sum (r_{ij} - r_0)^2 & \text{if } r_{ij} < r_0 \text{ and } i \neq j, \\ 0 & \text{if } r_{ij} \geq r_0 \end{cases} \tag{6.3}$$

where r_0 is the minimum approach distance, r_{ij} the distance between atoms i and j including symmetry equivalent positions, and the summation is over all atoms of type j. The penalty is zero when atoms are far apart but rises rapidly if atoms get closer than r_0.

A penalty function suitable for keeping a series of n atoms (e.g., an aromatic ring) flat can be expressed as:

$$flat = \frac{6}{n(n-1)(n-2)} \sum_{i=1}^{n} \sum_{j=i+1}^{n} \sum_{k=j+1}^{n} \left(|\mathbf{b}_i \times \mathbf{b}_j \cdot \mathbf{b}_k| - tol \right)^2 \tag{6.4}$$

$$\text{if } |\mathbf{b}_i \times \mathbf{b}_j \cdot \mathbf{b}_k| > tol,$$

where tol is the allowed deviation from the least squares plane of the atoms before a penalty is applied and $\mathbf{b}$ are Cartesian unit length vectors between the sites and the geometric center of the n sites.

More sophisticated electrostatic potentials for ionic compounds (Figure 6.1), like the Lennard-Jones or Born–Mayer potentials can be used as penalty functions by

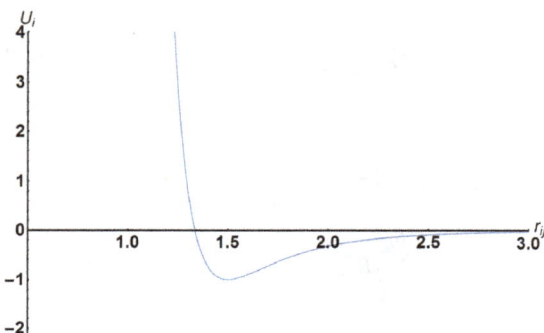

Figure 6.1: Schematic electrostatic potential as a combination of a repulsion term and an attractive Coulomb term, which can be used as penalty function.

applying what TOPAS calls the General Real Space (GRS) interaction (Coelho & Cheary, 1997). For a particular site i, the electrostatic potential U_i contains a Coulomb term C_i and a repulsive term R_i and is written as:

$$U_i = C_i + R_i, \tag{6.5}$$

where

$$C_i = \frac{e^2}{4\pi\varepsilon_0} \sum_{j=1 \neq i}^{n} \frac{Q_i Q_j}{r_{ij}} \tag{6.6}$$

and

$$R_i = \sum_{j=1 \neq i}^{n} \frac{B_{ij}}{r_{ij}^m} \tag{6.7}$$

for Lennard-Jones and

$$R_i = \sum_{j=1 \neq i}^{n} c_{ij} e^{-dr_{ij}} \tag{6.8}$$

for Born–Mayer, respectively. Here ε_0 is the permittivity of free space, e is the charge of an electron, Q_i and Q_j are the ionic charges of atoms i and j in units of e, r_{ij} is the distance between atoms i and j including symmetry equivalent positions, and the summation is over all atoms to infinity. The repulsive constants B_{ij}, m, c_{ij} and d are characteristic of the atomic species and their surroundings. The penalty has the form shown in Figure 6.1 and will guide atoms toward a sensible interatomic separation during refinement.

We will see how these and other restraints are used in TOPAS in later sections of this chapter.

6.1 Constraints in TOPAS

Parameters can be constrained to other parameters via a linear or nonlinear equation of any complexity. Constraints are built into the refined model mathematically so that they have to be obeyed exactly.

An easy example is the definition of an overall displacement parameter for a group of atoms with an enhancement factor for hydrogen atoms:

```
prm B1 1.0 min 0 max 2
site C1 num_posns  8 x @  0.0` y @  0.0 z @  0.0 occ C  1 beq =B1;
site C2 num_posns  8 x @  0.0` y @  0.0 z @  0.0 occ C  1 beq =B1;
...
site H1 num_posns  8 x @  0.1  y @  0.1 z @  0.1 occ H  1 beq =1.5 B1;
```

Constraints are also frequently used for defining occupancies of mixed sites. In the following example the parameter *zr* is used in the equation "= 1–*zr*;" to define the Ti+4 site occupancy parameter[1]:

```
Site Zr x 0 y 0 z 0
        occ Zr+4 zr    1          min=0; max = 1; beq 0.5
        occ Ti+4 =     1-zr;                      beq 0.5
```

Many other constraints are built into the TOPAS macros. For example, the cell parameter macro *Cubic (@ 4.1)* writes a series of constraint equations that force the lattice parameters to obey $a = b = c$ during refinement.

It is possible to assign minimum and maximum values to any parameter to constrain the value to lie within a user-defined range.[2] For example, one might set a minimum and maximum value of ±0.1° on the zero point error of a well-aligned diffractometer to prevent a refinement diverging to unreasonable values in its early stages. Most of the TOPAS macros have this type of constraint built in. One example is the *Keep_Atom_Within_Box* macro that sets min and max values on fractional coordinates to force an atom to remain within a cubic volume element centered on its starting value during refinement. In the following example, the potassium site cannot move outside of a box with a length of $2 \times size$ (here 0.1 Å) around its starting position:

```
...
prm !size 0.1
site K1    x kx 0.25 y ky 0.25 z kz 0.25   occ K+1  ok 0.5  min 0 max 1  beq  B1
           Keep_Atom_Within_Box (size)
...
```

6.2 Restraints & penalties in TOPAS

Soft-restraints allow more flexibility than constraints (Section 6.1) or rigid bodies (Section 6.3). They are handled by the least squares in the same way as experimental observations (they are "extra observations"). TOPAS has two ways of doing this – using either *penalty* or *restraint* equations.

---

1 In TOPAS, equations always start with an equal sign and end in a semicolon.
2 Care must be taken if parameters hit their min/max limits at the end of a refinement as the model may not be fully converged and other parameters may be affected. TOPAS flags this in the OUT file. The origin of the problem should be investigated and, if necessary, parameters removed from the model.

Imagine a simple case where one wants to use some form of soft-restraint to keep a coordinate *x1* close to a particular expected value, for example, close to 0.137. In TOPAS one could use a penalty equation:

```
penalty = (x1 - 0.137)^2;
```

The parabolic form of the penalty means that it will increase the overall χ^2 (eq. (6.1)) by an increasing amount as x1 deviates from 0.137 regardless of whether its value is less than or greater than 0.137. The penalty will therefore guide the refinement toward a model with *x1* of 0.137 (low χ_P^2) provided this doesn't cause too great an increase in χ_0^2.

Alternatively one could use a restraint equation:

```
restraint = (x1 - 0.137);
```

Internally TOPAS will square the restraint equation. It therefore ends up being very similar to using the penalty equation. Don't get too confused by the terms "penalty" and "restraint."[3] Both methods try to steer the refinement toward a solution that is consistent with a non-diffraction based observation. In general penalty equations are more flexible as they're not automatically squared and can therefore take negative values (e.g., the potential of Figure 6.1). There can be small differences in how penalties and restraints feed into the minimization of χ^2, due to different weightings against the diffraction data. As a consequence the minimization pathway and final minimum may differ. One difference is that off-diagonal terms in the **A** matrix (see Section 2.7) are not calculated for penalties unless:

```
approximate_A
```

is defined. This means that you might get quicker minimization using restraint equations. If penalties have a significant contribution to χ^2 than using *approximate_A* may give faster convergence as the off diagonal terms are then approximated by the BFGS method.

Penalties typically aren't included in esd calculations (restraints are) though you can override this with:

```
do_errors_include_penalties
```

---

3 For historical reasons even TOPAS syntax uses the terms inconsistently. For example *only_penalties* will apply to both penalty and restraint equations; macros like *Distance_Restrain* typically use a penalty equation.

TOPAS uses a separate **A** matrix for the data, penalties and restraints. It chooses how to scale penalties or restraints against the data (i.e., how closely each is obeyed; the terms K_P and K_R in eq. (6.2)) by considering the relative magnitudes of the inverse error terms in each matrix. The overall relative importance of penalties and restraints to data can be increased using the following keyword, which sets the value of K_1 in eq 6.2:

```
penalties_weighting_K1 1
```

Normally the default value of 1 is appropriate. By using larger values of K_1 penalties and restraints will be more closely obeyed and they will start to mimic constraints.

If one has several penalty equations their individual relative weighting can be changed by using an equation like:

```
penalty = w*(x1 - 0.137)^2;
```

The value w might be different for a soft-restraint on a bond angle compared to a bond distance. Typically one would weight by $1/\sigma^2$ where σ is the standard uncertainty on the quantity. The relative weighting for soft-restraints on distances ($\sigma = 0.01$) and angles ($\sigma = 1$) might therefore be 10000 to 1.

To instruct the minimization procedure to minimize on penalty (and/or restraint) functions only use:

```
only_penalties
```

This *only_penalties* switch is used either when there is no observed data or when one wants to temporarily ignore the data. With sufficient restraints a structural model can be refined without data, analogous to the DLS method of Baerlocher (Baerlocher, 1978). Note that parameters that are not dependent on the penalties are not refined.

Penalties can also be a function of the iteration number and can be turned off or decreased in importance once a certain iteration number is reached using the reserved parameter *Cycle_Iter* that returns the current iteration within a cycle with counting starting at zero. In the following example, the penalty is only applied for the first 10 iterations of the current refinement cycle:

```
penalty = If(Cycle_Iter < 9, (x1 - 0.137)^2, 0);
```

In general, the overall R_{wp} should not increase significantly when restraints are applied to a model. If it does, it is a sign of deficiencies in the model and/or the measurement. The possibilities for user defined penalties and constrains are essentially infinite. For convenience there are many predefined restraint macros, some of which are described in the following section.

6.2.1 Distance, angle and flatten restraints

Common macros in TOPAS that apply penalty equations include anti bumping, bond lengths, bond angles and flatness restraints:

```
AI_Anti_Bump(…)
Distance_Restrain(…)
Angle_Restrain(…)
Flatten(…)
```

As an example, let's consider how each of them might be used to control the geometry of an aromatic cyclopentadienyl ring consisting of 5 carbon atoms C1…C5 (Figure 6.2). It can be assumed that the bond lengths between two carbon atoms will be 1.4 Å and the angle between three carbon atoms will be close to 108°. The molecule is assumed to be flat with a maximum deviation from the plane of 0.01 Å. The positional parameters in the TOPAS input file are:

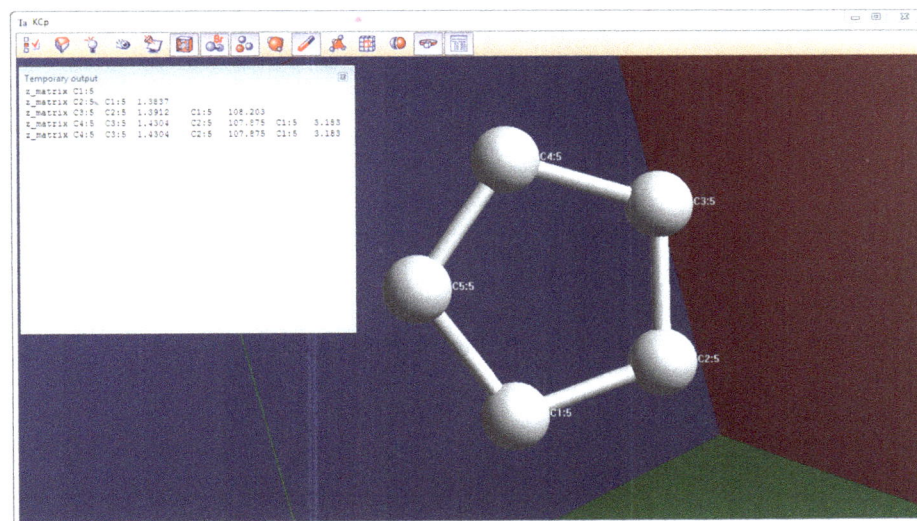

Figure 6.2: Aromatic cyclopentadienyl ring consisting of 5 carbon atoms C1…C5. The insert shows a Z-matrix description as discussed later.

```
site C1  x @  0.31646` y @  0.22174` z @  0.36153` occ C  1 beq 2
site C2  x @  0.31488` y @  0.35432` z @  0.32264` occ C  1 beq 2
site C3  x @  0.18779` y @  0.40617` z @  0.34644` occ C  1 beq 2
site C4  x @  0.10624` y @  0.29993` z @  0.39525` occ C  1 beq 2
site C5  x @  0.18537` y @  0.18763` z @  0.40500` occ C  1 beq 2
```

When setting up restraints, care must be taken to restrain the correct atoms as the list of atomic sites might contain atoms from different parts of the unit cell that are in different molecules. An inadvertent restraint on an inter- rather than intra-molecular C–C bond will cause catastrophic problems during refinement! To facilitate this, the command:

```
append_bond_lengths
```

---

writes bond lengths, angles and torsion angles to the OUT file after refinement (which could be after 0 refinement cycles (*iters* 0) so that a model is unchanged). In the listing each atom is uniquely identified by four numbers corresponding to the symmetry operation used to generate it, and fractional offsets of the cell in which the atom sits relative to the pivot atom. If all four numbers are zero they can be omitted, otherwise they have to be specified for each atom in the restraint. Here the five sites listed in the asymmetric unit are all part of the same molecule and the output looks like:

```
{
C1:0   C2:0      0   0   0   1.38365
       C5:0      0   0   0   1.42564  108.178
       C3:0      0   0   0   2.24781   72.209   36.011
       C4:0      0   0   0   2.26421   36.962   35.247   72.941
...
}
```

The most basic restraint would be to prevent atoms getting too close using the *AI_Anti_Bump* macro between the five carbon sites. The relative weight given to the penalty function is set to one and we penalizee approaches below 1.2 Å:

```
prm !bump_dist     1.2
prm !bump_weight   1
AI_Anti_Bump(C*, C*, bump_dist, bump_weight)
```

The closer the atoms are, the higher the penalty becomes. The *AI_Anti_Bump* macro includes the penalty function given in eq (6.1), and its expression in TOPAS syntax can be found in the TOPAS.INC file. Since anti-bumping restraints force atoms to stay away from each other, they are mainly used in structure determination using global optimization methods (see Chapter 7). Applying the restraint only for the first few iterations of a refinement cycle can also be beneficial.

A more powerful approach is to use the *Distance_Restrain* macro to restrain the bond length between the carbon atoms C1 and C2 to 1.4 Å, with a tolerance of 0.02 Å and a relative weight of 100.[4] The tolerance means a penalty is only applied if the refined distance lies outside the range 1.38–1.42 Å:

```
prm !bond_length    1.4
prm !bond_tol      0.02
prm !bond_weight    100
Distance_Restrain(C1 0 0 0 0  C2 0 0 0 0,  bond_length, 1.384, bond_tol, bond_weight)
```

The actual value of the bond length (1.384 Å) is automatically added to the OUT file after the first refinement run is completed. The penalty function used here is a parabola [$(r − 1.4)^2$ for $r < 1.38$ or $r > 1.42$], which means that the penalty increases symmetrically for shorter and longer distances.

Similarly, the *Angle_Restrain* macro can be used to restrain the bond angle between the carbon atoms C1, C2 and C5 to 108° with an angle tolerance of 2° and a relative weight of 10 by a similar parabolic function:

```
prm !angle         108
prm !angle_tol       2
prm !angle_weight   10
Angle_Restrain(C5 0 0 0 0 C1 0 0 0 0 C2 0 0 0 0, angle, 108.18037`,angle_tol, angle_ weight)
```

Finally, restraining the five carbon atoms C1...C5 to a flat plane with a tolerance of 0.01 Å and a relative weight of 1000 can be performed by the *Flatten* macro:

```
prm !flatten_tol 0.01
prm !flatten_weight 1000
Flatten(C1 C2 C3 C4 C5, 0.0107887918`, flatten_tol, flatten_weight)
```

which uses the penalty function defined in eq. (6.4). Note that more than three sites must be used with this macro.

The relevant part of the corresponding TOPAS INP file for restraining the entire five-ring using *Distance_Restrain, Angle_Restrain* and *Flatten* macros is given below:

```
penalties_weighting_K1 1
prm !bond_tol        0.02
prm !bond_weight      10
```

---

4 The symmetry label and fractional offsets of all carbon sites in this example are zero. Although the four zeros could be omitted, they are given here for completeness.

```
prm !angle_tol        2
prm !angle_weight     1
prm !flatten_tol        0.01
prm !flatten_weight 1000
Distance_Restrain(C1 0 0 0 0 C2 0 0 0 0, 1.4, 1.38365`, bond_tol, bond_weight)
Distance_Restrain(C2 0 0 0 0 C3 0 0 0 0, 1.4, 1.39125`, bond_tol, bond_weight)
Distance_Restrain(C3 0 0 0 0 C4 0 0 0 0, 1.4, 1.43036`, bond_tol, bond_weight)
Distance_Restrain(C4 0 0 0 0 C5 0 0 0 0, 1.4, 1.37359`, bond_tol, bond_weight)
Distance_Restrain(C5 0 0 0 0 C1 0 0 0 0, 1.4, 1.42564`, bond_tol, bond_weight)
Angle_Restrain(C1 0 0 0 0 C2 0 0 0 0 C3 0 0 0 0, 108, 108.19957`, angle_tol, angle_ weight)
Angle_Restrain(C2 0 0 0 0 C3 0 0 0 0 C4 0 0 0 0, 108, 107.87748`, angle_tol, angle_ weight)
Angle_Restrain(C3 0 0 0 0 C4 0 0 0 0 C5 0 0 0 0, 108, 107.68999`, angle_tol, angle_ weight)
Angle_Restrain(C4 0 0 0 0 C5 0 0 0 0 C1 0 0 0 0, 108, 107.95641`, angle_tol, angle_ weight)
Angle_Restrain(C5 0 0 0 0 C1 0 0 0 0 C2 0 0 0 0, 108, 108.18037`, angle_tol, angle_ weight)
Flatten(C1 C2 C3 C4 C5, 0.0107887918`, flatten_tol, flatten_weight)
```

6.2.2 Electrostatic potentials

Penalty equations can take any form – that is, they can be more complex than the sum of differences squared that appear in restraint equations. For example the equations in *grs_interaction* (which calculates a Lennard-Jones or Born–Mayer potential according to eqs. (6.5)–(6.8) can't be written as the sum of differences between observed and calculated parameters. In TOPAS, these are called using the predefined macros:

```
Grs_Interaction(…)
Grs_BornMayer(…)
```

The following example forces chemically sensible coordination environments between an aluminum and several oxygen sites. Charges have been set to +3 and −2 for aluminum and oxygen, respectively. The expected bond length is 2.6 Å between oxygen sites and 1.8 Å between aluminum and oxygen sites. First, the Lennard-Jones potential (eq. (6.7)) for the repulsion term is used:

```
prm !val_charge_Al 3
prm !val_charge_O -2
prm !dist_Al_O      1.8
prm !dist_O_O       2.6
prm !expo           5
Grs_Interaction(O*, O*, val_charge_O, val_charge_O, oo, dist_O_O, expo) penalty = oo;
Grs_Interaction(Al, O*, val_charge_Al, val_charge_O, alo, dist_Al_O, expo) penalty = alo;
```

with the exponent of the repulsion part set to five.[5] Alternatively, the Born–Mayer equation (eq. (6.8)) for the repulsion term can be used:

```
prm !val_charge_Al 3
prm !val_charge_O -2
prm !dist_Al_O    1.8
prm !dist_O_O     2.6
prm !const        3
Grs_BornMayer(O*, O*, val_charge_O, val_charge_O, oo, dist_O_O,  const) penalty = oo;
Grs_BornMayer(Al, O*, val_charge_Al, val_charge_O, alo, dist_Al_O, const) penalty = alo;
```

with the constant for the repulsion part set to three.

6.3 Rigid bodies

In many structures, groups of atoms (molecules or coordination polyhedra) have strong local bonding and form a more or less rigid unit. Rather than refining the constituent atoms independently it may be better to define them as a single rigid body. Typical examples are the cyclopentadienyl anion or the benzene ring. Rigid bodies have been a common tool in single crystal X-ray diffraction analysis for more than 50 years (Scheringer, 1963), and are especially valuable when the quality of the data is low, the ratio of observations to parameters is low and/or the structure is very complicated (e.g., proteins). There are several general advantages for using rigid bodies, which are probably even more advantageous with powder data than single crystal (Dinnebier, 1999):

- Since the group is forced to shift as a complete unit, meaningless changes in internal geometry cannot occur.
- The number of refined parameters can be drastically reduced, allowing them to be determined with much higher accuracy. This is in particular useful in the case of powder data, where the ratio of the number of independent observations (Bragg intensities) to refineable parameters is typically low.
- The range of convergence to the correct structure is much larger than in normal refinement.
- Hydrogen atoms can be included in the refinement process at an early stage. Only their relative positions with respect to the other atoms are needed (a "riding model").

---

5 TOPAS allows the use of the wild card character "*" and the negation character "!" to simplify the creation of lists of atom identifiers.

– Thermal parameters can be defined to describe the group as a whole. The use of *TLS* matrices allows anisotropic refinement of the translational and librational parts of the temperature factor with relatively few parameters.
– Rigid bodies can allow you to refine individual atomic positions even if disorder is present. By using rigid bodies, it can be possible to model disorder even with powder data (e.g., Behrens et al., 2008).

6.3.1 Definition of a rigid body

A rigid group of atoms can be set up using a variety of internal reference coordinate systems. The most common are fractional, Cartesian, spherical or Z-Matrix formalisms. TOPAS supports each of these and even allows a mixture of them to be used within a single INP file.

Whichever coordinate system is used, there must be a one-to-one match between the sites defined in the rigid body (e.g., *point_for_sites*) and available atomic sites in the *str* section of the INP file. The positions in the atomic site list can be set to (0,0,0) as they are updated by TOPAS during refinement based on the degrees of freedom of the rigid body. It is the responsibility of the user to check for correct site symmetries and, if necessary, to restrict rigid body rotations and/or translations and to set the fractional occupancy accordingly.

The graphical rigid body editor in TOPAS is of great help in setting up and checking rigid bodies.

6.3.2 Cartesian coordinates

Atoms of a rigid body can be defined in the Cartesian coordinate system $\mathbf{I} = \{\mathbf{i}, \mathbf{j}, \mathbf{k}\}$ with unity axis length (usually one Angstrom) or in the equivalent fractional coordinate system $\mathbf{D} = \{\mathbf{a}, \mathbf{b}, \mathbf{c}\}$. There is an infinite number of ways of defining the natural basis of a crystal in terms of a Cartesian basis. In TOPAS, the conversion from fractional to Cartesian coordinates in terms of the lattice vectors $\mathbf{a}$, $\mathbf{b}$ and $\mathbf{c}$ is as follows: (1) x-axis in the same direction as the *a* lattice parameter $\mathbf{x} \parallel \mathbf{a}$; (2) z-axis perpendicular to the *a-b* plane $\mathbf{z} \parallel (\mathbf{b} \times \mathbf{a}) \parallel \mathbf{c}^*$; (3) y-axis in the direction defined by the cross product of *a* and *c* $\mathbf{y} \parallel \mathbf{a} \times (\mathbf{b} \times \mathbf{a})$ (Fig. 6.3).

The conversion matrix $\mathbf{M}$ to convert the natural crystallographic coordinate system $\mathbf{D}$ into the Cartesian coordinate system of the crystal $\mathbf{I}$ and vice versa:

$$\mathbf{I} = \mathbf{M}\mathbf{D} \text{ and } \mathbf{D} = \mathbf{M}^{-1}\mathbf{I}, s \tag{6.9}$$

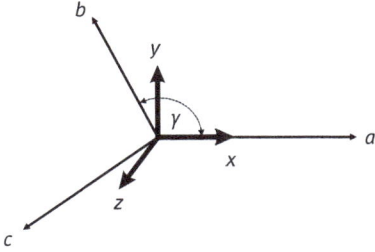

Figure 6.3: Orthonormalization of the crystallographic coordinate system **D** into a Cartesian coordinate system **I**.

is thus given by:

$$
\mathbf{M} = \begin{pmatrix} 1 & 0 & 0 \\ \cos\gamma & \sin\gamma & 0 \\ \cos\beta & \dfrac{\cos\alpha - \cos\beta\cos\gamma}{\sin\gamma} & \sqrt{1 - \cos^2\beta - \left(\dfrac{\cos\alpha - \cos\beta\cos\gamma}{\sin\gamma}\right)^2} \end{pmatrix}. \tag{6.10}
$$

In TOPAS, a point in Cartesian space is denoted by the parameters ux, uy, uz. The origin of the coordinate system is the basepoint of the rigid body and is normally the first *point_for_site* defined. For practical reasons, in particular to ensure similar movement of atoms at similar distances from the origin, the center of gravity of the rigid body is a good origin choice. The origin doesn't have to be at an atomic position, and in these cases it can be convenient to place a dummy atom with zero occupancy at the origin.

There are an infinite number of ways of setting up a rigid body all of which have pros and cons. It is usually advantageous to build a rigid body in a way that allows direct refinement of the parameters (e.g., bond lengths and angles) of most interest.

As a first example of a rigid body in Cartesian coordinates, a regular octahedron of ZrO_6 with zirconium in the center can be defined in terms of a bond length r as (Figure 6.4).

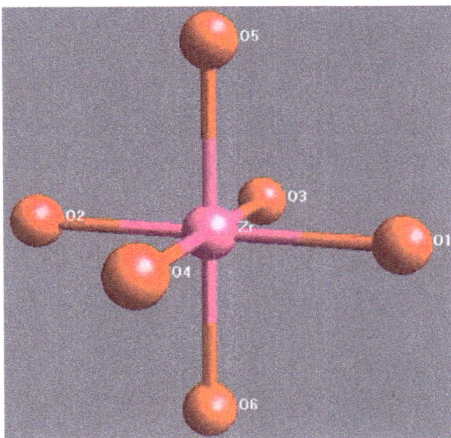

Figure 6.4: Rigid body of a regular ZrO_6 octahedron.

```
prm r 2 min 1.8 max 2.2
rigid
   point_for_site Zr              ' The origin; ux, uy, uz not specified as default to 0
   point_for_site O1 ux =  r;
   point_for_site O2 ux = -r;
   point_for_site O3 uy =  r;
   point_for_site O4 uy = -r;
   point_for_site O5 uz =  r;
   point_for_site O6 uz = -r;
```

The Cartesian coordinates, *ux, uy* and *uz* must be defined for each *point_for_site* unless they are zero. The isotropic Zr–O bond length *r* is set as a refinable internal degree of freedom in the limit between $2.0 \le r \le 2.2$ Å.

If the octahedron is located on one or more symmetry elements, fewer atoms are needed in the rigid body definition as the rest are created by space-group symmetry. If the central Zr atom lies on a center of inversion, it is sufficient to define only the three ligands O1, O3 and O5. Alternatively, the fractional occupancy of a particular atomic site can be scaled appropriately; in this case sites O1 to O6 could be set to occupancy 0.5.

More flexibility can be built into the ZrO_6 group by making the three main axes non-equidistant. This also distorts the 90° angle between three oxygen atoms in the base plane, simulating compressive strain:

```
prm r1 2 min 1.8 max 2.4
prm r2 2 min 1.8 max 2.4
prm r3 2 min 1.8 max 2.4
rigid
   point_for_site Zr
   point_for_site O1 ux =  r1;
   point_for_site O2 ux = -r1;
   point_for_site O3 uy =  r2;
   point_for_site O4 uy = -r2;
   point_for_site O5 uz =  r3;
   point_for_site O6 uz = -r3;
```

To turn the octahedron into a tetragonal or even orthorhombic bispyramid, thus keeping the angle between three oxygen atoms in the base plane rectangular, a different definition of the rigid body is more useful:

```
prm angle 45 min 0 max 90
prm !s45 = Sin(Deg angle);
prm !c45 = Cos(Deg angle);

prm r1 1.8 min 1.8 max 2.4
prm r2 2.2 min 1.8 max 2.4
```

```
prm r3 2.4 min 1.8 max 2.4
rigid
   point_for_site Zr
   point_for_site O1 ux =  s45 r1;   uy = c45 r1;
   point_for_site O2 ux = -s45 r1;   uy =  c45 r1;
   point_for_site O3 ux =  s45 r2;   uy = -c45 r2;
   point_for_site O4 ux = -s45 r2;   uy = -c45 r2;
   point_for_site O5 uz =  r3;
   point_for_site O6 uz = -r3;
```

Using this approach, more and more degrees of freedom can be introduced. For example, the polyhedron can be further distorted by changing the 45° angle.

The maximum number of degrees of freedom of the rigid body is always three times the number of atoms, $3n$, or 21 in our case. A perfect octahedron has six degrees of freedom (three translational and three rotational) resulting in 15 fewer refinable parameters. Refining one overall length leads to seven degrees of freedom. There are 9 degrees of freedom for an orthorhombic bipyramid and 10 degrees of freedom if one angle in the equatorial plane is variable.

A general definition of a tetrahedron that can be distorted to a tetragonal bisphenoid by changing the tetrahedral angle between the ligands (Figure 6.5) is given below:

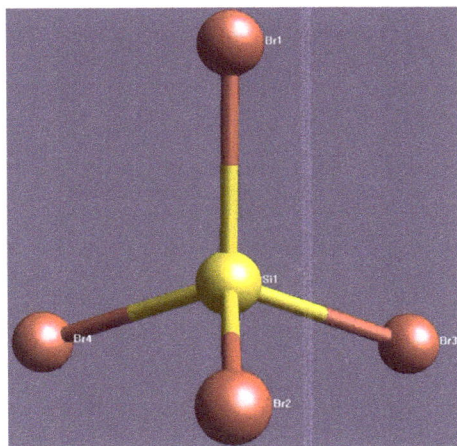

Figure 6.5: Rigid body description of a regular $SiBr_4$ tetrahedron.

```
prm r  2.17460 min 2.0 max 2.4
prm a 109.4712 ' = 2 Arccos(1/Sqrt(3));
prm s = Sin(Deg a/2);
prm c = Cos(Deg a/2);
```

```
rigid
    point_for_site Si1
    point_for_site Br1 ux =  s r; uy =   c r;
    point_for_site Br1 ux = -s r; uy =   c r;
    point_for_site Br1             uy = -c r; uz =  s r;
    point_for_site Br1             uy = -c r; uz = -s r;
```

A more complicated example is *para*-hydroxybenzoate that consists of a central benzene ring with hydroxy and carboxylate groups in the 1 and 4 positions. These can both be twisted adding two additional internal degrees of freedom to the rigid body (Figure 6.6). The most logical choice for the origin of the rigid body is in the center of the benzene ring. Although not necessary, a dummy atom (e.g., called "X") with zero occupancy could be introduced at this position. Note that it needs to be defined both in the *point_for_site* section of the rigid body and the *site* section of the *str*.

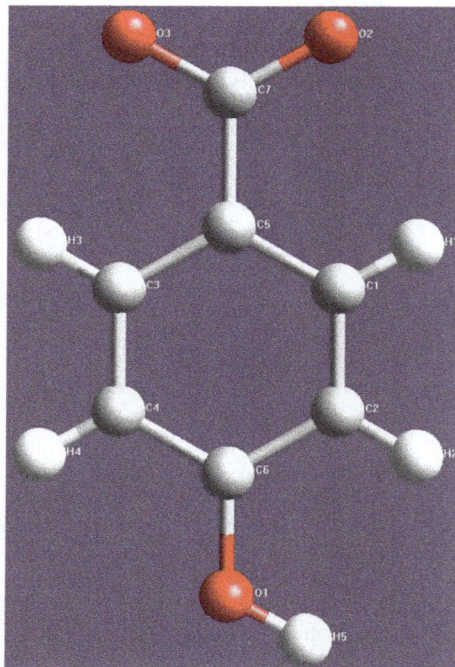

Figure 6.6: Rigid body of a *para*-hydroxybenzoate molecule.

```
prm !b_COH         1.377 min 1.33 max 1.40
prm !b_CC_aroma    1.392 min 1.37 max 1.42
prm !b_CC_single   1.540 min 1.50 max 1.60
```

```
prm !b_CH            1.000 min 0.95 max 1.05
prm !b_CO_aroma      1.281 min 1.25 max 1.30

prm !s30 = 0.5;
prm !c30 = Sqrt(3) .5;

rigid
   point_for_site X
   point_for_site C1 ux =   b_CC_aroma c30;        uy =   b_CC_aroma s30;
   point_for_site C2 ux =   b_CC_aroma c30;        uy =  -b_CC_aroma s30;
   point_for_site C3 ux =  -b_CC_aroma c30;        uy =   b_CC_aroma s30;
   point_for_site C4 ux =  -b_CC_aroma c30;        uy =  -b_CC_aroma s30;
   point_for_site C5                               uy =   b_CC_aroma;
   point_for_site C6                               uy =  -b_CC_aroma;

   point_for_site H1 ux =  (b_CC_aroma + b_CH) c30; uy =  (b_CC_aroma + b_CH) s30;
   point_for_site H2 ux =  (b_CC_aroma + b_CH) c30; uy = -(b_CC_aroma + b_CH) s30;
   point_for_site H3 ux = -(b_CC_aroma + b_CH) c30; uy =  (b_CC_aroma + b_CH) s30;
   point_for_site H4 ux = -(b_CC_aroma + b_CH) c30; uy = -(b_CC_aroma + b_CH) s30;

   point_for_site O1                               uy = -(b_CC_aroma + b_COH);
   point_for_site H5 ux =  b_CH c30;               uy = -(b_CC_aroma + b_COH) - (b_CH s30);
   point_for_site C7                               uy =  (b_CC_aroma + b_CC_single);
   point_for_site O2 ux =  b_CO_aroma c30;         uy =  (b_CC_aroma + b_CC_single) + (b_CO_aroma s30);
   point_for_site O3 ux = -b_CO_aroma c30;         uy =  (b_CC_aroma + b_CC_single) + (b_CO_aroma s30);
```

The two internal torsions of the CO_2^- and the OH group can be defined by rotations around a vector defined by two points. Take the carboxylate-group as an example, the atoms O2 and O3 are rotated around a vector defined by the atoms C5 and C7. Technically this is done by first translating the starting position of the rotation vector to the origin together with the atoms to be rotated. Afterwards, the rotation is performed around the rotation vector. Finally, all rotated atoms are translated back to their original position. In TOPAS notation:

```
Translate_point_amount(C5, -) operate_on_points "O2 O3"
rotate @ 0.0
Rotation_vector_from_points(C5, C7) operate_on_points "O2 O3"
Translate_point_amount(C5) operate_on_points "O2 O3"
```

or much more simply by the corresponding macro:

```
Rotate_about_points(@   0.0, C5, C7, "O2 O3")
```

6.3.3 Fractional coordinates

Setting up a rigid body in fractional coordinates has the advantage that the coordinates of an existing CIF file can be copied one to one to the rigid body. Any internal degrees of freedom can then be added. A typical application would be a molecular crystal structure for which a CIF file exists from low-temperature analysis that is to be refined at higher temperature using powder data allowing determination of changes of location, rotation and torsion angles of the molecules.

In TOPAS, a point in space with fractional coordinates is given by the parameters *ua, ub* and *uc*. Starting location and orientation of the rigid body are predetermined by the coordinates. Torsions can be defined and refined as in the example above.

In the following example, the fractional coordinates of a $C-CF_3$ group are copied into the rigid body:

```
rigid
    point_for_site C1    ua  0.2151  ub  1.0328  uc  1.0323
    point_for_site C2    ua  0.2302  ub  0.9579  uc  0.9878
    point_for_site F1    ua  0.2917  ub  1.0360  uc  1.0743
    point_for_site F2    ua  0.1324  ub  1.0038  uc  1.0592
    point_for_site F3    ua  0.1841  ub  1.1079  uc  1.0137
    ...
```

In order to refine the rotation of the three fluorine atoms F1, F2 and F3 around the C1-C2 axis, the following code must be added:

```
Rotate_about_points(@  0,  C2,  C1,  "F1 F2 F3")
```

6.3.4 Spherical coordinates

In some cases, such as molecules like C_{60}, it can be beneficial to define a rigid body in spherical coordinates. In order to introduce spherical coordinates in the rigid body definition, they must be transformed to Cartesian coordinates in the definition.

A simple rigid body, consisting of only one atom would look like[6]:

```
prm radius 1.0 min 0
prm theta 10   min 0 max 180
prm phi    10  min 0 max 360
```

---

[6] In TOPAS, the following constants are predefined: $Pi = \pi$, $Deg = 2\pi/360$, $Deg\_on\_2 = \pi/360$, $Rad = 360/2\pi$.

```
rigid
   point_for_site Ca ux = radius Sin(Deg theta) Cos(Deg phi);
                   uy = radius Sin(Deg theta) Sin(Deg phi);
                   uz = radius Cos(Deg theta);
```

or better defined as rotations:

```
...
rigid
   point_for_site Ca
   rotate theta qc 1 radius
   rotate phi   qa = Sin(Deg theta); qb = -Cos(Deg theta); radius
```

It is recommended to write a macro to facilitate the conversion.

6.3.5 Using internal rotations and translations to create a rigid body

Finally one can use tricks such as duplications, translations and rotations of sites with built-in TOPAS macros to rapidly build very complex molecules without complex trigonometry. As an example, let's consider how to build a complex molecule like the $C_{10}H_8$ molecule that consists of two hinged benzene rings as shown in Figure 6.7.

First, a benzene ring is formed by duplicating a first point that is iteratively rotated by 60° around the z-axis:

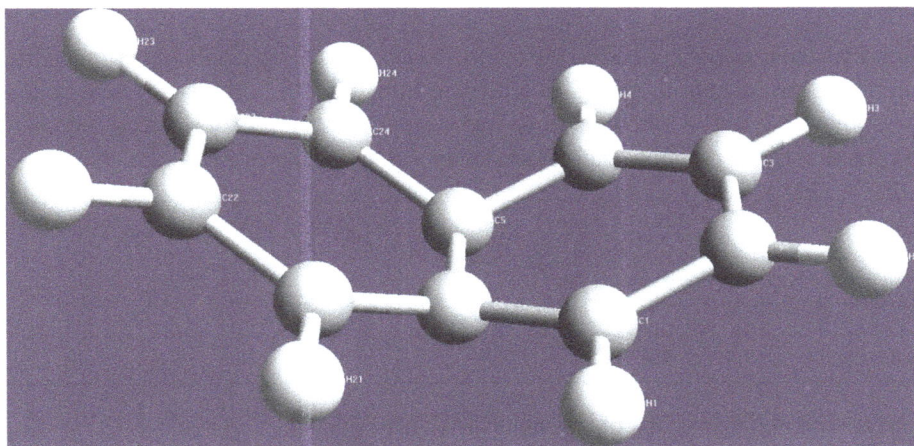

Figure 6.7: Rigid body of the final bent $C_{10}H_8$ molecule.

```
prm r1 1.3
prm r2 1.08
rigid
   point_for_site C1 ux = r1;
   Duplicate_rotate_z(C2, C1, 60)
   Duplicate_rotate_z(C3, C2, 60)
   Duplicate_rotate_z(C4, C3, 60)
   Duplicate_rotate_z(C5, C4, 60)
   Duplicate_rotate_z(C6, C5, 60)
   ...
```

This results in a regular benzene ring without hydrogen atoms (Figure 6.8).

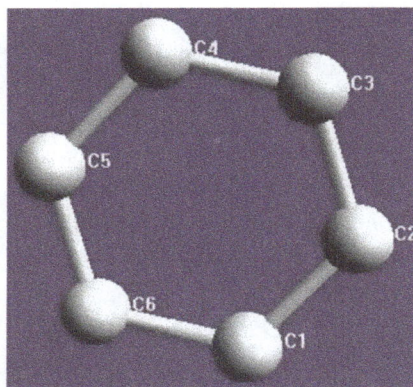

Figure 6.8: Rigid body of a benzene ring without hydrogen atoms.

Four hydrogen atoms bonded to the carbon atoms C1...C4 are created by duplicating one hydrogen atom at the position of C1 and translating it in the x-direction by the C–H bond length. The following three hydrogen atoms are then formed by duplicating the first hydrogen atom that is iteratively rotated by 60° around the z-axis:

```
...
Duplicate_Point(H1, C1)
translate tx = r2; operate_on_points H1
Duplicate_rotate_z(H2, H1, 60)
Duplicate_rotate_z(H3, H2, 60)
Duplicate_rotate_z(H4, H3, 60)
```

This results in the first half of the molecule (Figure 6.9).

The other half of the molecule is created by duplicating the four C–H groups and rotating the duplicated atoms by 140° around the vector formed by the C5 and C6 atoms. This produces the final molecule that was shown in Figure 6.7:

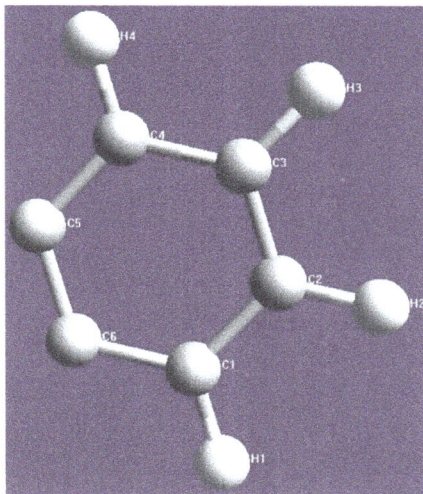

Figure 6.9: Rigid body of a benzene ring with four hydrogen atoms.

```
...
Duplicate_Point(C21, C1)
Duplicate_Point(C22, C2)
Duplicate_Point(C23, C3)
Duplicate_Point(C24, C4)
Duplicate_Point(H21, H1)
Duplicate_Point(H22, H2)
Duplicate_Point(H23, H3)
Duplicate_Point(H24, H4)
Rotate_about_points(140, C5, C6, "C21 C22 C23 C24 H21 H22 H23 H24")
```

6.3.6 Z-matrix

The Z-matrix notation is commonly used by chemists to define a group of atoms using an internal coordinate representation (Leach, 1996). Each atom in a molecule is defined in terms of its label, bond length, bond angle and dihedral angle relative to other atoms. The name arises because the Z-matrix always places the second atom in a molecule directly along the Z axis from the first atom, which defines the origin. The third atom is defined by a distance to the first or second atom and an angle between the three atoms. The fourth and all consecutive atoms are defined by a distance to one, an angle with two and a dihedral angle with three previously defined atoms. Owing to this flexible definition there are various different ways of setting up the Z-matrix. Particular care must be taken for cyclic compounds as the closing of the ring is not defined (there is no requirement to specify a geometric relationship between the last and the first atom), which sometimes leads to problems if individual bond lengths/angles are refined. A possible solution is the introduction of dummy atoms (with zero occupancy) at the center of the ring that can act as a kind of anchor.

Various tools are available that directly transform any molecule in CIF- or MOL-format to Z-matrix format. Z-matrices can always be converted to Cartesian coordinates and back, as the structural information contained is identical. In TOPAS, the conversion of Z-matrix coordinates to Cartesian is as follows: (1) the first atom, if defined using the *z_matrix* keyword, is placed at the origin; (2) the second atom is placed on the positive z-axis; (3) the third atom is placed in the *xz*-plane.

We can illustrate the Z-matrix syntax in TOPAS using the regular ZrO_6 octahedron discussed above in Z-matrix notation with the Zr atom located at the origin. In a general line each site is specified by a distance relative to the preceeding atom and two angles:

```
prm r 2 min 1.8 max 2.2
rigid
    z_matrix Zr
    z_matrix O1 Zr =r;
    z_matrix O2 Zr =r;  O1 90
    z_matrix O3 Zr =r;  O1 90   O2 90
    z_matrix O4 Zr =r;  O1 90   O2 180
    z_matrix O5 Zr =r;  O1 90   O2 270
    z_matrix O6 Zr =r;  O1 180 O2 0
```

A slightly more complicated example is a Fe_2O_7 double tetrahedron with the Fe–O–Fe bond angle and the twist between the two tetrahedra as internal degrees of freedom (Figure 6.10):

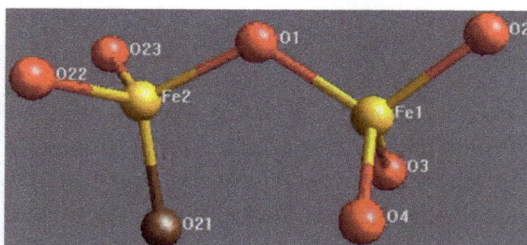

Figure 6.10: Rigid body description of a Fe_2O_7 double tetrahedron.

```
prm r        1.90
prm angle 120.00
prm twist    0.00
```

```
rigid
   z_matrix O1
   z_matrix Fe1 O1  = r;
   z_matrix O2  Fe1 = r; O1   109.5
   z_matrix O3  Fe1 = r; O2   109.5 O1   120
   z_matrix O4  Fe1 = r; O3   109.5 O2   240
   z_matrix Fe2 O1  = r; Fe1 = angle;O2  180
   z_matrix O21 Fe2 = r; O1   109.5 O2 = twist;
   z_matrix O22 Fe2 = r; O1   109.5 O2 = twist + 120;
   z_matrix O23 Fe2 = r; O1   109.5 O2 = twist + 240;
```

A third example is a rigid body for the cyclic pentamethyl-cyclopentadienyl $C_5(CH_3)_5^-$ anion, where three dummy atoms were defined to facilitate the definition of the inner and outer carbon atom positions (Figure 6.11).

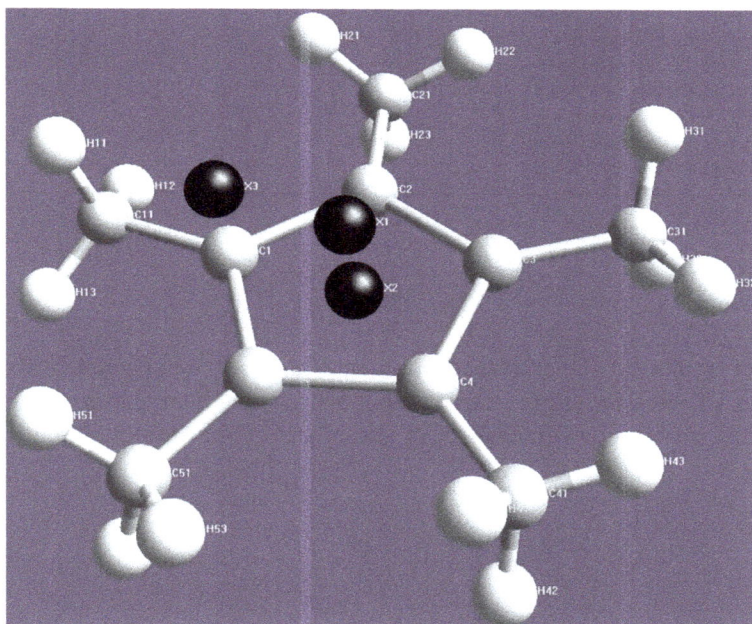

Figure 6.11: Rigid body of the pentamethyl-cyclopentadienyl $C_5(CH_3)_5^-$ anion and three dummy atoms X1, X2 and X3 in black.

```
rigid
   z_matrix X2
   z_matrix X1 X2 1
   z_matrix C1  X2  1.42 X1 90
```

```
z_matrix C2   X2   1.42 X1 90 C1 72
z_matrix C3   X2   1.42 X1 90 C1 = 2 72;
z_matrix C4   X2   1.42 X1 90 C1 = 3 72;
z_matrix C5   X2   1.42 X1 90 C1 = 4 72;
z_matrix X3   C1   1    X2 90 X1 0.0
z_matrix C11  C1   1.50 X3 90  X2 180.0
z_matrix C21  C2   1.50 C1 126 X2 180.0
z_matrix C31  C3   1.50 C2 126 X2 180.0
z_matrix C41  C4   1.50 C3 126 X2 180.0
z_matrix C51  C5   1.50 C4 126 X2 180.0
z_matrix H11  C11 1.05 C1 109.5 X3 0.0
z_matrix H12  C11 1.05 C1 109.5 X3 120.0
z_matrix H13  C11 1.05 C1 109.5 X3 240.0
z_matrix H21  C21 1.05 C2 109.5 X3 0.0
z_matrix H22  C21 1.05 C2 109.5 X3 120.0
z_matrix H23  C21 1.05 C2 109.5 X3 240.0
z_matrix H31  C31 1.05 C3 109.5 X3 0.0
z_matrix H32  C31 1.05 C3 109.5 X3 120.0
z_matrix H33  C31 1.05 C3 109.5 X3 240.0
z_matrix H41  C41 1.05 C4 109.5 X3 0.0
z_matrix H42  C41 1.05 C4 109.5 X3 120.0
z_matrix H43  C41 1.05 C4 109.5 X3 240.0
z_matrix H51  C51 1.05 C5 109.5 X3 0.0
z_matrix H52  C51 1.05 C5 109.5 X3 120.0
z_matrix H53  C51 1.05 C5 109.5 X3 240.0
```

6.3.7 External degrees of freedom of a rigid body

A rigid group of atoms can be positioned uniquely in space by specifying six external degrees of freedom: three translational parameters that define some reference point of the group with respect to the origin of the crystallographic coordinate system:

$$\mathbf{t} = \begin{pmatrix} t_a \\ t_b \\ t_c \end{pmatrix}, \tag{6.11}$$

and three angles that define its orientation with respect to the three crystallographic or Cartesian axes. The rotation angle is counter clockwise positive when looking toward the origin of the coordinate system:

$$\mathbf{R}_a(\omega) = \begin{pmatrix} 1 & 0 & 0 \\ 0 & \cos\omega & -\sin\omega \\ 0 & \sin\omega & \cos\omega \end{pmatrix}$$

$$\mathbf{R}_b(\omega) = \begin{pmatrix} \cos\omega & 0 & -\sin\omega \\ 0 & 1 & 0 \\ \sin\omega & 0 & \cos\omega \end{pmatrix} \tag{6.12}$$

$$\mathbf{R}_c(\omega) = \begin{pmatrix} \cos\omega & -\sin\omega & 0 \\ \sin\omega & \cos\omega & 0 \\ 0 & 0 & 1 \end{pmatrix}. \tag{6.12}$$

If the rigid body lies on a special position, some of these parameters will have fixed values. In general, the number of independent positional parameters for a group of n atoms in crystal space is therefore reduced from $3n$ to 6.

The conversion from a vector (e.g., atomic position) in Cartesian rigid body coordinates $\mathbf{s}$ to fractional crystallographic coordinates $\mathbf{u}$ is given by:

$$\mathbf{u} = \mathbf{M}^{-1}(\mathbf{R} \cdot \mathbf{s}) + \mathbf{t} \tag{6.13}$$

The rotation matrix $\mathbf{R}$ is the product of an arbitrary number of rotations around the crystallographic or Cartesian axes. In TOPAS notation, the rotation around the crystallographic a, b and c-axes (rx, ry and rz in degrees) of the rigid body is defined by:

```
prm rx 0;
prm ry 0;
prm rz 0;
rotate rx qa 1
rotate ry qb 1
rotate rz qc 1
```

while rotation around the Cartesian i, j and k-axes of the rigid body is defined by:

```
prm rx 0;
prm ry 0;
prm rz 0;
rotate rx qx 1
rotate ry qy 1
rotate rz qz 1
```

For rotations around the crystallographic axes a predefined macro exists:

```
prm rx 0;
prm ry 0;
prm rz 0;
Rotate_about_axes( rx, ry, rz)
```

The translation along the unit cell axes (tx, ty and tz in *fractional* coordinates) is:

```
prm tx 0;
prm ty 0;
prm tz 0;
translate ta tx tb ty tc tz
```

or via the predefined macro:

```
prm tx 0;
prm ty 0;
prm tz 0;
Translate( tx,  ty,  tz)
```

In general, it is not recommended to limit the rotations from 0 to 360° or the translations from 0 to 1. Doing so would require, for example, a parameter describing a −4° rotation to refine from 0 to 356°. This is less likely to occur by least squares than a 0 to −4° change. This is particularly important during structure determination as discussed in Chapter 7.

6.3.8 Finding a starting orientation

One of the most difficult problems when setting up a rigid body is to define its starting position and orientation. Sometimes the location of some or all of the atoms of the rigid body is approximately known from structure determination or related compounds. Although it is mathematically possible, it is usually impractical to hand-calculate starting values for translation, rotation and torsions. An easier approach is to subject the degrees of freedom of the rigid body to the global optimization method of simulated annealing to map the body onto the known coordinates. This is done by minimizing the difference between the known approximate fractional coordinates and those of the rigid body atoms. In the following example, the positions of the four atoms C1, C2, C3 and N1 of the rigid body should match as closely as possible the positions of the previously determined atoms CX1, CX2, CX3 and NX1. Since we're not refining against experimental data at this point, the *only_penalties* keyword must be used (see Chapter 7 for more explanation of this method).

```
Auto_T(2)
...
Rotate_about_axes(@ 0, @ 0, @ 0)
Translate(x1  0.61961`_0.00021,   @ -0.01596`_0.00084,  =x1; :  0.61961`_0.00021 )
Rotate_about_points(@  0, C1, C3, " C4")

Distance_Restrain(CX1 C1, 0.0, 0.25472, 0.0001, 10)
```

```
Distance_Restrain(CX2 C2, 0.0, 0.22346, 0.0001, 10)
Distance_Restrain(CX3 C3, 0.0, 0.24508, 0.0001, 10)
Distance_Restrain(NX1 N1, 0.0, 0.19113, 0.0001, 10)

only_penalties
```

6.3.9 Rotation of rigid bodies around an arbitrary axis

Quite often molecules need to be rotated around arbitrary axes in space, for example C_{60} around <111> axes. The dependence of such a rotation on rotations around the Cartesian axes x, y and z is usually non-linear. One option is to pre-rotate the entire rigid body in order to align the rotation axes with one of the Cartesian/crystallographic axes, perform the desired rotation around this axis and then back-rotate the rigid body into its original orientation. For example, rotating axis $\mathbf{r}$ to the $\mathbf{c}$-axis requires solving the equation

$$\frac{\mathbf{c}}{c} = \mathbf{R}_a(\omega_a)\mathbf{R}_b(\omega_b)\frac{\mathbf{r}}{r} \tag{6.14}$$

for ω_a and ω_b.

TOPAS offers an easier way of doing this. You simply define two atoms X1 and X2 on the rotation axis as dummy atoms (e.g., atom X1 at 0,0,0 and atom X2 at 1,1,1 for the <111> axis; $occ = 0$) and list all atoms to be rotated in quotation marks using the *Rotate_about_points* macro:

```
Rotate_about_points(@ 0,  X1, X2, "…")
```

6.3.10 TLS matrices

One significant advantage of using rigid bodies is that thermal parameters can be defined that refer to the group as a whole. In most Rietveld refinements it is sufficient to refine a single overall isotropic temperature factor. For high-quality powder data sets, it might be advantageous to try an anisotropic refinement of the thermal motion by means of so-called TLS matrices. A comprehensive explanation and mathematical treatment of TLS matrices is given by Willis and Pryor (1975) and by Downs (1992). Following Downs, their meaning can be easily understood by separating the displacement $\mathbf{u}$ of a rigid body into two parts, a translational component $\mathbf{t}$ and a librational component $\lambda \times \mathbf{r}$:

$$\mathbf{u} = \mathbf{t} + \lambda \times \mathbf{r} \tag{6.15}$$

While the translational component is the same for every part of the rigid body, the librational component of motion represents that part of the rigid body that is located at the end point of vector **r**. The vector λ is then defined as the direction of the rotational axis with its magnitude representing the magnitude of its arc of rotation. Both vectors originate from the same arbitrary origin. With respect to a Cartesian basis, the equation can be written as:

$$
\begin{pmatrix} u_x \\ u_y \\ u_z \end{pmatrix} = \begin{pmatrix} 1 & 0 & 0 & 0 & r_z & -r_y \\ 0 & 1 & 0 & -r_z & 0 & r_x \\ 0 & 0 & 1 & r_y & -r_x & 0 \end{pmatrix} \begin{pmatrix} t_x \\ t_y \\ t_z \\ \lambda_x \\ \lambda_y \\ \lambda_z \end{pmatrix} = (\mathbf{I} : \mathbf{A}) \begin{pmatrix} \mathbf{t} \\ \lambda \end{pmatrix}
\tag{6.16}
$$

The atomic displacement parameters, U, for a given atom located at position **r** can be obtained by taking the time average of the outer product $\otimes$ of the displacement vector **u** for that atom:

$$
\begin{aligned}
\mathbf{U} = \langle \mathbf{u} \otimes \mathbf{u} \rangle &= (\mathbf{t} + \lambda \times \mathbf{r}) \otimes (\mathbf{t} + \lambda \times \mathbf{r}) \\
&= \langle \mathbf{t} \otimes \mathbf{t} \rangle + (\mathbf{t} \otimes \lambda \times \mathbf{r}) + (\lambda \times \mathbf{r} \otimes \mathbf{t}) + (\lambda \times \mathbf{r}) \otimes (\lambda \times \mathbf{r}) \\
&= \langle \mathbf{t} \otimes \mathbf{t} \rangle + (\mathbf{t} \otimes \mathbf{Ar}) + (\mathbf{Ar} \otimes \mathbf{t}) + (\mathbf{A}\lambda) \otimes (\mathbf{A}\lambda) \\
&\equiv \mathbf{T} + \mathbf{AS} + \overline{\mathbf{SA}} + \mathbf{AL\overline{A}}
\end{aligned}
\tag{6.17}
$$

with the matrices **T** for the translational part, **L** for the librational part and **S** for the screw motion (mixing part between translation and libration).

The mathematics of TLS matrices is not trivial and their use requires some care to avoid divergence of the refinement. For practical purposes a few rules of thumb apply when using a single rigid body:

- If the center of the rigid body is also the center of gravity, the components of the **S** matrix (mixing term) can normally be set to zero.
- Refining only the diagonal components of the **T** matrix, constraining them to be equal, and fixing all elements of **L** and **S** to zero is the same as refining an overall isotropic temperature factor for the rigid body.
- Refining all independent elements of the **T** matrix, with **L** = **S** = 0, is the same as refining an overall anisotropic temperature factor for the rigid body.
- For flat molecules like benzene rings, it is often sufficient to refine only the components of the **T** matrix and the diagonal elements of the **L** matrix.
- If the rigid body is located on a symmetry element, some elements of the matrices have to be set to zero or constrained accordingly.

TOPAS macros for using TLS matrices can, for example, be found in Halasz & Dinnebier (2010). Figure 6.12 compares the *adps* from a single crystal refinement with those from a TLS model using powder data. In this example the center of the rigid body was also the center of mass (center of the shared C–C bond) and the **S** matrix was zero. Recently,

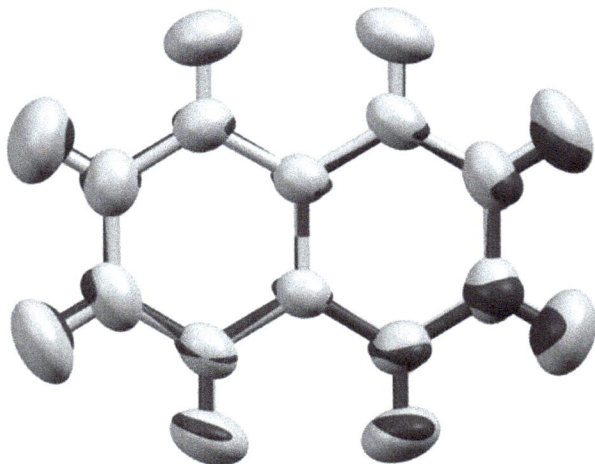

Figure 6.12: Representations of atomic anisotropic displacement parameters of naphthalene as calculated form molecular **T** and **L** matrices at 298 K. Overlap of ellipsoid from single crystal neutron diffraction (pale gray) and X-ray powder diffraction (dark gray) (from Halasz & Dinnebier, 2010).

routines have become available (e.g., those by Jacco van der Streek via the TOPAS wiki) to automatically generate TLS descriptions for TOPAS from .CIF files.

6.3.11 Example of a Rietveld refinement using rigid bodies

We now return to the Rietveld example of the double salt $Mg(H_2O)_6RbBr_3$ that we looked at in Section 2.10. We can use our knowledge of rigid bodies to greatly reduce the number of refined parameters and still get a fit equivalent to that in Figure 2.52.

From the crystal structure of $Mg(H_2O)_6RbBr_3$ (Figure 2.51) it is obvious that the main building blocks of the structure are the two $Mg(OH_2)_6$ and $RbBr_6$ octahedra (Figure 6.13), which have different degrees of freedom.

To define the rigid body of a flexible octahedron, the Z-matrix notation is particularly useful. Under the assumption that the center of symmetry and the angular frame are preserved, the $Mg(OH_2)_6$ octahedron (Figure 6.13) might be distorted with up to three different bond lengths turning it in an orthorhombic bisphenoid:

```
prm r1   2.10 min 2.0 max 2.2
prm r2   2.13 min 2.0 max 2.2
prm r3   2.08 min 2.0 max 2.2

rigid
   z_matrix Mg1
   z_matrix O1 Mg1 = r1;
```

```
z_matrix 02 Mg1 = r2; 01 = 90;
z_matrix 03 Mg1 = r3; 01 = 90; 02  90
z_matrix 04 Mg1 = r1; 01 180    02   0
z_matrix 05 Mg1 = r2; 02 180    03   0
z_matrix 06 Mg1 = r3; 03 180    01   0
```

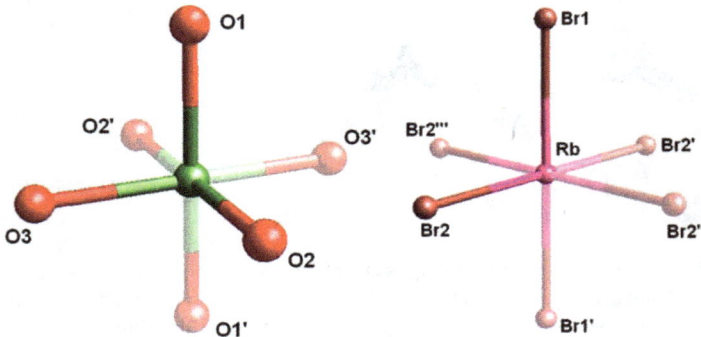

Figure 6.13: The two types of octahedra $Mg(OH_2)_6$ and $RbBr_6$ present in the crystal structure of Mg $(H_2O)_6RbBr_3$. Symmetry equivalent atoms are shaded and marked by primes.

Due to the symmetry of the LT-phase, only part of the rigid body needs to be defined as the remaining part is completed by symmetry (see Figure 6.13). Thus, the rigid body for the $Mg(OH_2)_6$ octahedron can be built from one central magnesium atom (Mg1) and three oxygen atoms (O1, O2 and O3). The central magnesium atom is located on a fixed position with inversion symmetry. Only the three possible rotations around the internal axes of the rigid body remain as external degrees of freedom. The internal degrees of freedom are the length of the three principal axes. Thus only six parameters are free to refine:

```
site Mg1 num_posns  4 x    0.50000 y    0.50000 z    0.50000 occ Mg+2  1    beq B1 1.8;
site O1   num_posns  8 x    0.58671 y    0.48270 z    0.63910 occ O-2   1    beq=B1;
site O2   num_posns  8 x    0.31996 y    0.39541 z    0.54806 occ O-2   1    beq=B1;
site O3   num_posns  8 x    0.40967 y    0.68389 z    0.53918 occ O-2   1    beq=B1;

prm r1   2.098 min 2.0 max 2.2
prm r2   2.131 min 2.0 max 2.2
prm r3   2.084 min 2.0 max 2.2
rigid
   z_matrix Mg1
   z_matrix O1 Mg1 =r1;
   z_matrix O2 Mg1 =r2; O1 90
   z_matrix O3 Mg1 =r3; O1 90  O2 90
```

```
Translate_point_amount(Mg1, -) operate_on_points "O*"
    rotate   15.877 qa 1 operate_on_points "O*"
    rotate   18.146 qb 1 operate_on_points "O*"
    rotate   30.729 qc 1 operate_on_points "O*"
Translate_point_amount(Mg1, +) operate_on_points "O*"

translate ta=1/2;
translate tb=1/2;
translate tc=1/2;
```

The rigid body for $RbBr_6$ can be built from one central rubidium atom (Rb1) and two bromine atoms (Br1 and Br2). Only one equatorial bromine atom is necessary, as the other three are created by symmetry. For numerical reasons, a very small deviation from the special position of the axial Br1 atom remains, even if all symmetry restrictions for the rigid body rotation and translation are obeyed. Therefore, the fractional site occupancy of Br1 must be set to ½. The y-coordinate of the central rubidium atom and the rotation around the c-axis of the rigid body remain as external degrees of freedom. The internal degrees of freedom are the equatorial and the axial Rb–Br lengths. Thus a total of 4 parameters can be refined:

```
site Rb1 num_posns  4  x  0.50000 y  -0.00225 z 0.75000 occ Rb+1  1   beq B2 1
site Br1 num_posns  8  x  0.50037 y  -0.00186 z 1.00018 occ Br-1 0.5 beq=B2;
site Br2 num_posns  8  x  0.25361 y  -0.25950 z 0.75012 occ Br-1  1   beq=B2;

prm r4  3.449 min 3.4 max 3.6
prm r5  3.476 min 3.4 max 3.6
rigid
    z_matrix Rb1
    z_matrix Br1 Rb1 =r4;
    z_matrix Br2 Rb1 =r5; Br1 90

    Translate_point_amount(Rb1, -) operate_on_points "Br*"
        rotate 0  qa 1           operate_on_points "Br*"
        rotate 0  qb 1           operate_on_points "Br*"
        rotate @  46.892 qc 1 operate_on_points "Br*"
    Translate_point_amount(Rb1, +) operate_on_points "Br*"

    translate ta=1/2;
    translate tb @ -0.00225
    translate tc=3/4;
```

Closer inspection of the crystal structure reveals that the parameter $r4$ (Rb1–Br1 bond length) is directly correlated to the length of the c-axis, which can also be deduced from the correlation matrix and the high standard deviation after preliminary refinement. The number of degrees of freedom can therefore be reduced by equating $r4$ to the length of the c-axis divided by four. In TOPAS notation this can be realized, for example, by the equation *prm r4=Get(c)/4;*.

Note that when rigid bodies are used, the refinement of the coordinates is exclusively controlled by the internal and external degrees of freedom of the rigid body and not by refinement of the individual atoms in the atom list, where all refinement flags on the coordinates must be turned off.

References

Baerlocher, C., Hepp, A., Meier, W.M. (1978): *DLS-76: A program for the simulation of crystal structures by geometric refinement.* Institute of Crystallography and Petrography, ETH Zurich, Switzerland.

Behrens, U., Dinnebier, R.E., Neander, S., Olbrich, F. (2008): *Solid-state structures of base-free rubidium and cesium pentamethylcyclopentadienides. Determination by high-resolution powder diffraction.* Organomet. 27, 5398–5400.

Coelho, A.A. Cheary, R. W. (1997): *A fast and simple method for calculating electrostatic potentials.* Comp. Phys. Comm. 104, 15–22.

Dinnebier, R.E. (1999): *Rigid bodies in powder diffraction. A practical guide.* Powder Diffraction 14, 84–92.

Downs, R.T. (1992): Librational displacements of silicate tetrahedra in response to temperature and pressure, *Ph.D. thesis*, Virginia Tech, Blacksburg VA 24061, USA (http://hdl.handle.net/10919/39442).

Halasz, I., Dinnebier, R.E. (2010): *Molecular motion by refinement of TLS matrices from high resolution laboratory powder diffraction data: implementation in the program TOPAS and application to crystalline naphthalene.* MatSci. Forum 651, 65–69.

Leach, A.R. (1996): *Molecular modelling: principles and applications*, Addison Wesley Longman Limited, Essex, England, 595 pages.

Scheringer, C. (1963): *Least-squares refinement with the minimum number of parameters for structures containing rigid-body groups of atoms.* Acta Cryst. 16, 546–550.

Willis, B.T.M., Pryor, A.W. (1975): *Thermal vibrations in crystallography*, Cambridge University Press, UK, 280 pages.

7 Solving crystal structures using the Rietveld method

7.1 Introduction

Structure determination from powder diffraction data can in principle be divided into algorithms working in reciprocal (diffraction) space (e.g., direct methods, Patterson methods, charge flipping) and in direct (crystal) space (e.g., simulated annealing, genetic algorithms). A comprehensive review on the different methods for structure determination from powders is given in the textbook by David et al. (2006). Since that text was published, the reciprocal space method of charge flipping (Oszlányi & Süto, 2004) has been successfully applied to powder diffraction data (Bärlocher et al., 2007; Coelho, 2007), and is available in TOPAS.

The focus of this book is on Rietveld refinement where the minimization algorithm is generally based on least squares with a limited radius of convergence. Nevertheless, there is no reason, other than computational efficiency, why the minimization algorithm used in Rietveld analysis can't serve as a robust global optimizer and this capability is now implemented in modern Rietveld codes. In this chapter, the global optimization capabilities of TOPAS, with a focus on structure determination from powder diffraction data, are described.

7.2 Local and global minima

Let us consider the monoclinic crystal structure ($P2_1$) of sodium-*para*-hydroxy-benzoate (Dinnebier et al., 1999). We select the x- and the z-coordinate coordinate of the single sodium atom as the only degrees of freedom. Figure 7.1 shows how the R_{Bragg} (Section 2.8) changes as a function of the two coordinates (resolution ≈ 0.01 Å). The global minimum is located at $x ≈ 0.947$, and $z ≈ 0.258$ and is only slightly lower than many of the local minima scattered over the complex hypersurface. There is therefore a risk of a local minimization routine, such as that in a traditional Rietveld least squares approach, getting trapped in one of the local minima. The situation for nontrivial examples with many parameters is even more complicated and gives a puckered hypersurface of the agreement factor (e.g., R_{Bragg}, R_{wp}, gof ...) with many local minima. In order to find the global minimum, the starting value must be either in the well around the global minimum, or the hypersurface must be probed in a more sophisticated way. Systematic grid search techniques are one approach, but are limited to very simple problems as the number of calculations necessary increases exponentially with the number of degrees of freedom.

https://doi.org/10.1515/9783110461381-007

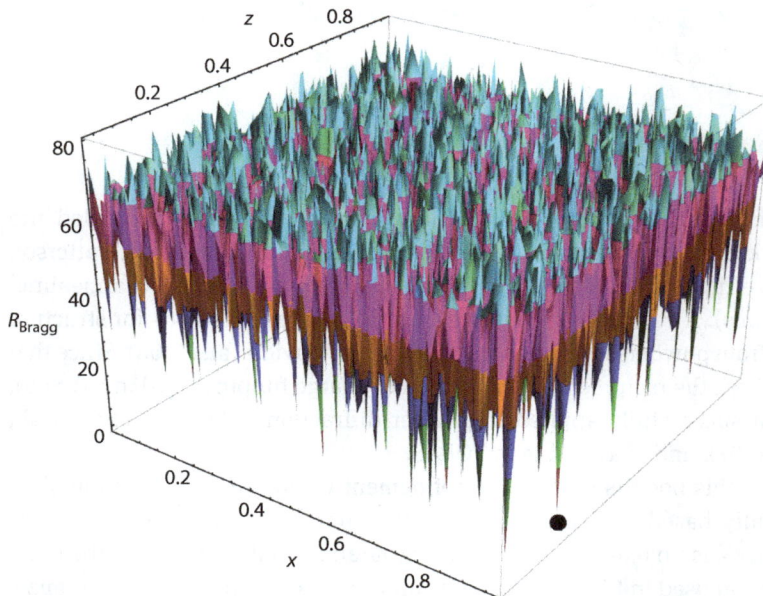

Figure 7.1: R_{Bragg} value as a function of the x and z coordinate of the sodium atom in sodium-para-hydroxybenzoate. All other parameters except for the linear scale factor were kept fixed. The global minimum is denoted by a black sphere.

TOPAS uses the global optimization method of simulated annealing (Newsam et al., 1992; Coelho, 2007). Just one macro is needed to switch between global optimization mode and refinement mode making it a powerful extension of the Rietveld method. In the following sections, the method of simulated annealing and its implementation in TOPAS is explained in some detail.

7.3 The method of simulated annealing

The most common and most easily implemented global optimizer, though one of the least efficient, is the Metropolis (Metropolis et al., 1953.) or *simulated annealing* (SA) algorithm that tests multiple models against the diffraction data. The most usual implementation is actually as a "regional" optimizer where the updates to parameters such as atomic position are constrained to be relatively close to the previous values in such a way that the algorithm makes a random walk through the parameter space. The algorithm can get out of a local minimum by "walking uphill" since changes to the parameters that produce a worse agreement may be accepted according to the Boltzmann criterion:

$$e^{\left(-\frac{\Delta\chi^2}{kT}\right)}$$ (7.1)

with the Boltzmann constant k. The temperature T in this expression is a fictitious temperature (i.e., it does not refer to any real temperature) and $\Delta\chi^2$ is the change in the agreement produced by the trial update. The temperature plays the role of tuning the probability of accepting a bad move. It is initially chosen to have a high value, giving a high probability of escaping a local minimum and allowing the algorithm to explore a large area of the parameter space. Later during the run, the temperature is lowered (Figure 7.3), trapping the solution into successively finer valleys in parameter space until it settles into (hopefully) the global minimum (Figure 7.1). The calculation of the cost function χ^2 can be based on the entire profile, or on integrated intensities. For the latter, the correlation between partially or fully overlapping reflections must be taken into account (as outlined in Figure 7.2).

A possible cost function based on integrated intensities can be defined as a double sum (David, 2004):

$$\chi^2 = \sum_{s1} \sum_{s2} \left[\left(I_{s1} - c|\mathbf{F}_{s1}|^2 \right) \left(V^{-1} \right)_{s1,\,s2} \left(I_{s2} - c|\mathbf{F}_{s2}|^2 \right) \right]$$ (7.2)

over all reflections **s1** and **s2**, with I_{s1} the observed intensity of reflection **s1**, $(V^{-1})_{s1,\,s2}$ the inverse of the correlation matrix reflecting the degree of overlap between the reflections and $c|\mathbf{F}_{s1}|^2$ the calculated intensity of reflection **s1**, and c is the scale factor.

Arbitrary parameters can be varied during SA. In the case of structure determination these typically include internal and external degrees of freedom (DOF) like translations (fractional coordinates or rigid body locations), rotations (Cartesian angles, Eulerian angles or quaternions, describing the orientation of molecular entities), torsion angles, fractional occupancies, temperature factors and so on. In some cases it is useful to restrict certain parameters like torsion angles to lie within reasonable values. Fractional or rigid body coordinates on general positions should not be restricted to stay within one unit cell since this could hinder free movement and, therefore, prevent the algorithm from finding the correct solution.

At the beginning, the first χ^2 is calculated from an initial, trial atomic[1] structure, containing all atoms/molecules that are thought to be located inside the unit cell, usually with the scale factor as the only variable. For structure determination all unknown fractional or rigid body coordinates can initially be set

---

1 A trial structure can contain all atoms on arbitrary or partly known positions, where the latter can be fixed or restrained.

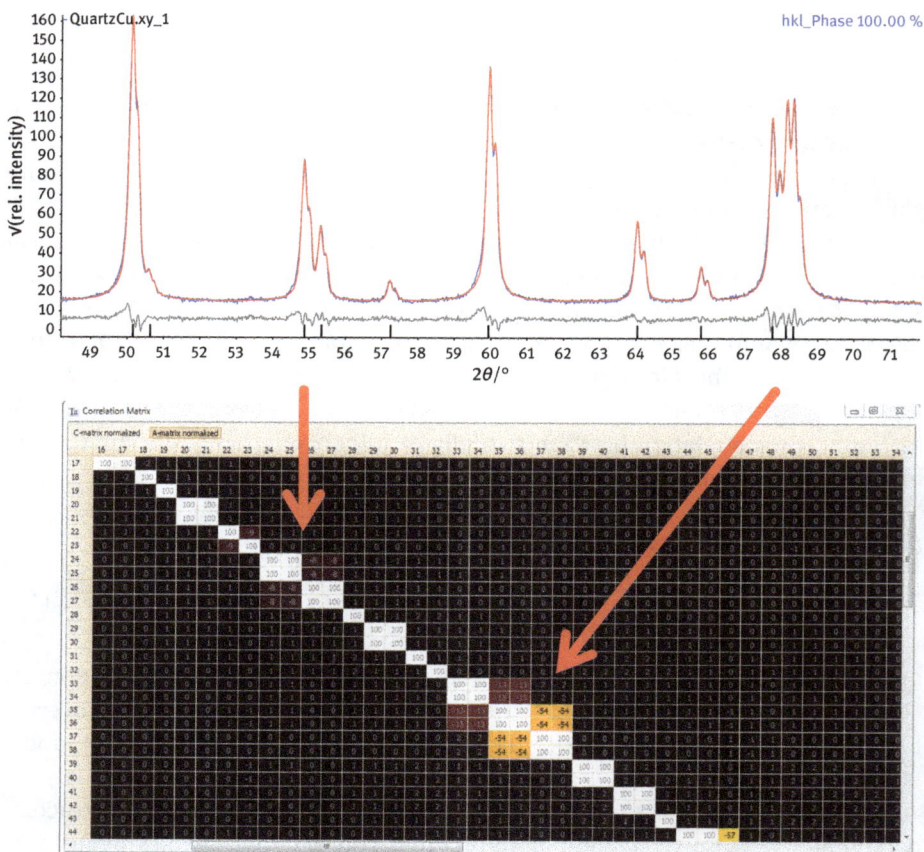

Figure 7.2: Part of a Pawley-fit of a laboratory diffraction pattern (Cu-K$_\alpha$ radiation) of quartz showing independent and partly overlapping reflections and the corresponding part of the correlation matrix. The arrows point to groups of two and three overlapping reflections. Colours represent the correlation between different variables on an artificial colour scale from black (0%) to white (100% correlated).

to the origin (0, 0, 0). After variation of some or all of the DOF's by the SA algorithm, a new χ^2_{new} is calculated. According to the difference

$$\Delta\chi^2 = \chi^2_{new} - \chi^2 \tag{7.3}$$

two possibilities exist. If $\Delta\chi^2 \leq 0$, the variation led to an improvement of the fit and is therefore automatically accepted. If $\Delta\chi^2 > 0$, the fit is worse but the parameter shifts are accepted with a probability according to eq. (7.1) that decreases with decreasing temperature (or time through the SA run). In other words, the entire real space is probed at the beginning of an SA where the "temperature" is high, whereas movements are much more hindered once the temperature decreases until the entire system "freezes." The actual temperature scheme and the type of movements might be much more sophisticated in the actual algorithm.

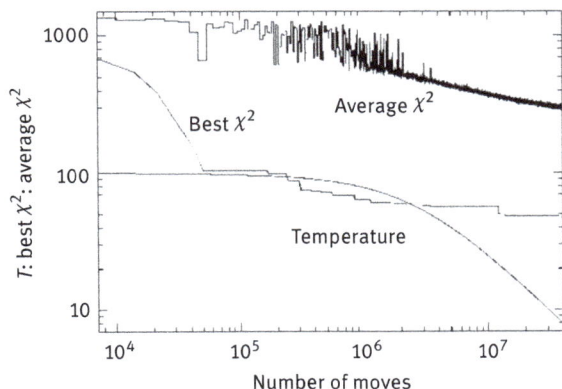

Figure 7.3: Average and best χ^2 (cost function) and "temperature" as a function of the number of moves during a simulated annealing run according to the classical definition (from Dinnebier and Müller, 2013; Copyright Wiley-VCH Verlag GmbH & Co. KGaA. Reproduced with permission).

Figure 7.3 shows a typical simulated annealing run for structure determination in which the χ^2 value falls dramatically in the first few thousand moves, indicating that the scattering is dominated by the positioning of heavier atoms or large molecules. Several million trial structures are usually generated before a minimum can be reached. At the end of the simulated annealing run, Rietveld refinement can be used to find the bottom of the global minimum valley more rapidly.

7.4 The simulated annealing algorithm in TOPAS

TOPAS typically uses a slightly different approach to "classical" SA in which Rietveld refinement is performed at each step of the process. As such each individual step is slower but fewer steps are required as each configuration will be least-squares refined to the bottom of the local minimum in χ^2. Simulated annealing is then equivalent to a continuation of the refinement after convergence has been reached using a specific "temperature" regime to scale parameter changes (Coelho, 2000). To instruct TOPAS to continue refinement after convergence, the keyword:

```
continue_after_convergence
```

is used. In the simplest approach, several actions can be performed before continuing.

(1) *randomize_on_errors* is a means of automatically randomizing parameters based on the approximate errors in the parameters as given by:

$$\Delta p_i = Q \; \text{Sign}(\text{Rand}(-1,1)) \sqrt{\frac{0.02T}{KA_{ii}}}, \quad K = \frac{1}{\sum_{m=1}^{M}\left(w_m y_{obs,m}^2\right)}, \tag{7.4}$$

where T is the current temperature and Q is a scaling factor determined such that convergence to a previous parameter configuration occurs 7.5% of the time on average.

(2) *rand_xyz* adds a vector $\mathbf{u} = (\Delta x, \Delta y, \Delta z)$ to each site described by the fractional coordinates (x, y, z), the direction of which is random and the magnitude in Å is: $|\mathbf{u}| = T$ *rand_xyz* where T is the current "temperature" that can be set by:

```
Temperature !E
```

Only fractional coordinates *xyz* that are flagged as independent parameters are randomized.

(3) *val_on_continue*[2] supplies a means of changing the parameter value after the refinement has converged in a user-controlled fashion. An example is given below:

```
site Pb1    x @  0.1    val_on_continue=0;
            y @  0.2    val_on_continue=0;
            z @  0.3    val_on_continue=0;
            occ Pb  1   beq 1
```

Here, the values of the fractional coordinates of a lead atom are reset to the origin before any new refinement cycle during an SA run.

The simplest simulated annealing could thus look like:

```
continue_after_convergence
randomize_on_errors
```

This does not include a temperature regime. Setting up an optimum temperature regime can be quite demanding. For most cases, the following macro can be used:

```
Auto_T(2)
```

---

2 When *val_on_continue* is defined then the corresponding parameter is not randomized according to *randomize_on_errors*.

which imposes a complex automated temperature program. It has shown to be adequate for a wide range of simulated annealing examples.

There is a huge variety of keywords in the TOPAS scripting language to manipulate and optimize the simulated annealing process. For example, some that speed up SA include:

```
chi2_convergence_criteria 1E-5 ' Convergence is determined when the change in χ² is
                                less than 1E-5 for three consecutive cycles
quick_refine 0.01               ' Removes parameters that influence χ² in a smaller
                                manner than 0.01 during a refinement cycle
```

An in-depth discussion of all possibilities is beyond the scope of this book and the prospective user is referred to the TOPAS technical reference.

For structure determination, one might consider a different weighting scheme with stronger weighting of the more intense reflections, like:

```
weighting = 1 / Sqrt(Max(Yobs, 1));
```

For inorganic crystal structures, the identification of special positions is useful during structure solution. This can be accomplished by a so-called "occupancy-merge" procedure as proposed by Favre-Nicolin and Cerny (2004). For example, if an atom refines to a position very close to a mirror plane it is likely that it actually lies on the mirror. Its occupancy would then need to be halved as the atom and its mirror image would be very close, effectively doubling the scattering from that site. In TOPAS this is handled by considering sites as spheres with a radius r and updating their occupancies if they approach more closely than $2r$. The occupancy is automatically adjusted to $1/(1 + \text{intersection fractional volumes})$ so that any number of sites can be merged. In the following example, special positions are identified when either two oxygen or two lead atoms approach each other within a distance less than the sum of their merging radii which are set to 0.7 Å:

```
occ_merge Pb*   occ_merge_radius .7
occ_merge O*    occ_merge_radius .7
```

An alternative to *occ_merge* is to refine the fractional occupancies during SA with a minimum limit of zero and a maximum limit equivalent to the scattering power of the strongest scatterer present. This increases the flexibility in structure determination but requires later manual reassignment of atomic species.

SA runs usually continue until the maximum number of iterations defined by *iters* is reached. Alternatively, the SA process can be stopped if e.g., the R_{wp} falls under a user-defined limit like:

```
iters = If(Get(r_wp) < 10, 0, 1000);
```

Usually no special algorithms are employed to prevent close contact of atoms or molecules during the global optimization procedure. In general these are not necessary, as the fit to the structure factors alone quickly moves the atoms or molecules to regions of the unit cell where they do not grossly overlap with neighbouring molecules.

It is common to use a Pawley/Le Bail fit prior to SA to fix quantities like cell parameters, peak shapes and background. This helps speed by reducing the number of iterations for each convergence. In many Pawley/Le Bail fits the background at high scattering angles, where lots of peak overlap occurs, correlates with the intensity of the reflections and leads to a poor (low) background estimation. This may prevent the SA from finding the correct crystal structure. It is therefore useful to visually inspect and, if necessary, to correct the background before SA starts.

7.4.1 Example of a structure determination with simulated annealing

In the following example, the crystal structure of the mixed-valent oxide Pb_3O_4 (space group $P4_2/mbc$, $a = 8.81$ Å, $c = 6.57$ Å) is determined by the global optimization method of simulated annealing.

The powder pattern of Pb_3O_4 was measured with a Bruker D8-Advance powder diffractometer (Cu-$K_{\alpha 1}$ radiation from a Ge(111)-Johannson primary beam monochromator) in Bragg–Brentano geometry on a flat plate low background single crystal sample holder at room temperature. The sample exhibits strong anisotropic line broadening due to microstrain that can be empirically handled by a spherical harmonics of fourth order applied to the Lorentzian *fwhm* (see Section 4.3.2). The INP file for the Pawley fit (Figure 7.4) is given below:

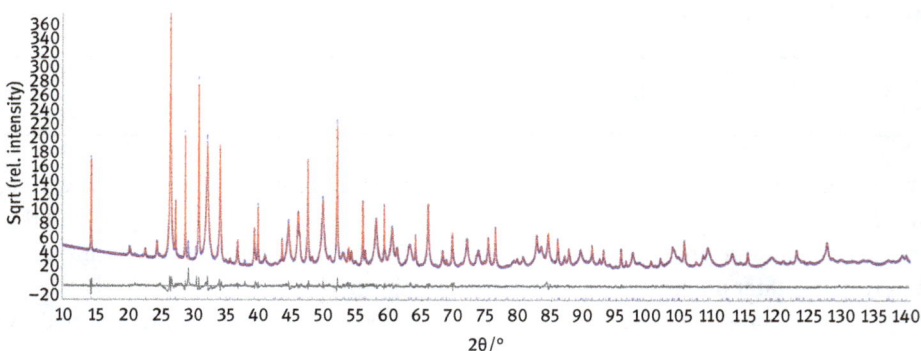

Figure 7.4: Pawley fit of powder diffraction data from Pb_3O_4.

```
iters 1000
xdd Pb3O4.raw
   bkg @  850.77 -581.44  611.07 -312.83 173.39 -81.07 1.37 3.36 38.58 54.48 50.61
   start_X 10    finish_X 140
   Specimen_Displacement(@, 0.00025`)
   LP_Factor( 27.3)
   Rs 217.5
   Simple_Axial_Model(, 1)
   Slit_Width( 0.1)
   Divergence( 1)
   lam
      ymin_on_ymax  0.00001
      la  1 lo  1.540596 lh  0.401844
   hkl_Is
      phase_name "Pb3O4 Pawley fit, Bruker D8 Advance, Bragg-Brentano"
      LVol_FWHM_CS_G_L(1, 142.03880`, 0.89, 151.56239`,CSG, 203.37399`,CSL, 602.52323`)
      prm p1  0.37733` min 0.0001
      spherical_harmonics_hkl sh1
      sh_order  4 load sh_Cij_prm {
          y00    !sh1_c00  1.00000
          y20    sh1_c20  -1.03954`
          y40    sh1_c40   0.16276`
          y44p   sh1_c44p  1.48377`
      }
      lor_fwhm = Max(0.0001, sh1 p1 Tan(Th));
      Tetragonal(@  8.813233, @  6.565034)
      space_group "P42/mbc"
      load hkl_m_d_th2 I
      {
      1   1   0   4    6.231897       14.20049    @  23.87964

      ...

      0   0   8   2    0.820629      139.65792    @  45.94778
      }
```

From the cell volume, the number of formula units per unit cell can be estimated to be $Z = 4$, equivalent to 12 lead and 16 oxygen atoms. With respect to the possible site symmetries, this could imply 1–4 oxygen and 2–3 unique lead positions. Here we assume 2 oxygen and 2 lead positions, but it might become necessary to probe other combinations.

To set up an SA run, the following changes to the INP file are performed:
- The macro *Auto_T(2)* is placed at the beginning of the INP file. All parameters with a refinement flag set on are now subject to global optimization.
- All refinement flags are turned off.
- Switch from WPPF to Rietveld mode by replacing *hkl_Is* by *str*.
- Change the weighting scheme to *weighting = 1/Sqrt(Max(Yobs, 1))*;
- Add a scale factor and turn its refinement flag on: *scale @ 0.0001*.

– Add the atomic sites, placing them arbitrarily at the origin and turning the
refinement flags of the positional parameters on:
  – *site Pb1 x @ 0 y @ 0 z @ 0 occ Pb 1 beq 1*
  – *site Pb2 x @ 0 y @ 0 z @ 0 occ Pb 1 beq 1*
  – *site O1 x @ 0 y @ 0 z @ 0 occ O 1 beq 1.5*
  – *site O2 x @ 0 y @ 0 z @ 0 occ O 1 beq 1.5*
– Identify special positions by merging occupancies between lead atoms and
between oxygen atoms that come closer than 0.7 Å:
  – *occ_merge Pb* occ_merge_radius .7*
  – *occ_merge O* occ_merge_radius .7*
  The updated input file after a successful SA run looks like:

```
iters 1000000
Auto_T(0.1)
xdd Pb3O4.raw
   bkg 850.77 -581.44 611.07 -312.83 173.39 -81.07 1.37 3.36 38.58 54.48 50.61
   start_X    10    finish_X  140
   Specimen_Displacement(, 0.00025)
   LP_Factor( 27.3)
   Rs 217.5
   Simple_Axial_Model(, 1)
   Slit_Width( 0.1)
   Divergence( 1)
   lam
      ymin_on_ymax  0.00001
      la  1 lo  1.540596 lh  0.401844

   weighting = 1 / Sqrt(Max(Yobs, 1));
   str
      phase_name "Pb3O4 SA, Bruker D8 Advance, Bragg-Brentano"
      LVol_FWHM_CS_G_L( 1, 142.03880, 0.89, 151.56239,!CSG, 203.37399,!CSL, 602.52323)
      ' Spherical harmonics of 4th order for anisotropic Lorentzian microstrain broadening
      prm !p1  0.37733 min 0.0001
         spherical_harmonics_hkl !sh1
         sh_order  4 load sh_Cij_prm {
            y00  !sh1_c00   1.00000
            y20  !sh1_c20  -1.04925`
            y40  !sh1_c40   0.00982`
            y44p !sh1_c44p  1.46390`
         }
         lor_fwhm = Max(0.0001, sh1 p1 Tan(Th));

      space_group "P42/mbc"
      Tetragonal(  8.813233,   6.565034)
      r_bragg  4.44686519
      scale @  0.000222785226
      site Pb1  x @  -0.16280  y @  0.14335  z @ -0.51714  occ Pb  0.56788  beq 1
      site Pb2  x @   0.00200  y @  0.49685  z @  0.26607  occ Pb  0.28815  beq 1
      site O1   x @   0.07062  y @  0.63909  z @  0.00085  occ O   0.50301  beq 1.5
```

```
site O2    x @  -0.19949  y @  0.68456  z @  2.24855   occ O   0.55529  beq 1.5
occ_merge Pb*  occ_merge_radius .7
occ_merge O*   occ_merge_radius .7
view_structure      ' Launches the structure viewer during SA
```

During an SA run, the R_{wp} is used as the indicator for the progress of the global optimization as shown in Figure 7.5. We see from this plot that a low R_{wp} solution was found around eight times during the SA run. TOPAS automatically returns to the lowest R_{wp} solution at the end of the run (note that the final point on Figure 7.5 is the lowest R_{wp}) and this is written to the output file.

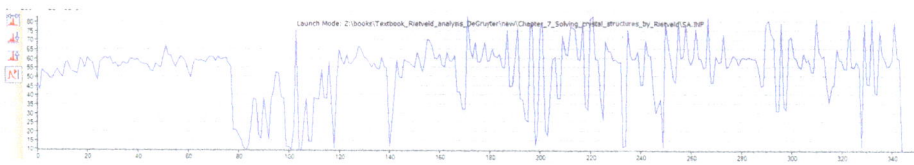

Figure 7.5: R_{wp} during the SA run on Pb_3O_4.

The best fit at the end of this process is shown in Figure 7.6, and the corresponding crystal structure of Pb_3O_4 is shown in Figure 7.7.

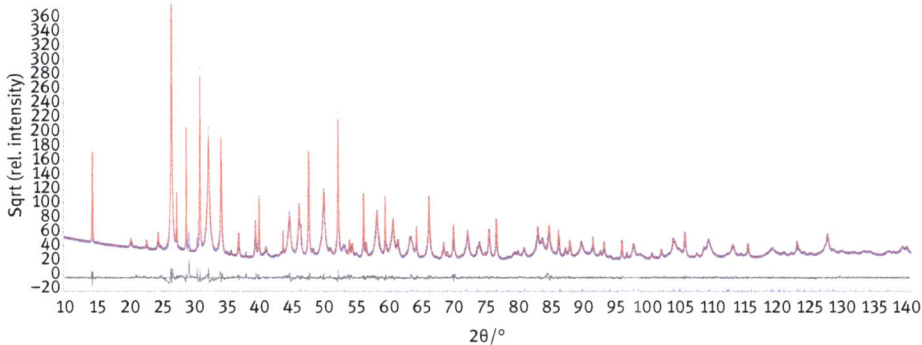

Figure 7.6: Best Rietveld fit during SA on Pb_3O_4.

By inspecting the structure visually and looking at the fractional coordinates and sites occupancies it's clear that atoms have refined close to special positions. These can be identified using, for example, the International Tables for Crystallography Vol. A (Aroyo, 2016):

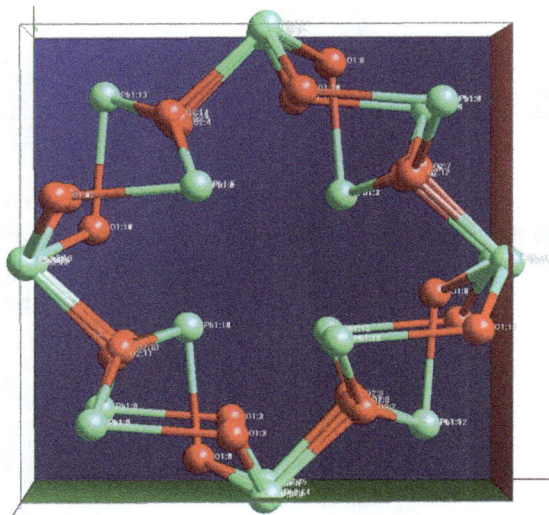

Figure 7.7: Crystal structure of Pb_3O_4 after an SA run. Although atoms are formally located on general positions, special positions can be easily seen by eye.

```
prm o2xx  -0.19949
site Pb1 num_posns 8 occ Pb  1 x  @  0.16280 y @  0.14335  z=1/2;  beq 1  '  8 h m..
site Pb2 num_posns 4 occ Pb  1 x=0             y=1/2;       z=1/4;  beq 1  '  4 d 2.22
site O1  num_posns 8 occ O   1 x @  0.07062 y @  0.63909 z=0;      beq 1.5'  8 h m..
site O2  num_posns 8 occ O   1 x =o2xx;        y =-o1xx+1/2; z=1/4; beq 1.5'  8 g ..2
```

Although the SA algorithm in TOPAS is extremely powerful, there is no guarantee that SA will find the global minimum. As a rule of thumb, crystal structures up to a complexity of 20 structural degrees of freedom can frequently be solved if good quality powder diffraction data with reflections up to at least $d \approx 1.75$ Å (better 1.2 Å) are available. The number of structural degrees of freedom can be greatly reduced by using rigid bodies and/or fixed atomic coordinates of partly known crystal structures (e.g., from charge flipping).

References

Aroyo, I.M. (ed.) (2016): *International tables for crystallography. Volume A: space-group symmetry.* IUCr Series, Kluwer Academic Publishers, Dordrecht, The Netherlands.

Baerlocher, C. McCusker, L.B., Palatinus, (2007): *Charge flipping combined with histogram matching to solve complex crystal structures from powder diffraction data.* Z. Kristallogr. l. 222, 47–53.

Coelho, A.A., (2000): *Structure solution by simulated annealing.* J. Appl. Cryst. 33, 899–908.

Coelho, A.A. (2007): *A charge-flipping algorithm incorporating the tangent formula for solving difficult structures.* Acta Cryst. A36, 400–406.

David, W.I.F (2004): *On the equivalence of the Rietveld method and the correlated integrated intensities method in powder diffraction*. J. Appl. Cryst. 37, 621–628.

David, W.I.F., Shankland, K., McCusker L.B., Baerlocher, Ch. (2006): *Structure determination from powder diffraction data*. IUCr Monographs on Crystallography, Oxford University Press, UK.

Dinnebier, R.E., Von Dreele, R., Stephens, P.W., Jelonek, S., Sieler, J. (1999). *Structure of sodium para-hydroxybenzoate, NaO_2C-C_6H_4OH by powder diffraction: application of a phenomenological model of anisotropic peak width*. J. Appl. Cryst. 32, 761–769.

Dinnebier, R.E., Müller, M. (2013): *Modern Rietveld refinement, a practical guide*. Chapter 2 in Modern Diffraction Methods, Mittemeijer, E. and Welzel, U. (eds.) Wiley-VCH Verlag GmbH & Co. KG Weinheim, Germany.

Favre-Nicolin, V. & Cerny, R. (2004): *Fox: modular approach to crystal structure determination from powder diffraction*. Mater. Sci. Forum 443–444, 35–38.

Metropolis, N., Rosenbluth, A.W., Rosenbluth, M.N., Teller, A.H. (1953): *Equation of state calculations by fast computing machines*. J. Chem. Phys. 21, 1087–1092.

Newsam, J.M., Deem, M.W., Freeman, C.M. (1992): *Direct space methods of structure solution from powder diffraction data*. NIST Special Publication 846 Accuracy in Powder Diffraction II, Prince, E. and Stalick,]. K. (eds.), 80–91.

Oszlányi, G., Süto A. (2004): *Ab initio structure solution by charge flipping*. Acta Cryst. A60, 134–141.

8 Symmetry mode refinements

8.1 Introduction

Many materials undergo phase transitions as a function of external variables such as temperature, pressure or changes in their chemical environment. Powder diffraction is a particularly powerful tool for studying these transitions as it is relatively easy to design cells for *in situ* or *operando* studies, and one doesn't have to worry about issues such as crystals shattering, which is a major difficulty in similar single crystal experiments.

We can broadly divide phase transitions into two classes: reconstructive and non-reconstructive. During a reconstructive transition there are large changes in the bonding and sufficient atomic rearrangement such that there is no simple relationship between the structures of the two phases. An example might be a CsCl structure transforming to rock salt under pressure. If one is analyzing powder diffraction data before and after the phase transition, the data need to be treated as two separate problems.

At a non-reconstructive phase transition the changes in structure are more subtle. They could involve small movements in atomic positions away from high-symmetry sites, such as the cooperative tilting of octahedra illustrated in Figure 8.1; they could involve metal atoms moving away from the center of an otherwise undistorted coordination polyhedron; they could involve atomic sites that are a 50:50 random mixture of two elements at high temperature ordering on cooling; or they could involve the ordering of magnetic moments as discussed in Chapter 9. Understanding this type of phase transition is an extremely important area of condensed matter science as they are often associated with changes in physical properties such as changes from insulator to metal to superconductor, paraelectric to ferroelectric, second harmonic generation inactive to active or paramagnet to anti/ferro/ferri-magnet.

Analyzing the powder diffraction pattern of a sample before and after a non-reconstructive phase transition can again be tackled as two separate problems. However, the close structural relationship between the two phases means that there are often more efficient ways of determining and understanding the structures of low-symmetry phases (often called *child* phases; more formally *hettotypes*) relative to their high-symmetry *parent* (*aristotype*). We discuss some of these methods in this chapter.

8.2 Symmetry lowering phase transitions

Non-reconstructive phase transitions often involve a material changing from a high-symmetry parent structure to a lower-symmetry child structure. The structures will

https://doi.org/10.1515/9783110461381-008

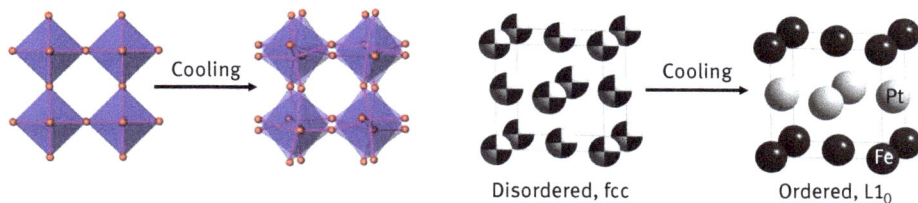

Figure 8.1: Left: a phase transition involving octahedral tilting in WO_3. Right: a phase transition involving site ordering in FePt (A1 (fcc) → $L1_0$ or CuAu type).

therefore have a group–subgroup relationship. High-symmetry forms are usually associated with high temperature (or low pressure) and low symmetry forms with low temperature (or high pressure), though there is no strict thermodynamic requirement for this. Materials therefore typically lose symmetry on cooling. We will use WO_3 as an example throughout this chapter. Even though it has never been observed experimentally, one can imagine a high-symmetry $Pm\bar{3}m$ form of WO_3 with perfectly regular WO_6 octahedra sharing corners with 180° W–O–W bond angles between them (Figure 8.1 left or Figure 8.2). The structure has W on the $1a$ Wyckoff site at (0, 0, 0) and O on $3d$ at (½, 0, 0) and therefore no refineable coordinates. On cooling, WO_3 undergoes a series of phase transitions that can be described as W moving away from the center of the WO_6 octahedra (a second order Jahn–Teller distortion) and tilts of the WO_6 octahedra around different crystallographic directions leading to non-180° W–O–W bond angles. The structures encountered on cooling are depicted in Figure 8.2. Even a material as compositionally simple as WO_3 can have a remarkable structural complexity, and seven different phases have been structurally characterized.

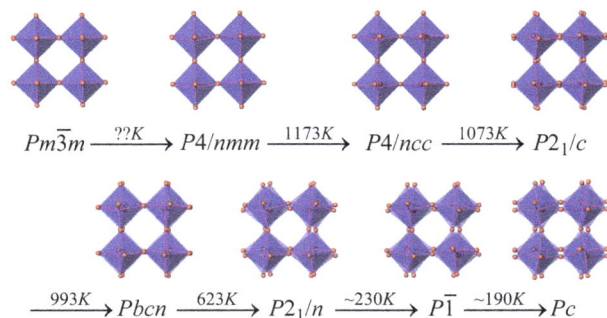

$Pm\bar{3}m \xrightarrow{??K} P4/nmm \xrightarrow{1173K} P4/ncc \xrightarrow{1073K} P2_1/c$

$\xrightarrow{993K} Pbcn \xrightarrow{623K} P2_1/n \xrightarrow{\sim 230K} P\bar{1} \xrightarrow{\sim 190K} Pc$

Figure 8.2: The phase transitions of WO_3. Blue WO_6 octahedra undergo different internal distortions and coupled rotations.

If we want to describe the structure of each of the phases of WO_3 using a conventional crystallographic approach, we need to determine: (1) the unit cell or basis of each phase; (2) the space-group symmetry; (3) the appropriate origin choice; (4) the fractional coordinates of each atom; and (5) decide which coordinates are free to change by symmetry. For simple cases we can use International Tables to help us. Under the entry for each space group there is a list given of "maximal isomorphic subgroups of lowest index" and "maximal non-isomorphic subgroups." Isomorphic subgroups are those with the same space group as the parent but a larger unit cell (so there are fewer symmetry elements per unit volume). The term *index* gives the ratio between the number of symmetry operations in the group and subgroup. The term *maximal* means that there are no other groups between the parent and child, and this is guaranteed by the term "lowest index." Non-isomorphic subgroups have a different space group. Under heading **I**, International Tables list translationengleiche subgroups (gleiche = German for same) that involve losses of rotational symmetry and under **II** klassengleiche subgroups, where translational symmetry elements are lost. Category **IIa** lists subgroups with the same (conventional) unit cell or basis, which arise by loss of centring, and **IIb** those with a larger unit cell.

For example, the highest symmetry characterized structure of WO_3 has space-group type $P4/nmm$ with a unit cell of $\sim\sqrt{2}a \times \sim\sqrt{2}a \times \sim a$ relative to the hypothetical cubic $Pm\bar{3}m$ parent (we will omit the "approximately equal to" signs and a from hereon). To describe the structure we need to specify a W at $(\frac{1}{4},\frac{1}{4}, z)$ $[z \approx 0]$ and an O at $(\frac{1}{4},\frac{1}{4}, z)$ $[z \approx \frac{1}{2}]$. There are therefore two free unit cell parameters and two free fractional coordinates. The distortion relative to the cubic phase is due to W moving away from the center of the WO_6 octahedron parallel to the c-axis. The next change that occurs on cooling is a tilt of octahedra around the c-axis that lowers the symmetry to $P4/ncc$ with a unit cell that is twice as large along c (i.e., $\sqrt{2} \times \sqrt{2} \times 2$). This space group is listed in International Tables as a maximal non-isomorphic subgroup of $P4/ncc$ of type **IIb**. We need to specify W at $(\frac{1}{4},\frac{1}{4}, z)$ $[z \approx \frac{1}{4}]$, O at $(x, -x, \frac{1}{4})$ $[x \approx 0]$ and O at $(\frac{1}{4},\frac{1}{4}, z)$ $[z \approx 0]$ to describe the structure. The change in the W z-coordinate from $\sim\frac{1}{4}$ to ~ 0 between the phases is due to an origin shift between the two space groups. Even for this relatively straightforward example there is significant work in relating the two structures and determining an appropriate structural description.

If we look at the (hypothetical) highest temperature transition from $Pm\bar{3}m$ to the observed $P4/nmm$ structure we hit a further complication: $P4/nmm$ is not listed as a maximal subgroup of $Pm\bar{3}m$ and we would have to describe the relationship between the two using intermediate groups. Even for a "simple" compound like WO_3 the possible transformations between different subgroups of the parent can become bewilderingly complex. In fact, if you work out all the possible subgroups between a $Pm\bar{3}m$ parent and a $2 \times 2 \times 2$ $P1$ child structure, there are 1427 possibilities!

8.3 Symmetry adapted distortion modes

8.3.1 Overview

Fortunately there are powerful group theory tools to help understand this type of structural distortion. These are sufficiently well-developed that they can be applied to structural problems without the user having a deep theoretical understanding of the methods. This is particularly true if one of the web-based tools such as the ISODISTORT software suite (http://stokes.byu.edu/iso/isodistort.html) or the Bilbao crystallographic server (http://www.cryst.ehu.es/) is used. In the following section, we will discuss the ISODISTORT symmetry adapted distortion mode approach implemented in TOPAS, and cover enough of its language to help understand its practical application.

The basic idea of the symmetry mode approach is that we describe a distorted child structure in terms of the parent structure plus the amplitudes of a family of symmetry-motivated distortions (Campbell, 2006 & 2007; Kerman, 2012). These distortions define the pattern and magnitude of atomic displacements and hence the symmetry of the material. We will find that this approach has a number of advantages:

1. We can often describe child structures with far fewer parameters than using traditional *xyz* coordinates. This can help us to extract detailed structural information from powder data.
2. The magnitude of the parameters we use will naturally be defined in the range 0 to ~4, where a large magnitude indicates a large structural change. This can help us focus on parameters that are most important in describing the phase transition.
3. Our new parameter set will allow us to rapidly explore different possible symmetries for child structures in a systematic way.
4. We can use well-developed, freely available, user-friendly computational infrastructure to take most of the hard work out of the process and help eliminate otherwise difficult-to-avoid errors.

The first of these advantages is probably the most powerful and most useful. It arises from the fact that the free energy change associated with any phase transition can be expressed in terms of the order parameters of the irreducible representations of the parent space-group symmetries (we'll explain the meaning of this sentence later; Landau and Lifshitz, 1970). Often, only a small number of these order parameters (frequently only one) dominate the free energy. These are the parameters we will use in the symmetry mode description, and we will therefore be refining parameters of direct energetic significance.

Before we delve into the details it might be useful, particularly for those from a chemistry background, to consider a simple molecular analogy to the process. H_2O is a bent molecule with C_{2v} symmetry and a 104.5° bond angle. If we wanted to describe a distorted H_2O molecule (e.g., one where the O–H bonds were of different lengths lowering the symmetry, or where the bond angle was different) we could specify xyz coordinates for each of the atoms. Alternatively, we could remember the vibrational properties of a H_2O molecule that are usually described using normal modes. In this approach we find that H_2O has a symmetric bond stretch that has symmetry label (irrep) a_1, an asymmetric bond stretch (b_2) and a bending mode (a_1). If we wanted to describe a statically distorted molecule we could do so in terms of the undistorted (parent) coordinates plus the amplitude of one or more frozen vibrational modes. Note that the a_1 symmetric stretch and a_1 bend would not lower the symmetry from C_{2v}, but the b_2 stretch would. a_1 is called the totally symmetric irreducible representation.

8.3.2 Definitions and terminology

The symmetry operations of a space group are conventionally described by a set of matrix operations called a *representation*. When these are expressed in their mathematically simplest block diagonal form this is called an *irreducible representation* or *irrep*. A representation has an *order* defined as the number of matrices and a *dimension* defined by the dimension of the matrices. A phase transition can always be associated with a k point or wave vector in reciprocal space. This k point describes where superlattice reflections will be seen following the phase transition. For example, a phase transition associated with k point ($\frac{1}{2}$, 0, 0) will give a child with extra ($h/2$, k, l) superlattice reflections or a doubled a-axis. For a given parent space group, specifying a k point defines a set of irreps. Each of these maps the symmetry elements onto a set of irreducible matrices. When a material undergoes a phase transition some of its symmetry elements are lost and some remain. Those that remain define what is called the *isotropy subgroup* or *distortion symmetry*. Since each irrep describes a set of parent symmetries that can be broken, they give us a "recipe" for lowering symmetry. The language used is that we superpose one (or more) irreps on the parent space group and this takes us to a specific isotropy subgroup. Within a given isotropy subgroup, many different distortions are possible due to the different values that the individual degrees of freedom can adopt.

One convenient way to think about any distortion is in terms of a *distortion vector*. For the simple case of a single atom moving away from a high-symmetry site in 1, 2 or 3 dimensions this is very intuitive. If we're thinking of a more complex case involving multiple atoms then we can use the same idea, but the distortion vector is harder to visualize as it lies in a higher dimensional "*distortion space*" (more formally the *carrier space* in which the matrices of the space group representation operate). The distortion vector is invariant under the symmetry operations of the isotropy subgroup. As with any vector, if we can define a basis for it, we can describe it in terms of vector components along each of the basis vector directions. In a traditional crystallographic approach we use parameters such as the unit cell, xyz fractional coordinates or site occupancies as the basis. In the symmetry mode approach we take advantage of the fact that we can use irreps, which are simply linear combinations of traditional crystallographic parameters, as a complete and unique basis for the distortion vector.

There's one final aspect of the distortion vector that it is useful to understand called the *order parameter direction* or *OPD*. An OPD is a specific direction (or subspace) of the generalized distortion space within which each distortion vector represents structures with the same symmetry. Remember that each different vector within this subspace represents a different specific distortion. A three dimensional irrep has its most general OPD expressed as $(a; b; c)$. This will give rise to a distortion symmetry known as the *kernel*. The kernel is the lowest distortion symmetry associated with an irrep. The same three-dimensional irrep may have a one-dimensional OPD where, for example, b and c happen to be zero or $b = -a$ expressed as $(a; 0; 0)$ or $(a; -a; 0)$, respectively. These will lead to an intermediate distortion symmetry that is a supergroup of the kernel and a subgroup of the parent. A simple analogy to this is to think of moving a single atom in a cubic structure away from $(0, 0, 0)$. If we move it in a general direction to (x, y, z), we will destroy certain symmetry elements; if we move it to (x, x, x), we would retain more symmetry. The variable parameters of an OPD are called *branches*.

In the kernel we can define a *distortion mode* as a vector component along one of the irrep basis vectors; in one of the higher symmetries it may be a linear combination of different irrep basis vectors. The ISODISTORT definition of an *order parameter* is a distortion vector along a specific OPD of a specific irrep at a specific k-point of the parent symmetry. The parameters we use to specify it are the individual mode amplitudes (one for each branch of the OPD) and these are therefore *order parameter components*.

We are now at the point where we've defined the various terms involved in a symmetry mode description and have found that we have a way of defining structural parameters (mode amplitudes) that relate directly to the order parameters used to express the free energy change at a phase transition. We will find in the examples below that these parameters often describe meaningful structural changes in a material such as rotations of polyhedral groups, recognizable distortions of polyhedra or Jahn–Teller distortions moving metals away from high-symmetry coordination environments. As such they can be extremely useful in describing structural changes.

8.3.3 Symmetry mode labels and TOPAS implementation

To recap, we've discussed that the recipe to define an isotropy subgroup or distortion symmetry is to choose a k point that defines a set of irreps. Selecting an irrep will produce a series of possible order parameter directions ranging from high to low (the kernel) distortion symmetries. Each of these produces a set of symmetry modes whose amplitudes define a specific structural distortion. The irrep chosen defines the distortion symmetry (it is called a *primary irrep*), though it will often give rise to additional irreps (called *secondary irreps*). These secondary irreps alone may not destroy sufficient symmetry to lead to the isotropy subgroup in question. When describing modes it is useful to give each a unique label. A small section of TOPAS INP code exemplifying these labels and showing how mode amplitudes relate to *xyz* fractional coordinates is given below for the room temperature WO_3 structure that we will discuss later. The code was automatically generated using the procedure in Section 8.4.1 on the ISODISTORT webserver:

```
'{{{mode definitions
   'etc
   prm   a7  1.12853 min  -4.00 max   4.00 'Pm-3m[1/2,1/2,1/2]R4+(a,0,b)[O:d]Eu(a)
   prm   a8  1.11797 min  -4.00 max   4.00 'Pm-3m[1/2,1/2,1/2]R4+(a,0,b)[O:d]Eu(b)
   'etc
'}}}
'{{{mode-amplitude to delta transformation
   'etc
   prm O_1_dx = +0.03125*a8 -  0.03125*a10 +  0.03125*a11 -  0.03125*a12;:  0.03494
   prm O_1_dy = -0.04419*a13 +  0.04419*a14;:  0.00000
   prm O_1_dz = +0.03125*a7 +  0.03125*a9 +  0.03125*a11 +  0.03125*a12;:  0.03527
   'etc
'}}}
'{{{distorted parameters
   'etc
   prm O_1_x = 1/4 + O_1_dx;:  0.28494
   prm O_1_y = 3/4 + O_1_dy;:  0.75000
   prm O_1_z = 3/4 + O_1_dz;:  0.78527
   'etc
'}}}
'{{{mode-dependent sites
   'etc
   site O_1 x = O_1_x;:0.28494 y = O_1_y;:0.75000 z = O_1_z;:0.78527 occ O 1 beq 0
   'etc
'}}}
```

At the top of the script two displacive symmetry modes are defined with TOPAS parameter names *a7* and *a8* and amplitudes 1.13 and 1.12 (the 22 other modes have been omitted for brevity). In the second section of the file, we see that *a8* feeds into parameter *O1_dx* that defines a displacement of this site in fractional coordinates. *a8* also feeds into *O2_dx*, *O3_dy* and *O4_dy* (omitted here). The next section has equations defining the final coordinates of each site in terms of the position derived from the undistorted parent plus the mode-dependent shifts. Finally, these values are used to specify the atomic sites. Most of the time these equations don't need to be read by the user, so they are normally hidden away in jEdit (https://community.dur.ac.uk/john.evans/topas_academic/jedit_main.htm) '{{{. . . '}}} folds by default.

After each mode amplitude we see the mode label in ISODISTORT language. This lists the parent symmetry ($Pm\bar{3}m$) the k-point (here ($\frac{1}{2}$, $\frac{1}{2}$, $\frac{1}{2}$) or the R-point of $Pm\bar{3}m$), the three-dimensional R_4^+ irrep using Miller and Love (1967) notation and the two-dimensional OPD [$(a, 0, b)$]. This is followed by the parent Wyckoff site of the atom in question and the spectroscopic label for the Wyckoff site point group irrep that induces the distortion. In some cases an additional order parameter number is appended (_1, _2) to distinguish different modes with the same local point group irrep. The final labels "a" and "b" distinguish the two branches of the two-dimensional order parameter.

8.3.4 Symmetry mode description of WO_3

We can see how these ideas work in practice using our WO_3 example. The first structural distortion on cooling WO_3 is an off-centring of the W inside the WO_6 octahedra. This gives rise to superstructure peaks in the powder pattern at ($h/2$, $k/2$, l) positions, corresponding to the M point of reciprocal space. The subsequent distortions involve tilting of octahedra and give rise to ($h/2$, $k/2$, $l/2$) type superstructure reflections corresponding to the R point.

If we take the parent structure and superpose an M point irrep then there are seven choices: M_1^+, M_2^+, M_3^+, M_4^+, M_5^+, M_3^- and M_5^-; each would lower symmetry in a different way. M_3^- is the appropriate choice (we discuss later how you would determine this experimentally) and there are then six choices for the OPD as in the table below. Each results in different degrees of freedom in the child resulting in a different space group, basis and origin combination. The first part of each line (e.g., P1, P2) is an OPD label and the other parts are as described above. The most general OPD is S1(a; b; c) that has a doubled cell and space group $I222$ (the kernel). This child has three different M_3^- modes and we can explore what each one does using the graphical tools in ISODISTORT.[1] We find that they displace W by different amounts along the a-, b- or c-axis of the parent cubic cell. If we choose identical amplitudes for each of the modes then W atoms move parallel to the three-fold axis of the parent cell and the symmetry

---

[1] Assuming you have some familiarity with ISODISTORT, you can explore these distortions using the default $SrTiO_3$ example: On the ISODISTORT home page click "Get started quickly with a cubic perovskite parent". On the "ISODISTORT: search" page select "M" as the k point in method 2 and click OK. On the pull down menu of the "ISODISTORT: irreducible representation" page select "M3−" and click OK. On the "ISODISTORT: order parameter direction" page select S1 and click OK. On the "ISODISTORT: distortion" page select "view distortion". Move the M3− slider bars for Ti sites. This mimics WO_3, which we can think of as a perovskite without an A site cation.

is higher. This mimics the third OPD choice P3(a; a; a) and corresponds to space-group symmetry $I\bar{4}3m$. This is an example of how the symmetry mode basis lets you easily explore different symmetries and how they are related. Similarly, two modes with equal amplitude move W along parent <110> directions corresponding to the second choice (P2(a; 0; a); space-group symmetry $I4/mcm$) and a single mode moves W along a single axis [P1(a;0;0), space group $P4/nmm$ with basis ({(1,1,0),(−1,1,0),(0,0,1)})]. This is the actual distortion observed above 1173 K with the √2 × √2 × 1 cell discussed above.

P1 (a;0;0) 129 P4/nmm, basis={(1,1,0),(-1,1,0),(0,0,1)}, origin=(0,1/2,0), s=2, i=6, k-active=
(1/2,1/2,0)
P2 (a;0;a) 140 I4/mcm, basis={(0,0,2),(2,0,0),(0,2,0)}, origin=(3/2,0,1/2), s=4, i=12, k-active=
(1/2,1/2,0),(0,1/2,1/2)
P3 (a;a;a) 217 I-43m, basis={(2,0,0),(0,2,0),(0,0,2)}, origin=(1/2,1/2,1/2), s=4, i=8, k-active=
(1/2,1/2,0),(1/2,0,1/2),(0,1/2,1/2)
C1 (a;0;b) 72 Ibam, basis={(0,0,2),(2,0,0),(0,2,0)}, origin=(1/2,0,1/2), s=4, i=24, k-active=
(1/2,1/2,0),(0,1/2,1/2)
C2 (a;b;a) 121 I-42m, basis={(0,0,2),(2,0,0),(0,2,0)}, origin=(1/2,1/2,1/2), s=4, i=24, k-active=
(1/2,1/2,0),(1/2,0,1/2),(0,1/2,1/2)
S1 (a;b;c) 23 I222, basis={(2,0,0),(0,2,0),(0,0,2)}, origin=(1/2,1/2,1/2), s=4, i=48, k-active=
(1/2,1/2,0),(1/2,0,1/2),(0,1/2,1/2)

The next two distortions that occur on cooling involve rotations of the WO_6 octahedra. If we superpose R_4^+ distortions on the parent we get the following OPD choices:

P1 (a;0;0) 140 I4/mcm, basis={(1,1,0),(-1,1,0),(0,0,2)}, origin=(0,0,0), s=2, i=6, k-active=
(1/2,1/2,1/2)
P2 (a;a;0) 74 Imma, basis={(1,0,1),(0,2,0),(-1,0,1)}, origin=(0,0,0), s=2, i=12, k-active=
(1/2,1/2,1/2)
P3 (a;a;a) 167 R-3c, basis={(-1,1,0),(0,-1,1),(2,2,2)}, origin=(0,0,0), s=2, i=8, k-active=
(1/2,1/2,1/2)
C1 (a;b;0) 12 C2/m, basis={(0,0,-2),(0,2,0),(1,0,1)}, origin=(0,1/2,1/2), s=2, i=24, k-active=
(1/2,1/2,1/2)
C2 (a;a;b) 15 C2/c, basis={(-1,2,-1),(-1,0,1),(1,0,1)}, origin=(0,1/2,1/2), s=2, i=24, k-active=
(1/2,1/2,1/2)
S1 (a;b;c) 2 P-1, basis={(0,1,1),(1,0,1),(1,1,0)}, origin=(0,0,0), s=2, i=48, k-active=
(1/2,1/2,1/2)

In a similar way to W displacements, an (a; b; c) OPD has three branches each of which rotates octahedra around a single <100> axis of the parent cell. An (a; a; a) OPD corresponds to rotations around the three-fold axis and (a; 0; a) to rotations around <110>.

We can then understand the $P4/ncc$ high temperature phase of WO_3 (Figure 8.2) in terms of superposing M_3^-(a; 0; 0) $\oplus R_4^+$(b; 0; 0) on the parent and the subsequent $P2_1/c$ phase in terms of M_3^-(a; 0; 0) $\oplus R_4^+$(c; b; b). We can think of these irreps as being primary and sufficient to define the symmetry of the child phase. When they are active other secondary irreps are also activated (X_4^-, X_5^-, M_5^-, R_3^+, R_5^+). These may or may not have significant amplitude in the distorted structure.

8.4 Examples of symmetry mode refinements

8.4.1 WO$_3$ 850 °C *P4/ncc* and 25 °C *P2$_1$/n* Rietveld refinements

The script below is an INP file for a symmetry mode refinement of laboratory powder diffraction data collected on *P4/ncc* WO$_3$ at 850 °C:

```
xdd d8_03901_850c_03.xy
   finish_X 90
   x_calculation_step = Yobs_dx_at(Xo); convolution_step 4
   bkg @ 209.298187 -54.7924369 13.8094009 -1.59425645 8.68555976 1.26939144
   LP_Factor(!th2_monochromator, 0)
   CuKa2(0.0001)
   Specimen_Displacement(height,-0.06930)
   Simple_Axial_Model(axial, 8.50263)

   str
      space_group P4/ncc:2 'transformPp a,b,c;0,0,0
      a lpa   5.280882
      b lpa   5.280882
      c lpc   7.847197
      al     90.00000
      be     90.00000
      ga     90.00000
      scale @  0.000242874444
      r_bragg  11.0005319
      TCHZ_Peak_Type(pku,-0.007,pkv,0.009,pkw,-0.00437,!pkz,0.0,pky,0.141,!pkx,0.0)
'{{{mode definitions
      prm a1   0.52070 min -2.00 max 2.00 'Pm-3m[1/2,1/2,0]M3-(a;0;0)[W:a:dsp] T1u(a)
      prm a2   0.01301 min -2.00 max 2.00 'Pm-3m[1/2,1/2,0]M3-(a;0;0)[O:d:dsp] A2u(a)
      prm a3   0.75857 min -2.83 max 2.83 'Pm-3m[1/2,1/2,1/2]R4+(a,0,0)[O:d:dsp] Eu(a)
'}}}
'{{{mode-amplitude to delta transformation
      prm W_1_dz = +0.06649*a1;:  0.03462
      prm O_1_dx = +0.04702*a3;:  0.03567
      prm O_2_dz = +0.06649*a2;:  0.00087
'}}}
'{{{distorted parameters
      prm !W_1_x = 1/4;:  0.25000
      prm !W_1_y = 1/4;:  0.25000
      prm  W_1_z = 3/4 + W_1_dz;:  0.78462
      prm  O_1_x = 1/2 + O_1_dx;:  0.53567
      prm  O_1_y = 1/2 - O_1_dx;:  0.46433
```

```
        prm !O_1_z = 1/4;: 0.25000
        prm !O_2_x = 1/4;: 0.25000
        prm !O_2_y = 1/4;: 0.25000
        prm O_2_z = 0 + O_2_dz;: 0.00087
        prm !W_1_occ = 1;: 1.00000
        prm !O_1_occ = 1;: 1.00000
        prm !O_2_occ = 1;: 1.00000
'}}}
'{{{mode-dependent sites
        site W_1 x = W_1_x; y = W_1_y; z = W_1_z; occ W = W_1_occ; beq bval -0.99268
        site O_1 x = O_1_x; y = O_1_y; z = O_1_z; occ O = O_1_occ; beq bval -0.99268
        site O_2 x = O_2_x; y = O_2_y; z = O_2_z; occ O = O_2_occ; beq bval -0.99268
'}}}
```

This is a relatively straightforward case in which there are three symmetry mode amplitudes ($a1$–$a3$) that can be refined. As discussed above, $a1$ and $a2$ describe distortions of the WO_6 octahedra and $a3$ describes a tilt of the octahedra around the parent c-axis. The mode amplitudes are converted to fractional atomic coordinates by the equations given. The plots in Figure 8.3 show Rietveld fits using three different models. The lowest plot shows the fit if no mode amplitudes are refined and cell parameters are exactly related to their cubic values. We see that this model fails to produce the observed peak splittings and position shifts due to the distortion to tetragonal symmetry. Interestingly, the only peak in the correct position is the (202) at ~41° 2θ – you might like to think why this is! This fit is 100% equivalent to fitting the parent $Pm\bar{3}m$ model to the data, and is another example of how a symmetry mode refinement lets us easily simulate fitting a higher-symmetry model.

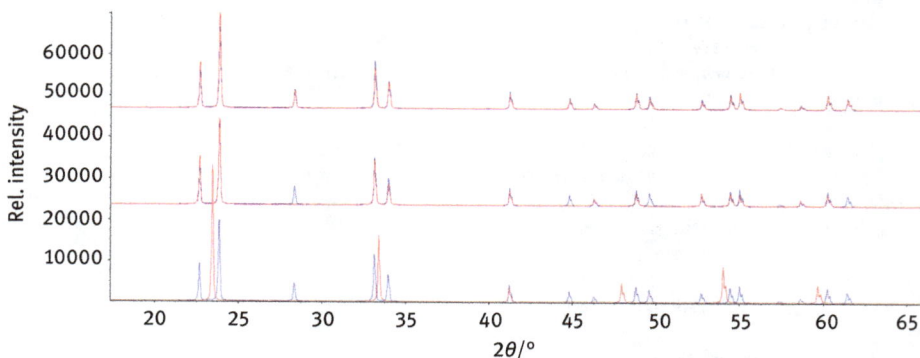

Figure 8.3: Rietveld fits to data collected on $P4/ncc$ WO_3 at 850 °C using the three models discussed in the text.

The middle plot shows the fit when the cell parameters are allowed to distort but mode amplitudes are kept at zero. We see that most of the strong peaks are now correctly fitted, but several superstructure reflections have zero calculated intensity.

This model is equivalent to fitting a $P4/mmm$ model with a $1 \times 1 \times 1$ unit cell (one of the maximal non-isomorphic subgroups of $Pm\bar{3}m$). The top fit shows the effect of refining the three mode amplitudes. We now get an excellent fit to the diffraction data.

You might notice from the TOPAS script that the temperature factors have refined to negative values, which makes no physical sense. This is because the data were collected in Bragg–Brentano geometry on a small sprinkled sample that was not larger than the illuminated area at all angles, and not densely packed. Hence, the criteria for validity of the usual 2θ-dependent intensity scaling (see Section 2.2) were not fulfilled. As a result, there are small systematic errors in the peak intensities as a function of 2θ that are "mopped up" by the negative temperature factor.

This example demonstrates advantages 2–4 of the symmetry modes that we introduced in Section 8.3.1. It doesn't, however, give any reduction in the number of parameters used (advantage 1): in the symmetry mode description we have three refineable mode amplitudes and in a conventional description we would have three xyz parameters; all of them are necessary to describe the structure.

We can, however, demonstrate parameter reduction using the more complex room temperature $P2_1/n$ structure. The lower plot in Figure 8.4 shows a Rietveld fit to a 30 °C data set using a conventional crystallographic description with 24 xyz parameters (R_{wp} = 9.0%) and the upper plot shows a fit using just five symmetry mode amplitudes (R_{wp} = 9.2%). We can see that the five-parameter fit is essentially as good as a 24-parameter fit. The five modes are sufficient to describe the key distortions of the material and are: tungsten X_5^- and M_3^- (W off-center displacements in different directions), $2 \times$ oxygen R_4^+ and a $1 \times$ oxygen M_3^+ (octahedral rotations). Even if we try and do simultaneous fitting of X-ray and neutron data (which is more sensitive to O displacements) we get a very good fit with five parameters and an excellent fit with seven. We'll discuss these combined refinements in more detail in the following sections.

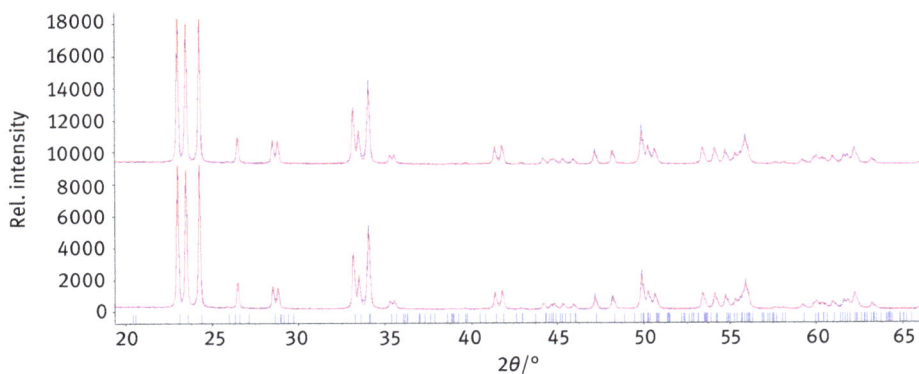

Figure 8.4: Rietveld fits to $P2_1/n$ WO_3 data collected at 30 °C using (lower plot) a 24-parameter conventional xyz refinement and (upper plot) a five-parameter symmetry mode refinement.

8.4.2 Using a genetic algorithm (GA) to identify symmetry

In the previous section, we discussed using symmetry modes in cases where we've already known the symmetry of the material and which distortions are the most important. How would we determine this in unknown cases?

One way of demonstrating what's possible is to forget for a moment that we know that the true symmetry of room temperature WO_3 is $P2_1/n$, and think whether it's possible to solve the structure with no symmetry assumptions. If WO_3 had a $2 \times 2 \times 2$ unit cell and $P1$ symmetry, there would be 96 possible mode amplitudes. If we could detect which of these are actually necessary to describe the diffraction data, we would essentially simultaneously determine the space group of the material (via which branches of the OPD are needed) and the specific distortions of the structure (from the amplitudes of the modes). It turns out that there are several ways to do this. One is to use a Genetic Algorithm or GA. In this approach we can flag whether modes in a structural model are active ("turned on" or 1) or inactive ("turned off" or 0) with a string of 96 1s and 0s that we can think of like a piece of genetic code (001011000 ...). We then test the model against the diffraction data by performing a refinement in which the selected mode amplitudes (*prms a1 to a96* in TOPAS syntax) are allowed to refine and recording the R_{wp} of each model. We can do this for a "population" of several different models and rank the "fitness" of each according to the R_{wp}. We can then apply some of the rules of Darwinian selection to our population of structures to produce a new offspring population. The rules might include "survival of the fittest" (deleting some of the highest-R_{wp} structures from the population); "mating" some of the better models (e.g., taking half of the string of 0s and 1s from one model and half from another to make a new structure) or "mutating" (randomly changing some of the 1s to 0 and vice-versa). We can then refine each member of this offspring population to obtain its R_{wp}, and keep repeating the process until we get a good fit to the diffraction data. If we add a small penalty to the R_{wp} fitness criterion that scales with the number of modes turned on, we can rapidly find structures that are simple in terms of their number of degrees of freedom, but still provide a good fit to the diffraction data.

There are simple PYTHON scripts published for performing this procedure, and within a few minutes on a normal desktop computer we find that for WO_3 a seven-mode model will give an excellent simultaneous fit to both X-ray and neutron data. From the resulting model we can use an algorithm such as FINDSYMM (http://stokes.byu.edu/iso/findsym.php) to identify that the true symmetry of this $P1$ symmetry mode description is $P2_1/n$. The method can be explored via an online tutorial at: http://community.dur.ac.uk/john.evans/topas_work shop/tutorial_GA_wo3.htm.

8.4.3 Using inclusion runs to identify symmetry

We can also tackle the problem of identifying the important modes in WO_3 in a way that is conceptually simpler (Lewis, 2018). Figure 8.5 shows the Rietveld R_{wp} from a simple "experiment" in which each of the 96 possible modes amplitudes was refined individually and the one that gave the lowest R_{wp} fit to X-ray and neutron data selected (here it was an oxygen R_4^+ mode describing an octahedral tilt that reduced R_{wp} from 30.1 with no mode amplitudes refined to 27.0%). We then simultaneously refined this mode amplitude along with each of the other 95 in turn (i.e., a series of two-parameter fits). This process was repeated until all the 96 mode amplitudes were included in the fit. As such, the plot in Figure 8.5 shows the lowest R_{wp} achievable using this recipe as a function of the number of modes in the model. We see that R_{wp} reaches low values when just seven modes are used and that improvements beyond this are much less marked. Reassuringly, these modes turn out to be the same ones identified by the GA and uniquely define the structure as $P2_1/n$. We also find that this plot is fully "reversible" and we see an identical behavior of R_{wp} versus the number of modes if we start with 96 modes on and sequentially eliminate the ones with the lowest impact on R_{wp}.

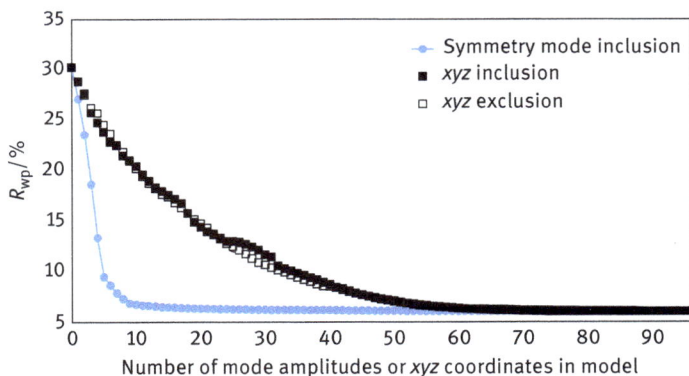

Figure 8.5: R_{wp} values obtained as a function of the number of structural degrees of freedom using either symmetry modes or *xyz* parameters. Values from simultaneous fit to X-ray and neutron data.

Figure 8.5 also includes data from a similar "experiment" using a traditional *xyz* parameter set instead of mode amplitudes (with equivalent constraints to limit coordinate shifts to those in the mode analysis). We again see that R_{wp} reduces as more parameters are included in the model, but that we need far more parameters to achieve an equivalent fit to the data. If we repeat this experiment or perform the reverse mode exclusion procedure, we always obtain similar plots to those in Figure 8.5, though the precise shape of the plot can change as the differentiation between the contributions of each parameter to R_{wp} is less marked. It is therefore much easier to make a small "mistake" in the sequence of parameter choice and follow a different pathway across the R-factor hypersurface.

Figure 8.5 provides a quantitative feel for how much more efficient symmetry mode refinements can be in terms of the number of parameters employed.

8.4.4 Exhaustive searching using a subgroup tree

A final way to tackle the WO_3 problem is to use a feature of ISODISTORT that produces a subgroup tree containing an exhaustive list of all the possible child structures. If one considers a parent structure (here $Pm\bar{3}m$ WO_3 in a $1 \times 1 \times 1$ cell) and a low-symmetry base structure with sufficient degrees of freedom to fully explain the diffraction data (for example a $2 \times 2 \times 2$ description in $P1$), it is possible to derive all the possible subgroups between the parent and base child. In this case there are 1427 of them. Each of these candidate structures can be tested against the diffraction data to determine its R_{wp}, giving a comprehensive "map" of the R_{wp} surface for all possible models. It is then relatively simple to make informed decisions about the "best" structural model, where best typically means the simplest description that describes all the important features of the diffraction data. This approach has been described in the literature (Lewis et al, 2016) and there are online tutorials describing how to do it in practice (http://community.dur.ac.uk/john.evans/topas_workshop/tutorial_exhaustive_symmetry.htm).

8.4.5 Other order parameters – occupancies, rotational modes and magnetism

In this chapter we've focussed on examples where the order parameters are symmetry modes that describe atomic displacements and have the tensor properties of microscopic (because they act on atoms) polar vectors. It's possible to use a similar language to describe site occupancies (microscopic scalar parameters), magnetic ordering (microscopic axial vectors or pseudo-vectors) or the rotational properties of rigid molecular groups (microscopic axial vectors). Some of these ideas are explored in Chapter 9 on magnetic Rietveld refinement.

8.5 Example of a Rietveld refinement using symmetry modes

We'll finish with the Rietveld example of the double salt $Mg(H_2O)_6RbBr_3$ that we looked at previously in Sections 2.10 and 6.3.11. In our earlier analyses on this material we didn't discuss the fact that it undergoes a phase transition at high temperature. With our knowledge of symmetry modes we can develop a simple description of the low-temperature (LT) structure relative to the parent high temperature (HT) structure using only a small number of parameters.

If the $Mg(OH_2)_6$ unit of the room temperature phase of $Mg(H_2O)_6RbBr_3$ was replaced by an atom A, the resulting $ARbBr_3$ structure would be analogous to a $\sqrt{2} \times \sqrt{2} \times 2$ supercell of a distorted cubic perovskite with $a = 6.94189$ Å, and Rb on the B site. The phase transition takes $Mg(H_2O)_6RbBr_3$ to a cubic ($Pm\bar{3}m$) HT structure analogous to an ideal perovskite. The HT structure exhibits four-fold orientational disorder of the $Mg(OH_2)_6$ octahedral tilts as illustrated in the right-hand panel of Figure 8.6. These become orientationally ordered in the LT structure, so that the tilts alternate along the long axis of the supercell.

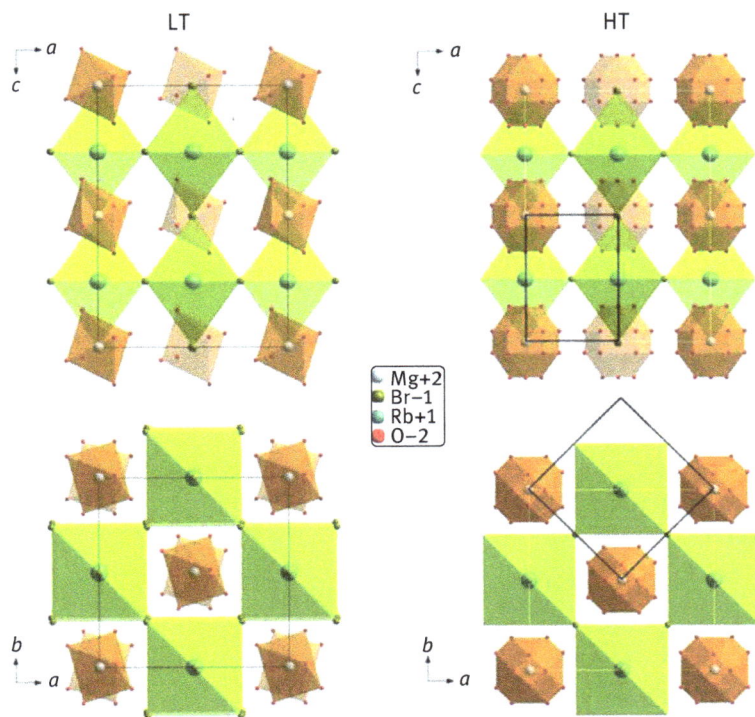

Figure 8.6: Comparison of the LT (left) and HT (right) crystal structures of $Mg(H_2O)_6RbBr_3$ as viewed along the b-axis (top) and the c-axis (bottom) showing green $RbBr_6$ and brown $Mg(OH_2)_6$ octahedra (from Dinnebier et al., 2008).

In the following paragraphs we will go through a cookbook recipe for a symmetry mode refinement of LT-$Mg(H_2O)_6RbBr_3$. We will again use the ISODISTORT (Campbell, 2006) package to calculate the symmetry modes for us, and to automatically produce a TOPAS STR file. We will reference the LT structure to an idealized hypothetical cubic parent structure in which the $Mg(OH_2)_6$ octahedra are ordered and aligned parallel to the cell axes, though the actual HT structure is orientationally disordered (Figure 8.6).

Firstly, the *ordered* high-symmetry parent structure should be uploaded to the ISODISTORT server from a CIF-file. ISODISTORT includes a viewer (Figure 8.7) to check the structure has read in correctly.

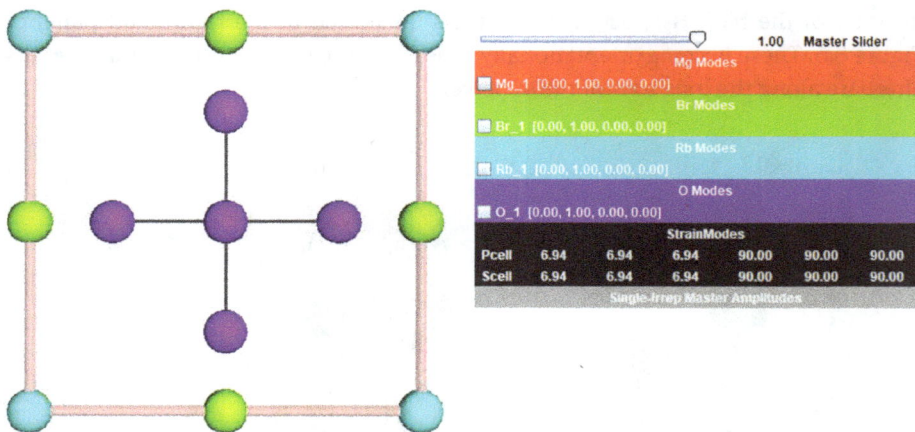

We can use ISODISTORT method 3 *"Search over arbitrary k points for specified space group and lattice"* to create a description of the LT *C2/c* structure. In the *"Select space group symmetry box"* we specifiy "15 C2-c C2h-6." We then choose the basis for the child structure as:

$$
\begin{pmatrix} a \\ b \\ c \end{pmatrix}_{child} = \begin{pmatrix} 1 & -1 & 0 \\ 1 & 1 & 0 \\ 0 & 0 & 2 \end{pmatrix} \begin{pmatrix} a \\ b \\ c \end{pmatrix}_{parent}
\tag{8.1}
$$

and click on OK. On the next screen we select the second choice, *"15 C2/c, basis = {(1,1,0),(−1,1,0),(0,0,2)}, origin=(0,0,1/2), s = 2, I = 24,"* and click OK. On the following page we can either select "View distortion" to visualize the structure in the child setting (Figure 8.8), "TOPAS.STR" to write out an STR file in TOPAS format or "View diffraction" to visualize the calculated powder pattern of the material. In the "View distortion" option you can adjust the values of the horizontal slider bars to investigate the effect of each symmetry mode on the structure. For example, you should see that changing the amplitudes of the O Γ_4^+ or X_1^- modes gives rise to rotations of the $Mg(OH_2)_6$ octahedra. In the "View diffraction" window you can see how these modes influence the powder diffraction data.

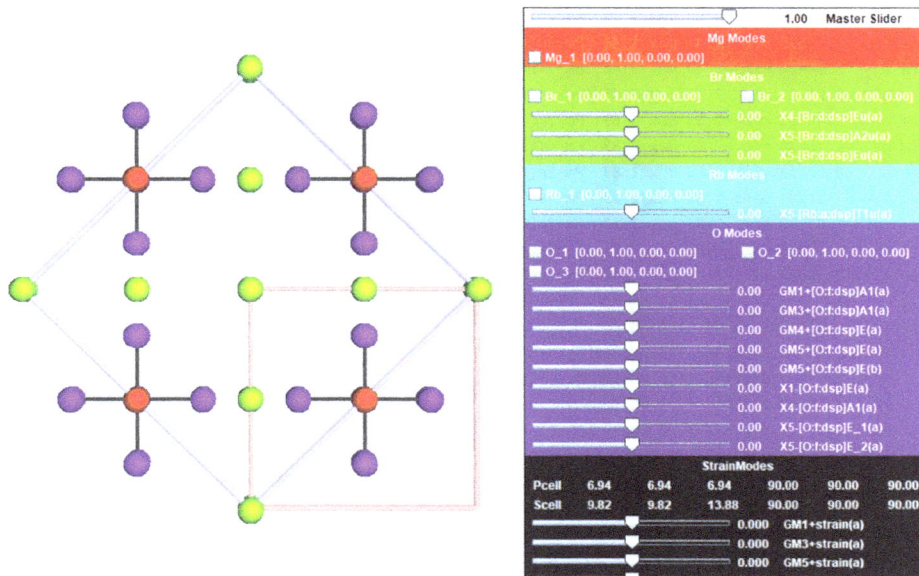

Figure 8.8: Screenshot of ISODISTORT showing the parent and child structures of $Mg(H_2O)_6RbBr_3$ and all displacive and strain modes allowed by symmetry. The parent cell is in pink and the child cell in purple.

For our purposes we can write out a TOPAS.STR file to allow us to refine the LT structure relative to the HT parent. The contents of TOPAS.STR replace the normal structural coordinates in the INP file that was given in Section 2.10. We find that there are 13 possible displacive mode amplitudes which correspond to the 13 free fractional coordinates in a conventional description. After refinement the relevant part of the INP file reads:

```
str
   phase_name double_salt
   scale @  8.50799022e-005
   a  @  9.641327
   b  @  9.865327
   c  @  13.786095
   be @  90.08790
   CS_L(@, 802.58724)
   CS_G(@, 819.13894)
   Strain_G(@, 0.12267)
   space_group C12/c1

'{{{mode definitions
prm a1  -0.03397 min -2.00 max 2.00 'Pm-3m[0,1/2,0]X4-(0;0;a)[Br:d:dsp] Eu(a)
prm a2   0.07955 min -2.00 max 2.00 'Pm-3m[0,1/2,0]X5-(0,0;0,0;a,-a)[Br:d:dsp] A2u(a)
```

```
prm a3   -0.18402 min -2.00 max 2.00 'Pm-3m[0,1/2,0]X5-(0,0;0,0;a,-a)[Br:d:dsp] Eu(a)
prm a4    0.03595 min -1.41 max 1.41 'Pm-3m[0,1/2,0]X5-(0,0;0,0;a,-a)[Rb:a:dsp] T1u(a)
prm a5    0.34429 min -3.46 max 3.46 'Pm-3m[0,0,0]GM1+(a)[O:f:dsp] A1(a)
prm a6    0.01731 min -2.45 max 2.45 'Pm-3m[0,0,0]GM3+(a,0)[O:f:dsp] A1(a)
prm a7    2.36162 min -2.83 max 2.83 'Pm-3m[0,0,0]GM4+(a,-a,0)[O:f:dsp] E(a)
prm a8    0.15038 min -2.83 max 2.83 'Pm-3m[0,0,0]GM5+(a,b,b)[O:f:dsp] E(a)
prm a9    0.02173 min -2.83 max 2.83 'Pm-3m[0,0,0]GM5+(a,b,b)[O:f:dsp] E(b)
prm a10  -1.61761 min -2.83 max 2.83 'Pm-3m[0,1/2,0]X1-(0;0;a)[O:f:dsp] E(a)
prm a11   0.07180 min -2.83 max 2.83 'Pm-3m[0,1/2,0]X4-(0;0;a)[O:f:dsp] A1(a)
prm a12   0.17459 min -2.83 max 2.83 'Pm-3m[0,1/2,0]X5-(0,0;0,0;a,-a)[O:f:dsp] E_1(a)
prm a13   0.33838 min -2.00 max 2.00 'Pm-3m[0,1/2,0]X5-(0,0;0,0;a,-a)[O:f:dsp] E_2(a)
'}}}
'{{{mode-amplitude to delta transformation
prm Br_1_dx = +0.03601*a2 + 0.03601*a3;: -0.00376
prm Br_1_dy = -0.03601*a2 + 0.03601*a3;: -0.00949
prm Br_1_dz = -0.03601*a1;: 0.00122
prm Rb_1_dy = +0.07203*a4;: 0.00259
prm O_1_dx = +0.02079*a5 + 0.01470*a6 + 0.02547*a8 + 0.02547*a10 - 0.02547*a11;: -0.03179
prm O_1_dy = -0.02079*a5 - 0.01470*a6 + 0.02547*a8 + 0.02547*a10 + 0.02547*a11;: -0.04295
prm O_1_dz = +0.01801*a7 + 0.01801*a9 + 0.02547*a12;: 0.04737
prm O_2_dx = +0.02079*a5 + 0.01470*a6 + 0.02547*a8 - 0.02547*a10 + 0.02547*a11;: 0.05427
prm O_2_dy = -0.02079*a5 - 0.01470*a6 + 0.02547*a8 - 0.02547*a10 - 0.02547*a11;: 0.03579
prm O_2_dz = +0.01801*a7 + 0.01801*a9 - 0.02547*a12;: 0.03848
prm O_3_dx = -0.03601*a7 + 0.03601*a9;: -0.08426
prm O_3_dy = -0.05093*a13;: -0.01723
prm O_3_dz = +0.02079*a5 - 0.02940*a6;: 0.00665
'}}}
'{{{distorted parameters
prm !Mg_1_x = 0;:  0.00000
prm !Mg_1_y = 1/2;:  0.50000
prm !Mg_1_z = 0;:  0.00000
prm  Br_1_x = 3/4 + Br_1_dx;:  0.74624
prm  Br_1_y = 1/4 + Br_1_dy;:  0.24051
prm  Br_1_z = 3/4 + Br_1_dz;:  0.75122
prm !Br_2_x = 0;:  0.00000
prm !Br_2_y = 0;:  0.00000
prm !Br_2_z = 0;:  0.00000
prm !Rb_1_x = 0;:  0.00000
prm  Rb_1_y = 0 + Rb_1_dy;:  0.00259
prm !Rb_1_z = 1/4;:  0.25000
prm  O_1_x  = 0.35500 + O_1_dx;:  0.32321
prm  O_1_y  = 0.14500 + O_1_dy;:  0.10205
prm  O_1_z  = 0 + O_1_dz;:  0.04737
prm  O_2_x  = 0.35500 + O_2_dx;:  0.40927
prm  O_2_y  = 0.14500 + O_2_dy;:  0.18079
prm  O_2_z  = 1/2 + O_2_dz;:  0.53848
prm  O_3_x  = 1/2 + O_3_dx;:  0.41574
prm  O_3_y  = 0 + O_3_dy;: -0.01723
prm  O_3_z  = 0.85500 + O_3_dz;:  0.86165
'}}}
```

```
'{{{mode-dependent sites
site Mg_1 x = Mg_1_x; y = Mg_1_y; z = Mg_1_z; occ Mg 1 beq bm  2.19023
site Rb_1 x = Rb_1_x; y = Rb_1_y; z = Rb_1_z; occ Rb 1 beq =bm;
site Br_1 x = Br_1_x; y = Br_1_y; z = Br_1_z; occ Br 1 beq bbr  1.67475
site Br_2 x = Br_2_x; y = Br_2_y; z = Br_2_z; occ Br 1 beq =bbr;
site O_1  x = O_1_x;  y = O_1_y;  z = O_1_z;  occ O 1 beq bo   0.07540
site O_2  x = O_2_x;  y = O_2_y;  z = O_2_z;  occ O 1 beq =bo;
site O_3  x = O_3_x;  y = O_3_y;  z = O_3_z;  occ O 1 beq =bo;
'}}}
```

From the TOPAS script above we can see that two modes have particularly large amplitude: $a7$ (O Γ_4^+) and $a10$ (O X_1^-). These are therefore the most important in describing the differences between the parent and child structures. We can visualize the effect of all the refined mode amplitudes in ISODISTORT using the so-called method 4 *"Mode decomposition of a distorted structure by loading a distorted structure (child structure) from CIF file."* This gives an easy way of comparing the LT and HT structures. In the window shown in Figure 8.9 we select the child basis from a pull down list of options then click OK.

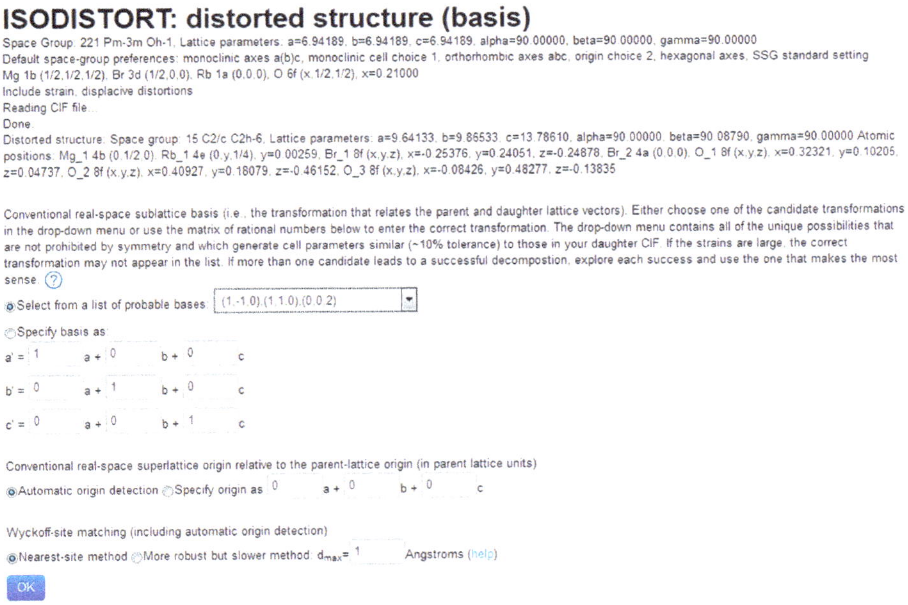

ISODISTORT: distorted structure (basis)

Space Group: 221 Pm-3m Oh-1. Lattice parameters: a=6.94189, b=6.94189, c=6.94189, alpha=90.00000, beta=90.00000, gamma=90.00000
Default space-group preferences: monoclinic axes a(b)c, monoclinic cell choice 1, orthorhombic axes abc, origin choice 2, hexagonal axes abc, SSG standard setting
Mg 1b (1/2,1/2,1/2), Rb 3d (1/2,0,0), Br 1a (0,0,0), O 6f (x,1/2,1/2), x=0.21000
Include strain, displacive distortions
Reading CIF file...
Done.
Distorted structure: Space group: 15 C2/c C2h-6. Lattice parameters: a=9.64133, b=9.85533, c=13.78610, alpha=90.00000, beta=90.08790, gamma=90.00000 Atomic positions: Mg_1 4b (0,1/2,0). Rb_1 4e (0,y,1/4). y=0.00259, Br_1 8f (x,y,z), x=-0.25376, y=0.24051, z=-0.24878, Br_2 4a (0,0,0), O_1 8f (x,y,z), x=0.32321, y=0.10205, z=0.04737, O_2 8f (x,y,z), x=0.40927, y=0.18079, z=-0.46152, O_3 8f (x,y,z), x=-0.08426, y=0.48277, z=-0.13835

Conventional real-space sublattice basis (i.e., the transformation that relates the parent and daughter lattice vectors). Either choose one of the candidate transformations in the drop-down menu or use the matrix of rational numbers below to enter the correct transformation. The drop-down menu contains all of the unique possibilities that are not prohibited by symmetry and which generate cell parameters similar (~10% tolerance) to those in your daughter CIF. If the strains are large, the correct transformation may not appear in the list. If more than one candidate leads to a successful decomposition, explore each success and use the one that makes the most sense. ⓘ

⦿ Select from a list of probable bases: (1,-1,0),(1,1,0),(0,0,2) ▾

⦿ Specify basis as:

a' = 1   a + 0   b + 0   c

b' = 0   a + 1   b + 0   c

c' = 0   a + 0   b + 1   c

Conventional real-space superlattice origin relative to the parent-lattice origin (in parent lattice units)
⦿ Automatic origin detection ⦿ Specify origin as 0   a + 0   b + 0   c

Wyckoff-site matching (including automatic origin detection)
⦿ Nearest-site method ⦿ More robust but slower method d_max = 1   Angstroms (help)

[OK]

Figure 8.9: Screenshot of ISODISTORT showing the Method 4 window for comparing the HT and LT structures.

The child structure can then be visualized as shown in Figure 8.10. By clicking "animate" in the ISODISTORT viewer we can play a simple movie between the parent and child structures. By pressing "z" on the keyboard we can zero all mode amplitudes; by pressing "r" we can return them all to the values refined for the LT structure.

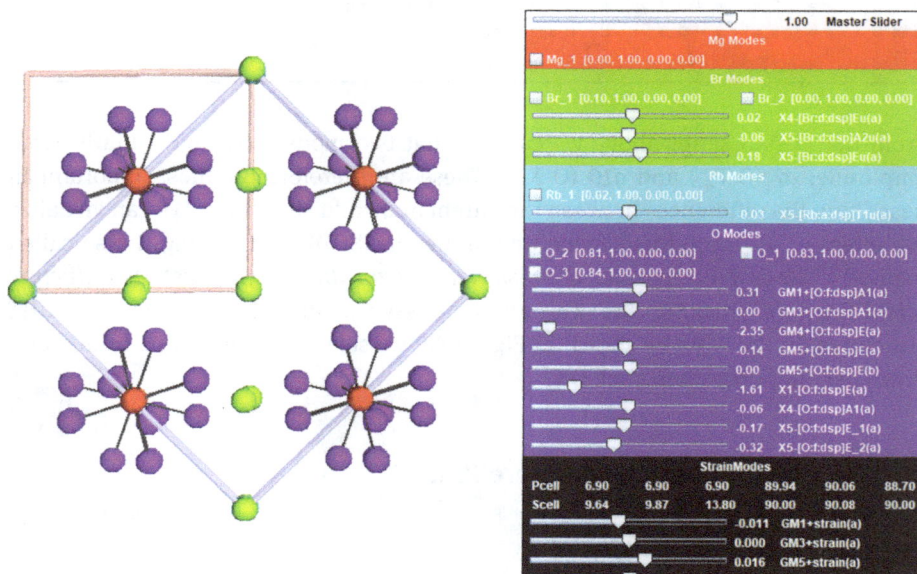

Figure 8.10: Screenshot from ISODISTORT showing the child structure of $Mg(H_2O)_6RbBr_3$ and the magnitudes of all displacive and strain modes allowed by symmetry.

It is worth noting that the distortions we called rotations above will in fact distort polyhedra to some extent. For example, the $Mg(OH_2)_6$ X_1^- rotations move oxygen atoms perpendicular to the Mg—O bonds so actually cause a slight expansion of the octahedra. To properly describe rotations of fully rigid bodies, a more sophisticated approach is needed using so-called rotational rigid body modes (Müller et al., 2014). Here a rotational axis of the rigid body is defined by a vector in space with the length of the vector describing the magnitude of rotation. This procedure is more complex and beyond the scope of this book. The interested reader is referred to the paper by Müller et al. (2014), which contains a description of the application of rigid body modes in TOPAS using our $Mg(H_2O)_6RbBr_3$ example, or the work of Liu et al. (2018).

References

Campbell, B.J., Stokes, H.T., Tanner D.E., Hatch, D.M. (2006): *ISODISPLACE: An internet tool for exploring structural distortions.* J. Appl. Cryst., 39, 607–614.

Campbell, B.J., Evans, J.S.O., Perselli, F., Stokes, H.T. (2007): *Rietveld refinement of structural distortion-mode amplitudes*. IUCr Comput. Commun. Newsletter, 8, 81–95.

Dinnebier, R.E., Liebold-Ribeiro, Y., Jansen, M. (2008): *The low and high temperature crystal structures of [Mg(H_2O_6)]XBr_3 double salts (X = Rb, Cs)*. Z. Anorg. Allg. Chem. 634, 1857–1862.

Kerman, S., Campbell, B.J., Satyavarapu, K.K., Stokes, H.T., Perselli, F., Evans, J.S.O (2012): *The superstructure determination of displacive distortions via symmetry-mode analysis*. Acta Cryst. A 68, 222–234.

Landau, L.D., Lifschitz, E.M. (1970): *Lehrbuch der theoretischen Physik V. Statistische Physik*. Akademie Verlag, Berlin, D.

Lewis, J.W., Payne, J.L., Evans, I.R., Stokes, H.T., Campbell, B.J., Evans, J.S.O. (2016): *An exhaustive symmetry approach to structure determination: phase transitions in $Bi_2Sn_2O_7$*. J. Am. Chem. Soc. 138, 8031–8042.

Lewis, J.W. (2018): *Symmetry methods for understanding structures of inorganic functional materials*. PhD thesis, Durham University, UK.

Liu, H., Zhang, W., Halasyamani, P.H., Stokes, H.T., Campbell, B.J., Evans, J.S.O., Evans, I.R. (2018): *Understanding the behavior of the above-room-temperature molecular ferroelectric 5-6-dichloro-2-methylbenzimidazole using symmetry adapted distortion mode analysis*. J. Am. Chem. Soc. 140, 13441–13448.

Miller, S.C., Love, W.F. (1967): *Tables of irreducible representations of space groups and co-representations of magnetic space groups*, Pruett Press, Boulder (USA).

Müller, M., Dinnebier, R.E., Dippel, A., Stokes, H.T., Campbell, B.J. (2014): *A symmetry-mode description of rigid-body-rotations in crystalline solids: a case study of Mg[H_2O]$_6RbBr_3$*. J. Appl. Cryst. 47, 532–538.

9 Magnetic refinements

9.1 Introduction

One significant advantage of neutron over X-ray (powder) diffraction is its sensitivity to the magnetic structures of materials. For magnetically ordered materials neutrons can therefore determine both the nuclear structure and the configuration of magnetic moments. The most general and powerful approach for describing magnetic structures uses the crystallographic cell to describe the nuclear structure and a Fourier series to describe magnetic moments, which often have a larger periodicity. This allows an elegant and unified description of both commensurate and incommensurate magnetic structures, which is described in detail elsewhere (Von Dreele & Rodriguez-Carvajal, 2008). The current TOPAS implementation of magnetic diffraction relies, however, on the use of a magnetic unit cell and is restricted to commensurate structures (though incommensurate structures can be approximated using large supercells). We will, therefore, restrict our discussion to this method. We give a brief introduction to magnetic scattering, introduce some of the key ideas of magnetic symmetry and Shubnikov groups, highlight some potential pitfalls in magnetic structure analysis and then discuss different ways to perform magnetic refinements in TOPAS.

9.2 Magnetic diffraction

Thermal neutrons produced by either reactor or spallation sources have wavelengths in the 1 to 4 Å range and are ideally suited for diffraction experiments. Even longer or shorter wavelengths (cold and hot neutrons) can be produced using moderators of appropriate temperature. Neutrons have a spin of ½ and an intrinsic magnetic moment of 1.9 nuclear magnetons, meaning that they will interact with the magnetic moments produced by unpaired electrons in materials. When these moments[1] order below a critical temperature[2] to give ferro-, ferri- or antiferromagnetic long range order we therefore see magnetic contributions to diffraction patterns. The magnitude of the effect is significant, and for many transition metal or rare earth containing samples magnetic and nuclear intensities can be comparable.

We will only consider the case of unpolarized neutrons where the beam contains all neutron spin orientations. In this situation the structure factor (ignoring Debye–Waller terms discussed in Section 2.2.3) can be expressed (Bacon, 1975) as the sum of two terms – one describing the nuclear scattering (see eq. (2.21)) and one the

---

[1] We will often use the term "spin" for brevity.
[2] The Curie temperature, T_C, for ferromagnets and Néel temperature, T_N for antiferromagnets.

https://doi.org/10.1515/9783110461381-009

magnetic. In cases where the nuclear and magnetic cells are different (see below), it's important that the two terms are appropriately scaled:

$$|F(\mathbf{s})|^2 = |F_{\text{nuc}}(\mathbf{s})|^2 + |F_{\text{mag}}(\mathbf{s})|^2$$

$$= \left| \sum_j \left(b_j e^{2\pi i \mathbf{s} \cdot \mathbf{x}_j} \right) \right|^2 + \left| \sum_j \left(\mathbf{q}_j p_j(s) e^{2\pi i \mathbf{s} \cdot \mathbf{x}_j} \right) \right|^2. \tag{9.1}$$

The second term, $|F_{\text{mag}}(\mathbf{s})|^2$, contains two quantities we haven't met before: $\mathbf{q}_j$ and p_j. $\mathbf{q}$ is the magnetic interaction vector that is defined as:

$$\mathbf{q} = \boldsymbol{\varepsilon}(\boldsymbol{\varepsilon} \cdot \mathbf{K}) - \mathbf{K}. \tag{9.2}$$

The various vectors in eq. (9.2) are illustrated in Figure 9.1. $\boldsymbol{\varepsilon}$ is a unit vector in the direction of the scattering vector $\mathbf{s}$ discussed in Section 1.1. It is perpendicular to the scattering plane and in the plane of the incident and diffracted beams. $\mathbf{K}$ is a unit vector in the direction of the spin.[3] From this definition we can deduce that $\mathbf{q}$ is in the plane of $\boldsymbol{\varepsilon}$ and $\mathbf{K}$ and perpendicular to $\boldsymbol{\varepsilon}$. Its magnitude is given by $\sin \alpha$ where α is the angle between the scattering vector and the spin:

$$|\mathbf{q}| = \sin \alpha. \tag{9.3}$$

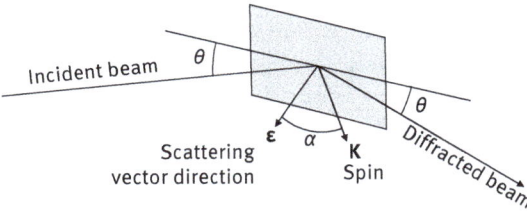

Figure 9.1: The relationship between the vectors given in eq. (9.2). After Bacon (1975).

We therefore see that magnetic intensities are sensitive to spin direction: when the spin is perpendicular to $\boldsymbol{\varepsilon}$, $|\mathbf{q}| = 1$, but when parallel $|\mathbf{q}| = 0$ meaning that no magnetic scattering is observed.

The quantity p is given by:

$$p = \frac{e^2 \gamma}{2 \cdot 4\pi\varepsilon_0 m_e c^2} 2Sf(s), \tag{9.4}$$

where γ is the magnetic moment of the neutron in nuclear magnetons, $e^2/m_e c^2$ the classical radius of the electron (e, charge on electron; m_e, mass of electron; c, speed of light) and S the spin quantum number. For a spin only ion in which the orbital moment

---

3 Note that the spin direction is opposite to the magnetic moment.

is quenched, $2S$ is equal to the magnetic moment in Bohr Magnetons (BM). In cases where spin and orbital contributions need to be considered, $2S$ is replaced by gJ where g is the Landé factor. The final term in eq. (9.4), $f(s)$, is the magnetic form factor. This is similar to the form factor familiar from X-ray diffraction (Section 2.2.1) and arises as the electrons that give rise to the magnetic moment are distributed over a volume of space comparable to the neutron wavelength. This is in marked contrast to nuclear scattering where the nuclei act as essentially point scatterers. Since the electrons responsible for magnetic moments are in high principal quantum number orbitals – the outer shells – magnetic neutron scattering factors fall off more rapidly with $\tilde{s} = s/2 = \sin\theta/\lambda$ than X-ray scattering factors. Taking Mn^{2+} as an example, at $\sin\theta/\lambda = 0.32$ (60° 2θ for $\lambda = 1.54$ Å) the X-ray and magnetic form factors have fallen to about 60% and 20% of their value at $\theta = 0$, respectively. The effect on magnetic intensities (which depend on f^2) is even more dramatic: in our Mn^{2+} example intensities will typically fall to 50% by $\sin\theta/\lambda = 0.15$ (~27° 2θ for $\lambda = 1.54$ Å).

Equation (9.4) also gives us a feel for the relative intensities of nuclear and magnetic scattering. Substituting values[4] we find $p = 0.54Sf \times 10^{-12}$ cm. For low $\sin\theta/\lambda$ and typical S values (up to 5/2 for transition metals), p will be the same order of magnitude as b_n. For example, for Fe^{3+} $b_n = 0.96 \times 10^{-12}$ cm and $p = 1.35 \times 10^{-12}$ cm at $\theta = 0$.

The form of eq. (9.1) also lets us understand where we will see magnetic peaks in the diffraction patterns of simple materials. For a simple ferromagnetic material like that in Figure 9.2a, $\mathbf{q}_j$ will be the same for all atoms and magnetic intensity will simply add to the nuclear peaks. For a simple antiferromagnetic material such as that shown in Figure 9.2b, $\mathbf{q}_j$ will alternate from + to − along the a-axis. The magnetic contributions at 2θ values corresponding to nuclear reflections will therefore be precisely zero. However, the larger magnetic unit cell means that there will be magnetic peaks at 2θ values corresponding to a doubled a-axis – (½, 0, 0), (½, 1, 0), (½, 1, 1) and so on using the nuclear cell. We therefore see nuclear-only and magnetic-only peaks in the neutron diffraction pattern. This type of magnetic ordering is therefore very easy to spot in a neutron diffraction experiment.

9.3 Magnetic symmetry

One significant difference between normal and magnetic crystallography is the need for an extended language to describe symmetry. In conventional crystallography one of the 230 space groups types is used to generate each atom in the unit cell from the asymmetric unit. In magnetic crystallography we need to do the same with the

---

4 $e = 1.602 \times 10^{-19}$ C, $\gamma = 1.9132$ nuclear magnetons (1 nuclear magneton is -9.662×10^{-27} JT^{-1}), $m = 9.1094 \times 10^{-31}$ kg, $c = 2.998 \times 10^8$ ms^{-1}; $\varepsilon_0 = 8.855 \times 10^{-31}$ $J^{-1}C^2m^{-1}$.

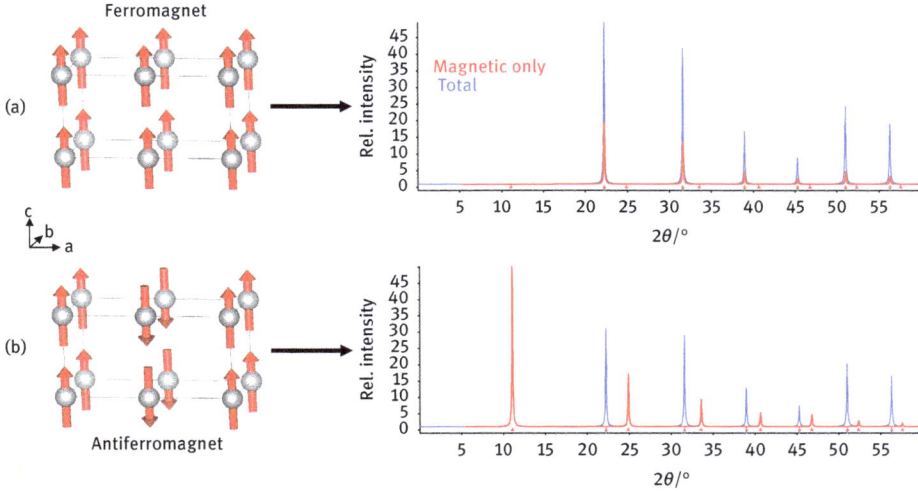

Figure 9.2: Neutron diffraction patterns from (a) a ferromagnetically ordered material: the red line shows the magnetic contribution to the diffraction pattern and the blue line the total diffraction pattern; and (b) an antiferromagnetically ordered material: each peak has either a magnetic or nuclear contribution. Note that the red magnetic intensities fall off rapidly with 2θ due to the magnetic form factor f.

magnetic moments or spin directions and must consider new symmetry operations that can, for example, change spins from "up" to "down." Much of the language needed to do this was developed for so-called two-color or black and white three dimensional space groups by Belov and others (Belov, 1955; Belov, 1957) and by Zamorzaev (Zamorzaev, 1953, 1957a, 1957b) building on ideas introduced by Heesch and Shubnikov (Heesch, 1930; Shubnikov, 1951). Some of the historical developments are covered by Wills (2017). These groups contain not only operations that rotate or translate atoms in space, but operations that can change their color from black to white. There are 1651 two-color space groups and they are often called *Shubnikov groups*. As described below, when we use them for magnetism we relate the change in color of a site (black to white) to the flip of a spin at that site (e.g., up to down).

Before we go into more detail, it's instructive to consider the effect of some simple symmetry elements on magnetic moments or spins. Figure 9.3 contrasts the effect of a mirror plane on two types of vector that we encounter in structural science. On the left we represent a polar vector, such as an electric dipole in a molecule, by an arrow with charges at either end. Here we see the "common sense" behavior that an arrow perpendicular to a mirror plane is flipped, whereas an arrow parallel to it retains the same orientation.

Magnetic dipoles show the opposite and initially counter-intuitive behavior. In the center of Figure 9.3 we see that an arrow perpendicular to the mirror plane remains pointing in the same direction, whereas one parallel to the plane is flipped.

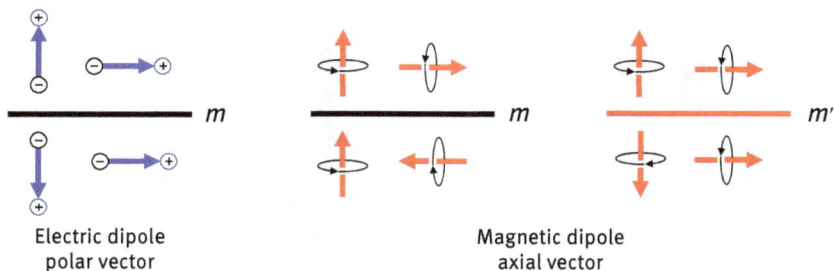

Figure 9.3: The action of a mirror plane on (left) a polar vector and (right) an axial vector such as a magnetic moment or spin. We can imagine the moment as being generated from a current loop.

We can understand this by considering the moment in a classical sense as being generated by a circulating current loop.[5] We therefore have to consider the effect of the symmetry operation on the direction of this current. In the ↑ case the mirror operation retains the current loop direction, whereas for → the current is reversed leading to the spin being flipped.[6] This type of vector is known as an *axial vector*. It's important to remember that the arrow we conventionally draw to depict a moment is a somewhat misleading depiction of its symmetry properties.

There will be cases where we need a mirror plane that also flips the direction of a ↑ spin. To do this, we have to introduce a new symmetry operator that is usually called *time reversal* or *spin reversal* and is labeled 1′. Physicists tend to favor the term time reversal, which we can interpret from Figure 9.3 in terms of the reverse in direction of the current loop giving rise to the spin – the electrons move backwards in time when their direction is changed. A symmetry element such as a mirror that combines reflection and spin reversal is called a primed element (*m*′); a normal mirror is said to be unprimed (*m*). Symmetry operations are often called "black" if primed and "white" if unprimed. Slightly confusingly, magnetic space group diagrams usually color symmetry elements red for primed and black for unprimed.

There are some concepts that arise from the introduction of time or spin reversal that we'll need to understand. First, we consider what happens to the 32 crystallographic point groups (which describe site symmetries in extended structures) when they're combined with time reversal. This process creates *magnetic point groups*. There are three ways in which this combination is done. In the first case, the 1′ operator is not present giving 32 magnetic point groups that are the same as the crystallographic point groups. They are called *trivial magnetic groups* or *monochrome/colorless groups* as all

---

5 Current is conventionally the flow of positive charge. The right-hand grip rule reminds us that the moment is in the direction of our thumb if our fingers are curled in the direction of current flow. Spin is in the opposite direction.

6 Magnetic moments or spins have the symmetry properties of a rotating cylinder: a center of inversion and a mirror plane perpendicular to their axis.

the operators are white. In the second case, 1' can be present as a separate element meaning every operator in the point group is both black and white. These 32 groups are therefore called *gray groups*. Finally, we can combine 1' with individual operations which leads to 58 *nontrivial* or *black-white point groups*. For example, ignoring the gray groups, point group $4/m$ will give rise to four magnetic point groups $4/m$, $4/m'$, $4'/m'$ and $4'/m$ – one trivial and three nontrivial.

Many magnetic point groups aren't possible for a magnetically ordered site in a crystal. First, the gray groups all contain 1' as an operator. This means that every site must contain both a moment and a reversed moment (↑ and ↓ giving ↕). This is only possible if the time-averaged moment on the site is zero, and gray groups will describe a paramagnetic site. Some trivial and black-white point groups also require zero moment. For example, in $4'/m$ a spin can't lie along the $4'$ axis as the operation would reverse its direction; equally if it were perpendicular to the $4'$ (in the mirror plane) its direction would be reversed by the mirror. $4'/m$ is therefore said to be a nonadmissible magnetic point group. In total there are 31 admissible magnetic point groups. These are listed along with the restrictions they impose on spin direction in Table 9.1.

Table 9.1: The 31 admissible magnetic point groups and the restrictions they impose on spin direction. After Cracknell, 1975.

Type	1			Type	3			Direction
1	$\bar{1}$							Any
				m'				Any in plane
				$2'$	$2'/m'$			$\perp 2'$
m				$mm'm'$	$m'm2'$			$\perp m$
2	$2/m$			$2'2'2$	$m'm'2$			$\parallel 2$
4	$\bar{4}$	$4/m$		$42'2'$	$4m'm'$	$\bar{4}2'm'$	$4/mm'm'$	$\parallel 4$ or $\bar{4}$
3	$\bar{3}$			$32'$	$3m'$	$\bar{3}m'$		$\parallel 3$ or $\bar{3}$
6	$\bar{6}$	$6/m$		$62'2'$	$6m'm'$	$\bar{6}m'2'$	$6/mm'm'$	$\parallel 6$ or $\bar{6}$

We can apply similar considerations to develop the magnetic space groups or Shubnikov groups. In doing so, it is useful to introduce a small amount of the formalism needed to understand the nomenclature used in the literature and in TOPAS input files. The 1651 magnetic space groups (we'll give them symbol **G**) are classified into four different types depending on how they are derived from one of the 230 crystallographic space groups (symbol **F**) as summarized in Table 9.2. For type 1 magnetic space groups none of the operators contain time reversal, and they are equivalent to the corresponding crystallographic space group. We can express this as **G = F**. A type 2 magnetic space group contains 1' as an operator and will have both an unprimed and a primed copy of each operator: **G = F + F1'**. As for magnetic point groups, these are called gray groups and describe paramagnetic materials in zero applied field.

Table 9.2: A summary of different magnetic space group types, and how they might be remembered by those of a similar age to the authors.

Type		G =	Description	Number	Comment
	1	F	colorless or monochrome	230	All operators unprimed
	2	F+F1′	gray or paramagnetic	230	Each operator primed and unprimed
	3	D+(F−D)1′	black-white first kind BW1	674	**D** equi-translation: translations not associated with spin reversal. Magnetic cell equals crystal cell.
	4	D+(F−D)1′	black-white second kind BW2	517	**D** equi-class: translations associated with spin reversal. Magnetic primitive cell larger than crystal primitive cell.

Types 3 and 4 magnetic space groups are slightly more complex. They can both be expressed as $G = D + (F − D)1′$ where D is a subgroup of index 2 of G; that is, D is a space group that contains half the symmetry operations of G. In words, this equation says that the magnetic space group G has half its symmetry operations unprimed (those in subgroup D) and the remaining half primed (those not in D). The distinction between type 3 and type 4 is in the type of subgroup chosen for D. In type 3, D is an equi-translation subgroup, meaning that the lattice translations don't involve spin reversal. Overall, exactly half the point operators are white and half are black so type 3 space groups will have a colored point group. In type 4, D is an equi-class subgroup, meaning that translational symmetry operations are either white or black and exactly half the lattice translations involve spin reversal. Type 4 magnetic space groups are therefore said to have a colored lattice, which means that the primitive magnetic cell will be larger than the primitive crystal cell.

The 1651 magnetic space groups are given labels similar to those used for crystallographic space groups. For type 1 the magnetic space group label is the same as the conventional label. For type 2 the symbol 1′ is appended to the label. For type 3 the generators of the magnetic point group that include time reversal are primed. For type 4 there are two common conventions: the BNS (Belov, Neronov, Smirnova) and OG (Opechowski and Guccione) schemes. In the BNS convention a subscript is appended to the first letter of the crystallographic space group symbol of D to indicate the type of colored lattice. For example, a primitive monoclinic lattice can give rise to P_a, P_b and C_a (see Figure 9.4). This is then followed by the point group generator symbols. No primes are needed as the combination of translation and point symmetries results in each point operator being present with and without time reversal (the magnetic point group is gray). In the OG convention the crystallographic space group of **F** is used for naming, and the equivalent labels are P_{2a}, P_{2b} and P_C. In this convention some of the point group generators need to be primed. As an example, the BNS symbol C_amma (67.509; TOPAS symbol C_amma) and OG symbol P_cmmm' (47.11.357) are the same magnetic space group. The BNS symbol is based on **D** ($Cmma$) and OG symbol on **F** ($Pmmm$). Each magnetic space group is given a serial number of the form $n.m$ (BNS) or $n.m.p$ (OG) and tables relating the different symbols are available online (e.g., at http://stokes.byu.edu/iso/magneticspacegroups.php). TOPAS uses the keyword mag_space_group along with the BNS number. These and the corresponding BNS symbol can be found in the file shubnikovgroups.txt.

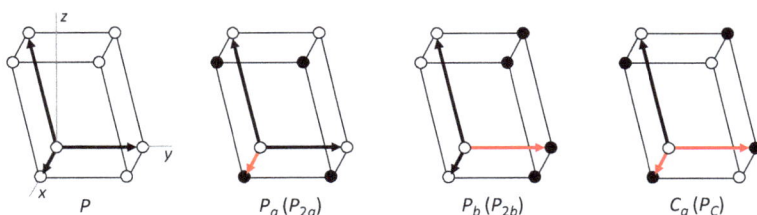

Figure 9.4: Magnetic lattices arising from a primitive monoclinic lattice. Labeled with BNS (OG) symbols. Lattice vectors coupled with time reversal are shown in red.

The crystallographic properties of magnetic space groups have been tabulated in an electronic book "Magnetic Space Groups" that has a similar layout to the regular International Tables. This is freely available online (Litvin, 2008). Part of a representative entry is shown in Figure 9.5 for magnetic space group $P_{2b}m'ma'$ (OG symbol; BNS symbol is P_amna serial 53.330). The figure on the top left shows the lattice diagram that has the conventional cell of **F**. The lattice vectors coupled with spin inversion (here b) are shown in red and those not coupled are shown in black. To the right of this are listed: the short international (Hermann–Maugin) symbol of the magnetic space group, the point group symbol and the crystal system, followed by the serial index and the long international symbol. The figure in the bottom right shows the symmetry elements with primed elements in red. The figure in the bottom left is the general position diagram. Positions at $+z$ are shown in red and $-z$ in blue. Magnetic moments are colored the same as the general positions they are associated with. In-plane moment directions are shown by an arrow and the + or – symbol at the end of the arrow indicates whether it is inclined above or below the z-plane. Note that the lower diagrams don't show the full unit cell for this type 4 magnetic space group. The full cell is, however, easy to visualize as the general position diagram is periodic in unprimed translations (here a and c) but has spins inverted in the primed translation (here b).

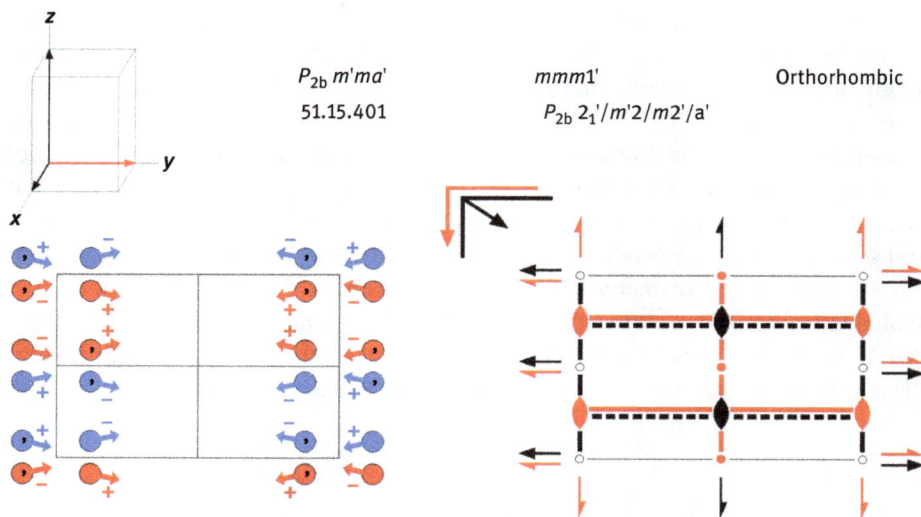

$P_{2b}\,m'ma'$

51.15.401

$mmm1'$

$P_{2b}\,2_1'/m'2/m2'/a'$

Orthorhombic

Multiplicity,
Wyckoff letter,
site symmetry

$(0,0,0)\,+$ $(0,1,0)'\,+$

16	l	1	(1) $x,y,z\,[u,v,w]$	(2) $\bar{x}+1/2,\bar{y},z\,[u,v,\bar{w}]$	(3) $\bar{x},y,\bar{z}\,[\bar{u},v,\bar{w}]$	(4) $x+1/2,\bar{y},\bar{z}\,[\bar{u},v,w]$
			(5) $\bar{x},\bar{y},\bar{z}\,[u,v,w]$	(6) $x+1/2,y,\bar{z}\,[u,v,w]$	(7) $x,\bar{y},z\,[\bar{u},v,\bar{w}]$	(8) $\bar{x}+1/2,y,z\,[\bar{u},v,w]$
8	k	$m'..$	$1/4,y,z\,[0,v,w]$	$1/4,\bar{y},z\,[0,v,\bar{w}]$	$3/4,y,\bar{z}\,[0,v,\bar{w}]$	$3/4,\bar{y},\bar{z}\,[0,v,w]$
8	j	$.m.$	$x,1/2,z\,[u,0,w]$	$\bar{x}+1/2,1/2,z\,[\bar{u},0,w]$	$\bar{x},1/2,\bar{z}\,[\bar{u},0,\bar{w}]$	$x+1/2,1/2,\bar{z}\,[u,0,\bar{w}]$
8	i	$.m.$	$x,0,z\,[0,v,0]$	$\bar{x}+1/2,0,z\,[0,v,0]$	$\bar{x},0,\bar{z}\,[0,v,0]$	$x+1/2,0,\bar{z}\,[0,v,0]$
8	h	$.2.$	$0,y,1/2\,[0,v,0]$	$1/2,\bar{y},1/2\,[0,v,0]$	$0,\bar{y},1/2\,[0,v,0]$	$1/2,y,1/2\,[0,v,0]$
8	g	$.2.$	$0,y,0\,[0,v,0]$	$1/2,\bar{y},0\,[0,v,0]$	$0,\bar{y},0\,[0,v,0]$	$1/2,y,0\,[0,v,0]$
4	f	$m'm'2$	$1/4,1/2,z\,[0,0,w]$	$3/4,1/2,\bar{z}\,[0,0,\bar{w}]$		
4	e	$m'm2'$	$1/4,0,z\,[0,v,0]$	$3/4,0,\bar{z}\,[0,v,0]$		
4	d	$.2/m'.$	$0,1/2,1/2\,[0,0,0]$	$1/2,1/2,1/2\,[0,0,0]$		
4	c	$.2/m.$	$0,0,1/2\,[0,v,0]$	$1/2,0,1/2\,[0,v,0]$		
4	b	$.2/m'.$	$0,1/2,0\,[0,0,0]$	$1/2,1/2,0\,[0,0,0]$		
4	a	$.2/m.$	$0,0,0\,[0,v,0]$	$1/2,0,0\,[0,v,0]$		

Figure 9.5: Extracts of an entry from "Magnetic Space Groups" by Litvin (2008). Reproduced with permission.

Other aspects of the tables are similar to conventional crystallographic tables. For each Wyckoff position the site symmetry is given for the first position quoted in a block of equivalent positions. For each position the symmetry-allowed magnetic moments are also given. For example, in $P_{2b}m'ma'$ Wyckoff site $4c$ has site symmetry .$2/m$. giving the two positions $0,0,\frac{1}{2}$ $[0,v,0]$ and $\frac{1}{2},0,\frac{1}{2}$ $[0,v,0]$ meaning that the moments are required to have equal magnitude (v) and lie along the b-axis for both sites. Equivalent information is included in electronic files for both OG and BNS conventions from the ISO-MAG resource (http://stokes.byu.edu/iso/magneticspa cegroups.php). The Bilbao crystallographic server (http://www.cryst.ehu.es/) also contains an excellent set of online interactive tools for exploring magnetic symmetry.

9.4 Ambiguities in magnetic structures

When performing any Rietveld analysis it is important to remember that the information content in a powder diffraction pattern can be low, and that results should be interpreted with caution. This is particularly true with magnetic data, where one usually relies on a relatively low number of reflections at low $\sin\theta/\lambda$, which are often collected at relatively low resolution, so prone to peak overlap.

There are also a number of cases where neutron diffraction simply can't distinguish between alternate magnetic models. For example, Figure 9.6 shows two very different magnetic structures. In each of them the magnetic cell is quadrupled in one direction relative to the nuclear cell. Figure 9.6a is a structure in which all sites have identical magnetic moments arranged in blocks of two-up then two-down (↑↑↓↓ ...). Figure 9.6b has zero moment on every second site and up/down ordering between magnetic sites (↑0↓0 ...). The magnetic (powder) diffraction pattern from these two arrangements differs only by a scale factor. If the moment in Figure 9.6b is $\sqrt{2}$ larger than in Figure 9.6a, the two configurations are indistinguishable.

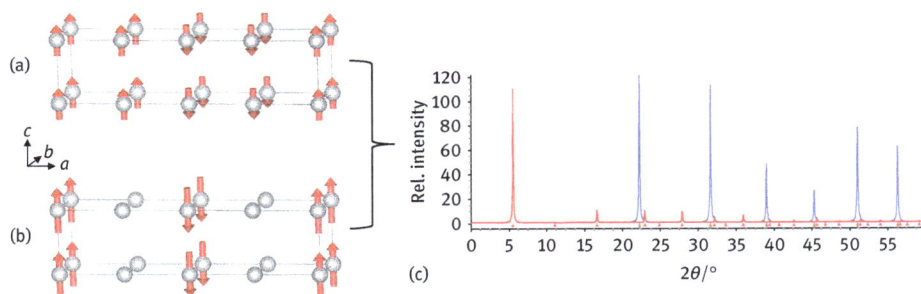

Figure 9.6: Two apparently different magnetic structures (a) and (b) give rise to identical nuclear and magnetic diffraction patterns. The magnetic diffraction is highlighted in red.

Problems also arise in magnetic structures that need more than one propagation vector to describe them (so-called multi-**k** structures). In this language the propagation vector **k** relates the reciprocal space cells of the nuclear and magnetic structures: in Figure 9.6b the relationship would be $\mathbf{k} = \frac{1}{4}\mathbf{a}^*$. In multi-**k** structures the spin configuration can't be unambiguously determined by diffraction data, though introducing symmetry constraints or restricting solutions to those with equal moments on each site can reduce the number of possible solutions.

The difficulties in determining magnetic structures are even greater with powder diffraction data on high-symmetry samples. Since an individual experimental peak may be the contribution of different symmetry-related reflections, information on the direction of moments is lost. As described by Shirane (Shirane, 1959), for collinear magnetic structures this means that the moment direction can't be determined for systems higher than orthorhombic. For cubic systems there is no information available on the moment direction. For tetragonal, rhombohedral and hexagonal systems only the angle of moments relative to the unique c-axis can be determined, and no information is available on their orientation in the ab-plane. Partial peak overlap means that similar ambiguities will arise when analyzing low-resolution powder data on orthorhombic, monoclinic and triclinic samples with only small metric distortions from higher symmetry.

9.5 LaMnO₃ TOPAS worked example

We will explore the various ways in which magnetic diffraction patterns can be analyzed in TOPAS using a simulated set of powder neutron diffraction data for the perovskite $LaMnO_3$ provided by John Evans and Branton Campbell. $LaMnO_3$ is of interest as it is the parent phase of the so-called Giant Magneto Resistive (GMR) perovskites that show a large change in their electrical conductivity under an applied magnetic field – an effect that can be exploited in data storage devices. The ideal perovskite structure of $LaMnO_3$ can be described as an infinite network of corner sharing $MnO_{6/2}$ octahedra with the La in 12-coordinate cuboctahedral coordination by O (Figure 9.7). The real structure undergoes two important distortions relative to this (see Chapter 8 for more discussion on this concept): a Jahn–Teller distortion driven by the d^4 configuration of Mn(III), which leads to octahedra with four shorter and two longer bonds, and cooperative tilts of the octahedra around the axes of the cubic cell (Glazer tilt system $a^+b^-b^-$) such that its space-group symmetry is $Pnma$ (the non-standard $Pbnm$ setting is frequently used in the literature). $LaMnO_3$ also undergoes magnetic ordering below its Néel temperature, T_N, of ~139 K and adopts the so-called A-type structure in which each Mn is ferromagnetically coupled to the four neighboring Mn sites in the ac-plane, but these planes couple antiferromagnetically along the b-axis (Moussa, 1996). The resulting magnetic space group is $Pn'ma'$, number 62.448 in the BNS convention. The magnetic moment on each Mn is around 4 Bohr magnetons (BM).

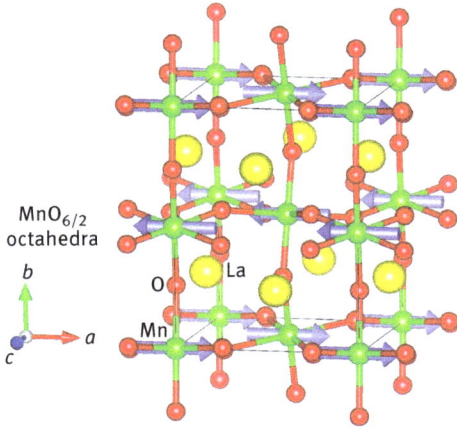

Figure 9.7: Magnetic structure of LaMnO₃.

9.5.1 Single phase nuclear and magnetic Shubnikov approach

The simplest way to perform a magnetic Rietveld refinement is to use a single phase for both nuclear and magnetic contributions and provide TOPAS with the magnetic spacegroup using the key word *mag_space_group*. This is specified as a serial number (here 62.448), and the corresponding symbol (*Pn'ma'*) is stored in the file shubnikovgroups.txt. Atoms that contribute to the magnetic diffraction are flagged with the keywords *mlx*, *mly* and *mlz*, which contain the components of the moment in Bohr Magnetons per Angstrom along each crystallographic axis. It is usually convenient to include the macro *MM_CrystalAxis_Display* that reports the overall moment in Bohr Magnetons.

A simple script for the magnetic refinement is given below. The top panel of Figure 9.8 shows the fit without any magnetic contribution (*mlx* = *mly* = *mlz* = 0) and the lower panel with the magnetic contribution. We can see that the magnetic contribution to the diffraction pattern is largest at low 2θ and decreases rapidly with 2θ due to the form factor.

```
xdd maglamno3.xy
   x_calculation_step = Yobs_dx_at(Xo); convolution_step 4
   bkg @  9.968 -0.002  0.018 -0.024 -0.004 -0.018
   lam ymin_on_ymax 0.0001 la 1.0 lo 1.54 lh 0.5
   neutron_data
   LP_Factor( 90)
   str
      phase_name "LaMnO3"
      mag_space_group 62.448
      r_bragg  0.289311571
      a lpa  5.747293
```

```
b lpb  7.693035
c lpc  5.536693
site La x @  0.04911 y 0.25        z @  0.49217 occ La 1 beq bval  1.00845
site Mn x 0           y 0          z 0            occ Mn 1 beq bval  1.00845
   mlx @ 0.67 mly @ 0.03 mlz @ 0.00 MM_CrystalAxis_Display( 3.87, 0.19, 0.00)
site O1 x @  0.48792 y 0.25        z @  0.57470 occ O  1 beq bval  1.00845
site O2 x @  0.30668 y @  0.03846  z @  0.22619 occ O  1 beq bval  1.00845
scale @  0.0100387974
CS_L(@, 49.82149)
```

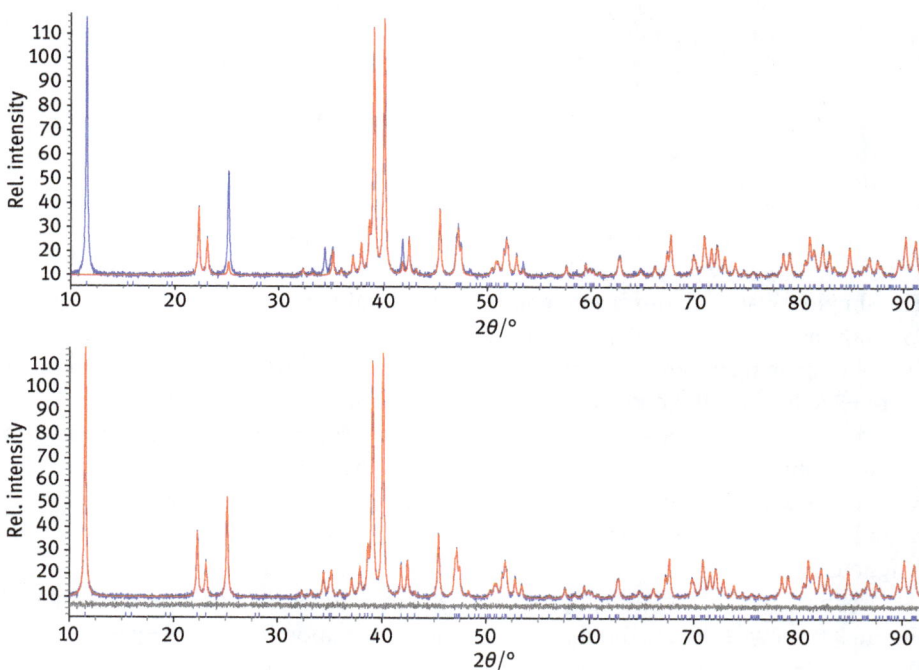

Figure 9.8: Rietveld fits to simulated neutron diffraction data of LaMnO$_3$ with (top) no magnetic scattering contribution and (bottom) magnetic scattering included.

9.5.2 Separate nuclear and magnetic phases

In many cases, magnetic ordering leads to a unit cell that is larger than the nuclear cell, though there are typically minimal changes in the nuclear structure on magnetic ordering. If we adopted the approach in the previous section, then we would need to describe coordinates of all the atoms in this larger cell during refinement (to reliably refine atomic coordinates, it would also be necessary to set up appropriate equations to describe their relationship to the paramagnetic structure). It is often more convenient to describe the magnetic diffraction using a separate magnetic-only phase that contains only the atoms that contribute to the magnetic diffraction. This can be done by including a second phase in the TOPAS INP file flagged with the keyword *mag_only_for_mag_sites* as shown below. In this case, since only the Mn atom carries a magnetic moment, and it lies on the special position (0, 0, 0), we could refine atomic coordinates on La and O sites in the nuclear-only structure alongside the Mn moment. With this approach the scale factor of the magnetic-only phase should be appropriately related to the nuclear phase. In this example, the nuclear and magnetic phases have the same unit cell and scales should be equated. In a general case, the scale of the magnetic-only phase should be $(V_{nuc}/V_{mag})^2$ times that of the nuclear phase (eq. (5.2)).

This approach has a secondary benefit in that the magnetic scattering can be displayed separately from the nuclear scattering as in Figure 9.9. This makes it easy to identify the different contributions to the overall diffraction pattern.

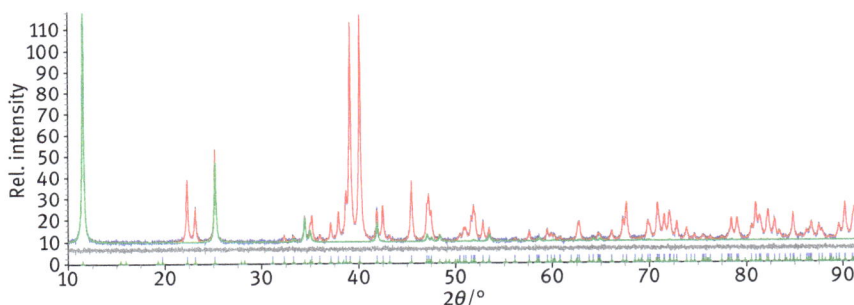

Figure 9.9: Rietveld fit using a second phase to describe magnetic scattering with the keyword *mag_only_for_mag_sites*. Magnetic peaks are highlighted in green.

```
str
    phase_name "LaMnO3_magnetic"
    mag_space_group 62.448
    mag_only_for_mag_sites
    a =lpa_nuc;:5.747294
    b =lpb_nuc;:7.693034
    c =lpc_nuc;:5.536694
    site Mn x 0 y 0 z 0 occ Mn 1 beq bval1  1.00974
        mlx @ 0.67 mly @ 0.02 mlz @ 0.00 MM_CrystalAxis_Display( 3.88, 0.18, 0.00)
    scale =scale_nuclear;:0.0100330659
    CS_L(@, 49.40491)
```

9.5.3 Symmetry modes approach

When the magnetic structure has a larger unit cell or lower symmetry than the nuclear structure, one has to derive a description of the nuclear structure in the magnetic cell and space group. It is also necessary to decide which magnetic moment components $mlx/mly/mlz$ are free to refine by symmetry. Both these processes can be time-consuming and it is very easy to make mistakes. Fortunately, we can use ideas very similar to those discussed in Chapter 8 to help us with this, and describe the magnetic structure using symmetry modes.

With our LaMnO$_3$ example we can use the *Pnma* nuclear structure as the parent structure of the magnetic phase transition. We can upload this to the ISODISTORT suite of software and specify that we want to consider magnetic distortions for the Mn site and atomic distortions for all sites. As the magnetic cell of LaMnO$_3$ is identical to the nuclear cell we can choose the gamma k-point, and select the $m\Gamma_4^+$ irrep. This gives rise to a single choice for the order parameter direction, which uniquely specifies the magnetic space group as 62.448, *Pn'ma'* (we discuss in Section 9.5.4 how to make this choice if you didn't know it). We can then write out a description of the magnetic structure in TOPAS STR format that, after refinement, has the form shown in the table below.

```
str
    phase_name "LaMnO3_magnetic"
    mag_space_group 62.448    'Pn'ma'
    a      5.74730
    b      7.69290
    c      5.53670
    al    90.00000
    be    90.00000
    ga    90.00000
    scale @  0.0100372179
    '{{{mode definitions
    prm !a1 0.00000 min -2.00 max 2.00 'Pnma[0,0,0]GM1+(a)[La:c:dsp] A'_1(a)
    prm !a2 0.00000 min -2.00 max 2.00 'Pnma[0,0,0]GM1+(a)[La:c:dsp] A'_2(a)
    prm !a3 0.00000 min -2.00 max 2.00 'Pnma[0,0,0]GM1+(a)[O1:c:dsp] A'_1(a)
    prm !a4 0.00000 min -2.00 max 2.00 'Pnma[0,0,0]GM1+(a)[O1:c:dsp] A'_2(a)
    prm !a5 0.00000 min -2.83 max 2.83 'Pnma[0,0,0]GM1+(a)[O2:d:dsp] A_1(a)
    prm !a6 0.00000 min -2.83 max 2.83 'Pnma[0,0,0]GM1+(a)[O2:d:dsp] A_2(a)
    prm !a7 0.00000 min -2.83 max 2.83 'Pnma[0,0,0]GM1+(a)[O2:d:dsp] A_3(a)

    prm mm1  7.74131 min -8.00 max 8.00 'Pnma[0,0,0]mGM4+(a)[Mn:a:mag]Ag_1(a)
    prm mm2  0.38675 min -8.00 max 8.00 'Pnma[0,0,0]mGM4+(a)[Mn:a:mag]Ag_2(a)
    prm mm3  0.00285 min -8.00 max 8.00 'Pnma[0,0,0]mGM4+(a)[Mn:a:mag]Ag_3(a)
    '}}}

    '{{{mode-amplitude to delta transformation
    prm  La_1_dx   = +  0.08700*a1;;  0.00000
    prm  La_1_dz   = +  0.09031*a2;;  0.00000
```

```
prm  O1_1_dx   = +  0.08700*a3;:  0.00000
prm  O1_1_dz   = +  0.09031*a4;:  0.00000
prm  O2_1_dx   = +  0.06152*a5;:  0.00000
prm  O2_1_dy   = +  0.04596*a6;:  0.00000
prm  O2_1_dz   = +  0.06386*a7;:  0.00000

prm  Mn_1_dmlx = +  0.08700*mm1;:  0.67349
prm  Mn_1_dmly = +  0.06499*mm2;:  0.02513
prm  Mn_1_dmlz = +  0.09031*mm3;:  0.00026
'}}}

'{{{distorted parameters
prm  La_1_x    =     0.04907 + La_1_dx;:  0.04907
prm !La_1_y    = 1/4;:  0.25000
prm  La_1_z    =     0.49220 + La_1_dz;:  0.49220
prm !Mn_1_x    = 0;:  0.00000
prm !Mn_1_y    = 0;:  0.00000
prm !Mn_1_z    = 0;:  0.00000
prm  O1_1_x    =     0.48787 + O1_1_dx;:  0.48787
prm !O1_1_y    = 1/4;:  0.25000
prm  O1_1_z    =     0.57474 + O1_1_dz;:  0.57474
prm  O2_1_x    =     0.30670 + O2_1_dx;:  0.30670
prm  O2_1_y    =     0.03846 + O2_1_dy;:  0.03846
prm  O2_1_z    =     0.22625 + O2_1_dz;:  0.22625

prm  Mn_1_mlx  = 0  + Mn_1_dmlx;:  0.67349
prm  Mn_1_mly  = 0  + Mn_1_dmly;:  0.02513
prm  Mn_1_mlz  = 0  + Mn_1_dmlz;:  0.00026

prm !La_1_occ  = 1;:  1.00000
prm !Mn_1_occ  = 1;:  1.00000
prm !O1_1_occ  = 1;:  1.00000
prm !O2_1_occ  = 1;:  1.00000
'}}}

'mode-dependent atoms
   site La_1   x = La_1_x;  y = La_1_y;  z = La_1_z;  occ La = La_1_occ;  beq bval1 1.0
   site Mn_1   x = Mn_1_x;  y = Mn_1_y;  z = Mn_1_z;  occ Mn = Mn_1_occ;  beq bval1 1.0
   mlx = Mn_1_mlx; mly = Mn_1_mly; mlz = Mn_1_mlz; MM_CrystalAxis_Display(3.87, 0.19, 0.00)
   site O1_1   x = O1_1_x;  y = O1_1_y;  z = O1_1_z;  occ O = O1_1_occ;  beq bval1  1.0
   site O2_1   x = O2_1_x;  y = O2_1_y;  z = O2_1_z;  occ O = O2_1_occ;  beq bval1  1.0
   CS_L(@, 49.82248)
```

The format of this structural description should be familiar from Chapter 8. The parameters $a1–a7$ describe displacive mode amplitudes. By refining them, the atomic coordinates will shift from the parent positions as allowed by the space group symmetry. Refining $a1–a7$ is equivalent to refining 7 independent xyz parameters in a conventional crystallographic description (the seven structural @ symbols in Section 9.5.1). The parameters $mm1–mm3$ describe magnetic mode amplitudes. The equations in the "mode-amplitude to delta transformation" section show how the changes in *a1–a7* and *mm1–mm3* determine the shifts of atomic coordinates and the change in magnetic moments from zero along the different crystallographic axes. The shifts in both coordinates and magnetic moments are then transformed to actual values in the "distorted parameters" section (actual = parent + shift) and the resulting parameters are used to describe the structure in the "mode-dependent sites" section. During most refinements many of these equations can be ignored and they are therefore hidden away in jEdit '{{{...'}}} folds by default.

In the $LaMnO_3$ example, we find that parameter *mm1* refines to a large value (7.74) and that it describes a magnetic moment of 3.87 BM along the *a*-axis. The fit is 100% equivalent to those shown in Figures 9.8 and 9.9. The values of *mm2* and *mm3* are close to zero and fixing them to zero has no impact on the quality of the fit. This tells us that although moments are allowed to have components along *b* and *c* by symmetry, the (simulated) data shows that these components are negligible.

9.5.4 Symmetry modes – determining the Shubnikov group

If you're working on a material like a perovskite, you may have a good idea of possible magnetic structures and their symmetry from literature analogues. It is then relatively straightforward to test these different models against your data. In other cases, however, you may have no idea of the possible magnetic structures and their symmetries. In these cases symmetry modes offer one possible way of solving the magnetic structure.

Let's assume that we have successfully identified a unit cell that will describe all the magnetic diffraction peaks and that it is a simple multiple (e.g., a $2a \times 1b \times 1c$) of the nuclear cell. We can then use ISODISTORT to produce a description of the magnetic structure in space group $P1$ (BNS number 1.1). We can then refine all of the allowed magnetic modes in TOPAS to see if it is possible to get a good fit to the magnetic diffraction. If we can, this tells us that this $P1$ model has sufficient degrees of freedom to describe the magnetic structure. It will, however, almost certainly have too many degrees of freedom and in fact be a subgroup description of the true structure. If we can choose which of the magnetic modes are actually required to fit the data, and which can be fixed at 0.0 without degrading the fit, then this will tell us about the true magnetic symmetry.

If we take the case of $LaMnO_3$, we can use a method analogous to that in the previous section to produce an STR description of the magnetic structure in magnetic space group 1.1. We find there are four unique Mn atoms in the unit cell, which means there are 12 possible magnetic modes (3 degrees of freedom on each Mn, described by modes *m1* to *m12*) that belong to four different irreps ($m\Gamma_1^+$, $m\Gamma_2^+$, $m\Gamma_3^+$ and $m\Gamma_4^+$). If we refine all of these mode amplitudes, we find that we get an excellent fit to the diffraction data with R_{wp} = 3.927%. However, on inspecting the output file we find that one of the amplitudes – *m10, Pnma[0,0,0]mGM4+(a)[Mn:a]Ag_1(a)* – is significantly larger than the others as in the table below.

```
prm m1   -0.12234  'Pnma[0,0,0]mGM1+(a)[Mn:a]Ag_1(a)
prm m2    0.02348  'Pnma[0,0,0]mGM1+(a)[Mn:a]Ag_2(a)
prm m3    1.93525  'Pnma[0,0,0]mGM1+(a)[Mn:a]Ag_3(a)
prm m4    0.08904  'Pnma[0,0,0]mGM2+(a)[Mn:a]Ag_1(a)
prm m5   -0.06189  'Pnma[0,0,0]mGM2+(a)[Mn:a]Ag_2(a)
prm m6   -0.64994  'Pnma[0,0,0]mGM2+(a)[Mn:a]Ag_3(a)
prm m7   -0.27857  'Pnma[0,0,0]mGM3+(a)[Mn:a]Ag_1(a)
prm m8    0.02419  'Pnma[0,0,0]mGM3+(a)[Mn:a]Ag_2(a)
prm m9    0.05751  'Pnma[0,0,0]mGM3+(a)[Mn:a]Ag_3(a)
prm m10  -7.50541  'Pnma[0,0,0]mGM4+(a)[Mn:a]Ag_1(a)
prm m11  -0.01403  'Pnma[0,0,0]mGM4+(a)[Mn:a]Ag_2(a)
prm m12  -0.31751  'Pnma[0,0,0]mGM4+(a)[Mn:a]Ag_3(a)
```

If we now reset the amplitude of all the magnetic modes to zero and fix them, then refine only the amplitude of *m10* we get a fit with R_{wp} = 3.92%, which is essentially identical to that with all the modes turned on. This tells us that the magnetic ordering can be described using just the $m\Gamma_4^+$ irrep. It is usual for magnetic ordering transitions to involve a single irrep. If we return to ISODISTORT, we can upload the parent structure, superpose the $m\Gamma_4^+$ irrep and find that there is a single choice of order parameter direction that leads to space group 62.448 *Pn'ma'*, and gives us the correct high symmetry description of the structure that we used in earlier sections. We have therefore managed to solve the structure and determine the magnetic space group.

For more complex examples, there are ways of automatically testing which modes are required. Some of these are explored in the online tutorial at http://community.dur.ac.uk/john.evans/topas_workshop/tutorial_GA_magnetic.htm.

References

Bacon, G.E. (1975): *Neutron diffraction*, Clarendon Press, Oxford, UK, 636 pages.

Belov, N.V., Neronova, N.N. & Smirnova, T.S. (1955): *The 1651 Shubnikov groups*. Trudy Inst. Kristallogr. Acad. SSSR 11, 33–67.

Belov, N.V., Neronova, N.N., Smirnova,T.S. (1957): *Shubnikov groups*. Sov. Phys. Crystallogr. 2, 311–312.

Cracknell, A.P. (1975): *Magnetism in crystalline materials*, Pergamon Press.

Heesch, H. (1930): *Über die vierdimensionalen Gruppen des dreidimensionalen Raumes*. Z. Kristallogr. 73, 325–345.

Litvin, D.B. (2008): *Tables of crystallographic properties of magnetic space groups*. Acta. Cryst. A64. 419–424; supplementary information PZ5052. http://www.bk.psu.edu/faculty/Litvin/download. html.

Moussa, F., Hennion, M., Rodriguez-Carvajal, Moudden, J.H., Pinsard, L., Revcolevschi, A. (1996): *Spin waves in the antiferromagnet perovskite LaMnO₃: a neutron-scattering study*. Phys. Rev. B 54, 15149–15155.

Opechowski, W., Guccione, R. (1965): *Magnetism*, edited by G.T. Rado and H. Suhl, Vol. 2A, ch.3, New York, Academic Press.

Shirane, G. (1959): *A note on the magnetic intensities of powder neutron diffraction.* Acta Cryst. 12, 282–285.

Shubnikov, A.V. (1951): *Symmetry and antisymmetry of finite figures.* In Russian, USSR Academy of Sciences, Moscow.

Von Dreele, R.G., Rodriguez-Carvajal, J. in Dinnebier, R., Billinge, S.J.L. (eds.) (2008): *Powder diffraction: theory and practice*, RSC publication, Cambridge (UK), 574 pages.

Wills, A.S. (2017): *A historical introduction to the symmetries of magnetic structures. Part 1. Early quantum theory, neutron powder diffraction and the coloured space groups.* Powder Diffraction, 32, 148–155.

Zamorzaev, A.M. (1953): *Dissertation*, Leningrad State University, Russia .

Zamorzaev, A.M. (1957a,b): *Generalisation of Federov groups.* Kristallografiya 2, 15–20 or Sov. Phys. Crystallogr. 2, 10–15.

10 Stacking disorder

10.1 Introduction

Many crystal structures contain two-dimensional layered structural motifs (atoms, ions or molecules) periodically repeated in the third dimension, giving the crystal a distinct stacking sequence. In many real crystals (including technologically important materials like battery electrodes and alloys), the stacking order can deviate from ideal periodicity. The stacking sequence becomes nonuniform, and a microstructure is created. Irregular stacking of layers is typically connected with the occurrence of what are called *stacking faults*, and can have a huge impact on the diffraction pattern. In this chapter we will discuss how this type of structural defect can be modeled in TOPAS.

10.2 Description of stacking orders

Two basic types of packings are possible for close-packed spheres: cubic close packing (*ccp*) and hexagonal close packing (*hcp*). Both packing types can be distinguished by their stacking sequence. In a *ccp*-structure that contains only one atom type (e.g., copper metal), copper metal, the layers are stacked perpendicular to the [111] direction in an $\alpha\beta\gamma$-stacking sequence, with metal position indicated by small Greek letters (Figure 10.1, left). A *hcp*-structure, (e.g., magnesium metal), exhibits layers stacked perpendicular to the [001] directions with an $\alpha\beta\alpha\beta$-stacking sequence (Figure 10.1, right).

A useful description of a stacking sequence can be achieved by using stacking vectors, which describe the transition from a given layer to the subsequent one (Figure 10.1 magenta and blue arrows). For convenience the unit cells (Figure 10.1 black lines), with lattice parameters a, b, c, α, β, γ are transformed into pseudo-orthorhombic or pseudo-trigonal unit cells, with lattice parameters a', b', c', $\alpha' = \beta' = 90°$, $\gamma' = 90°$ or $120°$, in which the layers are perpendicular to the c'-axis (Figure 10.1 dashed gray lines). The stacking vectors, **Si**, are given in fractional coordinates of the transformed unit cell. Each stacking vector consists of three components (eq. (10.1)): S_{xi}, S_{yi} and S_{zi}, with S_{xi} and S_{yi} describing the shift within the $a'b'$-plane and S_{zi} describing the shift parallel to c':

$$\mathbf{Si} = \begin{pmatrix} S_{xi} \\ S_{yi} \\ S_{zi} \end{pmatrix}. \tag{10.1}$$

A *ccp*-structure can be described by using only one stacking vector, **S1**, whereas a *hcp*-structure can be constructed by using two stacking vectors, **S1** and **S2**, that are repeated in an alternating fashion (Figure 10.1, magenta, blue):

https://doi.org/10.1515/9783110461381-010

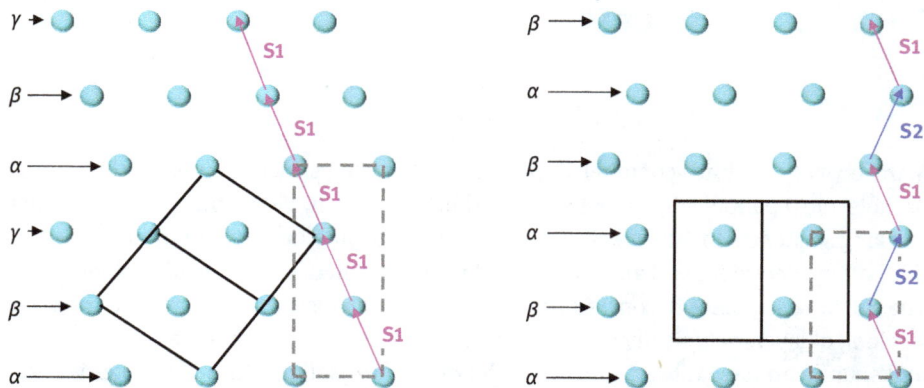

Figure 10.1: Stacking sequence in *ccp* (left) and *hcp*-packing (right). Unit cell edges are displayed in black and transformed unit cells in gray dashed lines. The stacking order is indicated by stacking vectors **S1** and **S2** describing layer to layer transitions. Positions of metal atoms are indicated by lowercase Greek letters.

$$ccp: \mathbf{S1} = \begin{pmatrix} 1/3 \\ -1/3 \\ 1/3 \end{pmatrix} \quad hcp: \mathbf{S1} = \begin{pmatrix} 1/3 \\ -1/3 \\ 1/2 \end{pmatrix}, \quad \mathbf{S2} = \begin{pmatrix} -1/3 \\ 1/3 \\ 1/2 \end{pmatrix}. \tag{10.2}$$

Many inorganic materials, like NaCl or MgO, also exhibit close packed struc-tures. In these structures layers of metal cations, denoted by small Greek letters and layers of anions, denoted by capital Roman letters, are arranged in an alternating fashion (Figure 10.2, left). Layered structures like $Mg(OH)_2$ (brucite) or $CdCl_2$ can be derived directly from close packings by removing half of the cation layers (Figure 10.2, right). As a consequence the stacking sequence trans-forms from $Ca B\gamma A\beta Ca B\gamma A\beta$ ... (NaCl-type) to $Ca B\square A\beta C\square B\gamma A\square$ ... ($CdCl_2$-type) with "$\square$" indicating a cation vacancy. Usually the vacancies are not explicitly shown and layers are indicated by round brackets. Therefore the stacking order in a $CdCl_2$-type lattice can be expressed as: $(Ca B)(A\beta C)(B\gamma A)$. Like close packed structures, layered structures exhibit several types of packings. In Figure 10.3 three examples are shown: C19-type ($CdCl_2$-type), C6-type (CdI_2/brucite-type) and CrOOH-type (sometimes denoted a 3R-type). There are many more basic stacking types in other layered compounds. For a good overview see, for example, Hulliger (Hulliger, 1976).

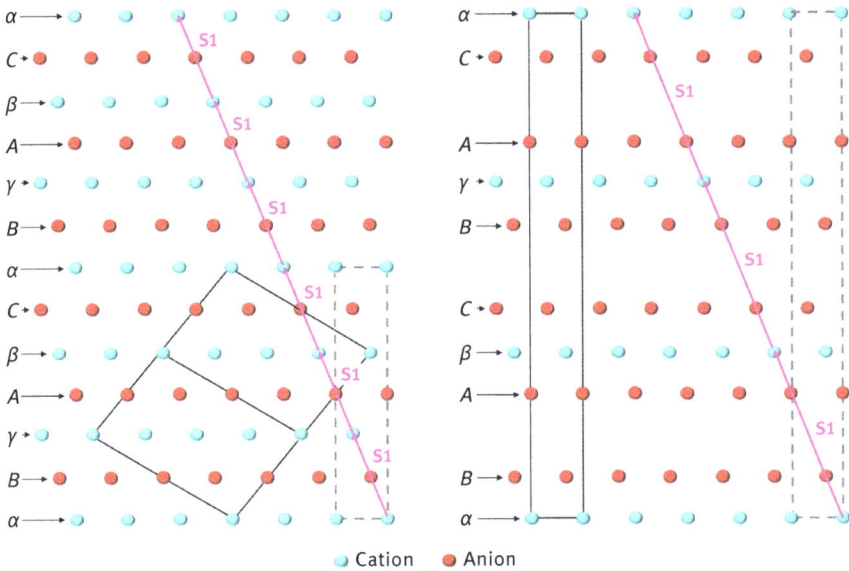

Figure 10.2: Stacking sequence in a binary MX *ccp*-structure (left) and in a related MX_2 structure. Unit cell edges are displayed as black solid lines, edges of transformed unit cells in gray dashed lines. The stacking order is indicated by stacking vectors **S1** describing layer to layer transitions. Positions of metal cations are indicated by lowercase Greek letters, anion positions are indicated by capital Roman letters.

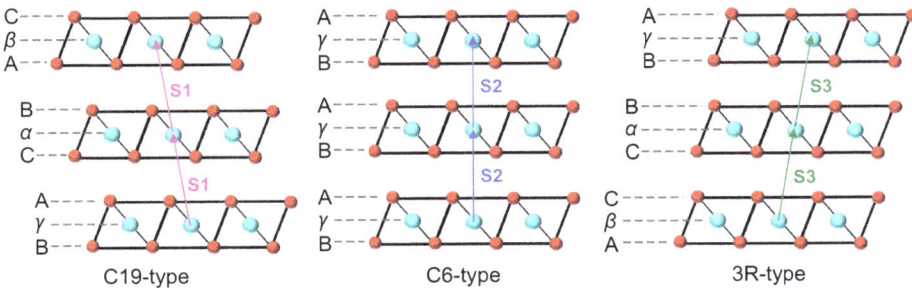

Figure 10.3: Examples of stacking types in layered structures. $MX_{6/3}$-octahedra with M = metal cation and X = anion are indicated by black solid lines.

10.3 Types of stacking faults

Several types of stacking faults can occur in a structure based on close packed spheres. In a *ccp*-structure the stacking order can have two different orientations: $\alpha\beta\gamma$ (stacking vector **S1**, Figure 10.4, top left) and $\gamma\beta\alpha$ (stacking vector **S2**). When two differently oriented crystals are intergrown, the stacking vector switches from **S1** to

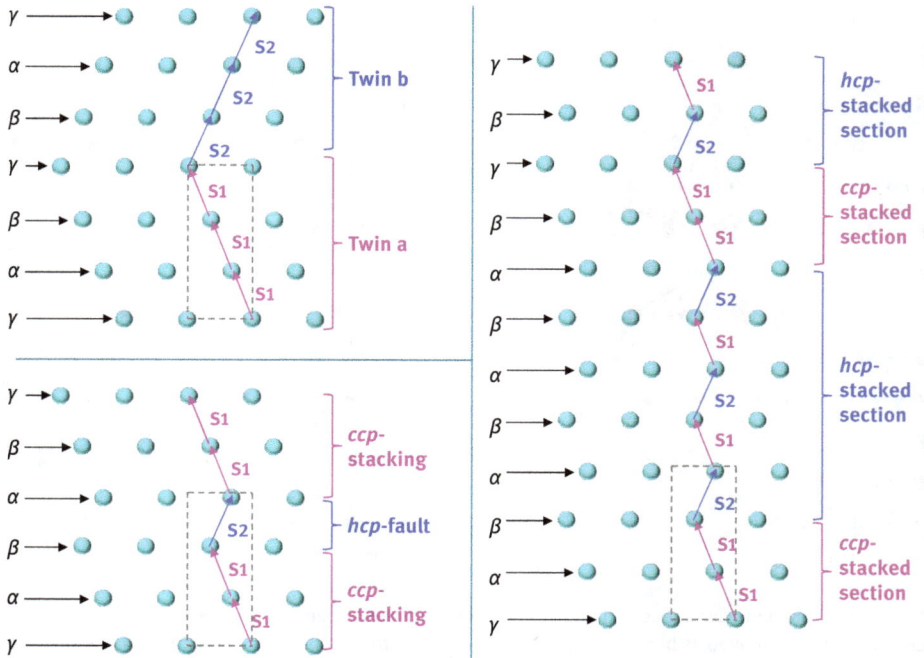

Figure 10.4: Types of stacking faults. Transitions between different stacking vectors leading to: twinning (top left), a single *hcp*-fault in a *ccp*-stacking (bottom left) and crystallographic intergrowth of *ccp*- and *hcp*-stacked sections (right).

S2 at the interface. This interface can be considered as a *stacking fault*. The switch of the stacking vectors creates a local *hcp*-like stacking, for example, *βγβ* (Figure 10.4, top left) in the stacking order. This type of intergrowth can be also described as *twinning*.

If the stacking vector switches from **S1** to **S2** and afterwards back to **S1**, then *ccp*-like packing is interrupted by a *hcp*-like transition (Figure 10. 4, bottom left). This can be considered as a *local stacking fault*.

Sometimes local faults exhibit a certain extension, for example, after several *ccp*-like layer-to-layer transitions more than one *hcp*-like transition occurs before the stacking order switches back to *ccp*-stacking (Figure 10.4, right). Hence distinct structural motifs are *crystallographically intergrown* and the interfaces between the intergrown sections can be considered as stacking faults.

In each of these cases the stacking vector can be described by a *finite* array of distinct vectors. For close packed spheres, this array contains only two vectors[1]:

---

1 $S_z = 1/3$ assumes c' is three times the layer spacing.

$$\mathbf{S} = \begin{pmatrix} S_x \\ S_y \\ S_z \end{pmatrix} \in \left\{ \mathbf{S1} = \begin{pmatrix} -1/3 \\ 1/3 \\ 1/3 \end{pmatrix}, \ \mathbf{S2} = \begin{pmatrix} 1/3 \\ -1/3 \\ 1/3 \end{pmatrix} \right\}. \tag{10.3}$$

These types of faulting can be found, for example, in highly alloyed TRIP steels (Martin et al., 2011).

The same considerations can be applied to layered structures. The main difference is that far more than two basic packing types exist. If, for example, the packing types presented in Figure 10.3 form a microstructure, the array that describes the stacking vector consists of three vectors:

$$\mathbf{S} = \begin{pmatrix} S_x \\ S_y \\ S_z \end{pmatrix} \in \left\{ \mathbf{S1} = \begin{pmatrix} -1/3 \\ 1/3 \\ 1/3 \end{pmatrix}, \ \mathbf{S2} = \begin{pmatrix} 0 \\ 0 \\ 1/3 \end{pmatrix}, \ \mathbf{S3} = \begin{pmatrix} 1/3 \\ -1/3 \\ 1/3 \end{pmatrix} \right\}. \tag{10.4}$$

If more packing types contribute to the microstructure of a layered compound, the array of possible stacking vectors will be larger. If the structure remains closely related to a close packing, then this array will always be finite.

In layered structures with a large interlayer spacing, weak interlayer interactions or with a high $r_{cation} : r_{anion}$ ratio, the stacking order of the layers often deviates from close packing. Hence layers can be randomly displaced from their ideal position by shifts within the layer plane or by rotation within the layer plane, leading to *turbostratic* like disorder (Figure 10.5). Accordingly a random component, Δ with Δ_x, Δ_y and Δ_z, can be added to the stacking vector. As turbostratic disorder doesn't affect the interlayer spacing, $\Delta_z = 0$:

$$\mathbf{S} = \mathbf{Si} + \mathbf{\Delta} = \begin{pmatrix} S_x \\ S_y \\ S_z \end{pmatrix} + \begin{pmatrix} \Delta_x \\ \Delta_y \\ 0 \end{pmatrix} \text{ with } 0 < \Delta_x, \ \Delta_y < 1 \text{ and } \Delta_x, \ \Delta_y \in \{\mathbb{R}\}. \tag{10.5}$$

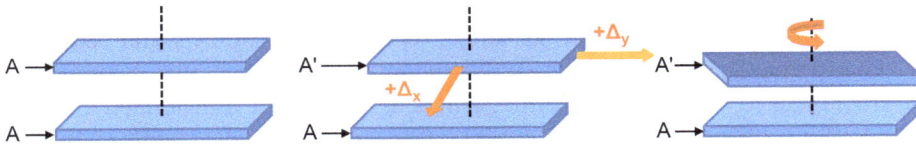

Figure 10.5: Ideal stacked layers (left); top layer randomly dislocated by a shift in the ab-plane (middle); and top layer randomly dislocated by a rotation in the ab-plane.

Since the x and y components can assume all real values between 0 and 1, the stacking is described by an *infinite* array of distinct vectors.

Some layered compounds can take up atoms, ions or molecules in-between the sheets in a process called intercalation. Intercalation can be inhomogeneous, that is,

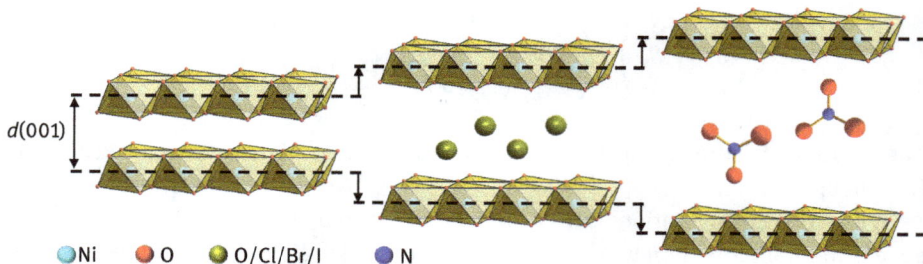

Figure 10.6: Disorder in α/β-Ni(OH)$_2$ by intercalation of ions and molecules ≡ interstratification.

some layers are intercalated and others are not, or homogenous. Intercalation is usually accompanied by an increase of the interlayer distance (Figure 10.6). When the distribution of intercalated particles is random or inhomogeneous then intercalation appears as a form of stacking faulting that is called *interstratification*. Interstratification affects the *z*-component of the stacking vector and often goes along with turbostratic-like disorder in materials such as brucite-type magnesium and nickel hydroxides (Radha et al., 2003; Ramesh et al., 2003; Ramesh et al., 2008).

10.4 Modeling the influence of stacking faults on diffraction

One of the most fundamental and significant theoretical works on the effects of stacking disorder on XRPD patterns was carried out by B.E. Warren (Warren, 1941). He considered complete random stacking (turbostratic disorder) in a layered structure with $\Delta_z = 0$. As rotations of layers (Figure 10.5, right) don't cause any additional reflections in an XRPD pattern, considering only random translations (Figure 10.5, right) is sufficient. In his work, Warren included only sharp 00l and diffuse hk0 reflections. The 2θ-dependent reflection intensities (i.e., the broadened reflections) were derived from the Debye formula. For this purpose, three vectors are introduced: vectors $\mathbf{a_1}$ and $\mathbf{a_2}$, describing the scheme of repetition in each layer and vector $\mathbf{a_3}$, which is perpendicular to $\mathbf{a_1}$ and $\mathbf{a_2}$ and therefore describes separation between the layers (Figure 10.7, red arrows). Lattice points describing the transformed unit cell with lattice parameters a', b', c', α', β' and γ' (see above) are indicated by m_1 and m_2 and an individual layer is denoted by the layer number m_3 (Figure 10.7, green numbers).

The position of an atom, n, in layer m_3 can be described by the vector $\mathbf{R_n}(m_3)$ that includes the basis vector, $\mathbf{r_n}$, that contains the fractional coordinates x', y', z', for the atoms n (Figure 10.7, blue balls and arrows):

$$\mathbf{R_n}(m_3) = m_1\,\mathbf{a_1} + m_2\,\mathbf{a_2} + m_3\,\mathbf{a_3} + \Delta x(m_3)\,\mathbf{a_1} + \Delta y(m_3)\,\mathbf{a_2} + \mathbf{r_n}. \qquad (10.6)$$

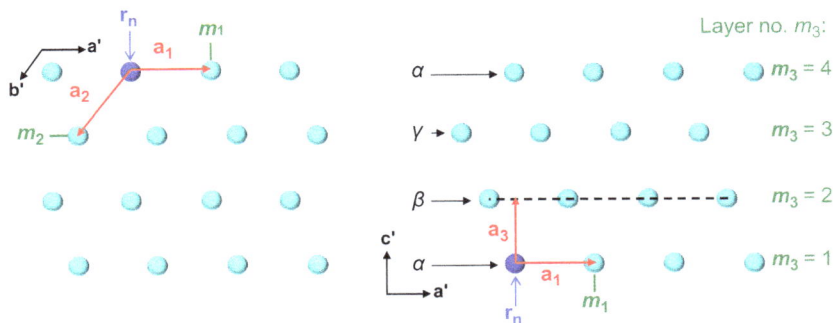

Figure 10.7: Description of the crystal structure of *ccp*-Cu in the notation used by Warren (Warren, 1941), m_1 and m_2 are lattice points, m_3 denotes the number of an individual layer, $\mathbf{a_1}$ and $\mathbf{a_2}$ are vectors describing the scheme of repetition in each layer, $\mathbf{a_3}$ is a vector describing the separation between the layers, $\mathbf{r_n}$ is the basis vector for atoms of type n.

The intensity for a particular scattering vector $\mathbf{s} = (hkl)$ is then given by:

$$
I_s = \left| \sum_n \left[f_n(s) e^{2\pi i \left(hx'_n + ky'_n + lz'_n \right)} \right] \right|^2
$$
$$
\cdot \frac{\sin^2(\pi N_1 a_1 h)}{\sin^2(\pi a_1 h)} \cdot \frac{\sin^2(\pi N_2 a_2 k)}{\sin^2(\pi a_2 k)} \tag{10.7}
$$
$$
\cdot \left| \sum_{m_3} \left[f_{m_3}(s) e^{2\pi i \left(h\Delta x(m_3)a_1 + k\Delta y(m_3)a_2 + lm_3 a_3 \right)} \right] \right|^2
$$

where $N_1 a_1$ and $N_2 a_2$ are the dimensions of the layers, which is assumed to be a parallelogram.

For $hk0$ reflections, the randomness does not play any role and the intensity is that of a crystalline reflection. The phase factors of these reflections are, however, completely random in the summation over m_3 resulting in incoherent scattering from the individual layers (Figure 10.8, left). For a two-dimensional layer the reciprocal lattice becomes a series of parallel lines perpendicular to the layer, called truncation rods (Warren, 1941) (Figure 10.8, middle). For stacking along $\mathbf{c^\star}$, these rods intersect the plane at the $hk0$ points. This leads to aniso-tropic broadening of the $hk0$ reflections. Warren demonstrated that these reflec-tions are always broadened toward higher diffraction angles leading to characteristic Warren-type triangular or saw-tooth peak shapes (Figure 10.8, right). The unaffected $00l$ reflections remain sharp.

One powerful tool to simulate and visualize the effects of stacking faults on diffraction patterns is the DIFFaX software package (Treacy et al., 1991). Simulations can be carried out in DIFFaX by using transformed unit cells of the type described above with layers stacked perpendicular to the c-axis. Faulted structures are generated

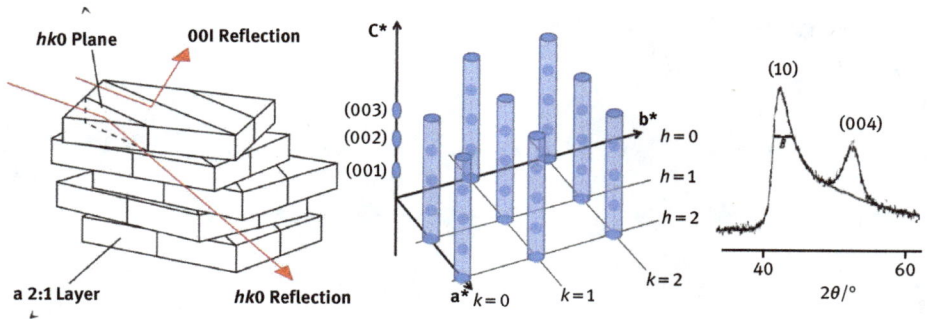

Figure 10.8: Left/middle: real and reciprocal space of a turbostratically disordered layer structure (Moore et al., 1997; Ufer et al., 2012). Right shows a small region of the diffraction pattern of carbon black (adapted from Briscoe and Warren, 1942).

by defining individual layers that can be chemically identical or different. Transitions from one layer to the next are described by using stacking vectors (e.g., eq. (10.1)) with associated probabilities. Patterns can be simulated using two fundamental program modes: *explicit* – an XRPD pattern from *one* pre-defined or random sequence of layers and stacking vectors is simulated; or *recursive* – an averaged diffraction pattern is calculated from *all* possible stacking sequences consistent with a *transition probability matrix*.

Transition probabilities used to describe the degree of faulting are expressed through a square probability matrix. In this matrix the elements P_{ij} describe the probability of layer i being followed by layer j. The description is often set up so the diagonal terms P_{ii} correspond to the probability of a faultless stacking and off-diagonal terms describe a stacking fault, though this doesn't have to be the case. An example of a transition probability matrix including three stacking vectors is given in Table 10.1:

Table 10.1: Example of a 3 × 3 transition probability matrix for a 3-layer system ($P_{n1} + P_{n2} + P_{n3} = 1$).

transition from↓/to→	layer 1	layer 2	layer 3
layer 1	P_{11}	P_{12}	P_{13}
layer 2	P_{21}	P_{22}	P_{23}
layer 3	P_{31}	P_{32}	P_{33}

When generating stacking sequences the position/type of the preceding layer is always taken into account. Hence, the transition probability matrix can be non-symmetric. The whole approach can be considered as *recursive*. As a statistical constraint, all transition probabilities within each row of the matrix must sum up to unity:

$$\sum_{j=1}^{n} P_{ij} = 1. \tag{10.8}$$

A demonstration of the two DIFFaX program modes is given in Figure 10.9. Diffraction profiles of polycrystalline copper with faultless *ccp*-stacking (black line) and a hexagonal intergrowth (colored lines) were simulated by applying a 2×2 transition probability matrix with $P_{12} = P_{22} = 0.02$ (implying $P_{11} = P_{21} = 0.98$) and stacking vectors as given in eq. (10.3). For the simulations, the *ccp* unit cell of copper was transformed to a hexagonal one with c' along the cubic [111] direction. Simulations in recursive mode using 500 layers show that the 003 reflection (111 in the original cubic unit cell) is slightly broadened and is less intense (Figure 10.9, green line) than a unfaulted sample. In addition, a weak feature appears at 40.6° 2θ that can be indexed as the 010 reflection of *hcp*-stacked copper, suggesting small sections of hexagonal intergrowths. When an equivalent simulation is carried out in the explicit mode with a single stack of 5000 layers (blue line) the pattern is essentially identical. This shows that the statistics of a 5000 layer stack is equivalent to the array of weighted stacking sequences that is modeled in the recursive approach. If, however, the number of layers is reduced from 5000 to 500, significant "ripples" appear in the pattern (red line). Each time a 500 layer stack is calculated, slightly different ripples appear due to the slightly different supercell

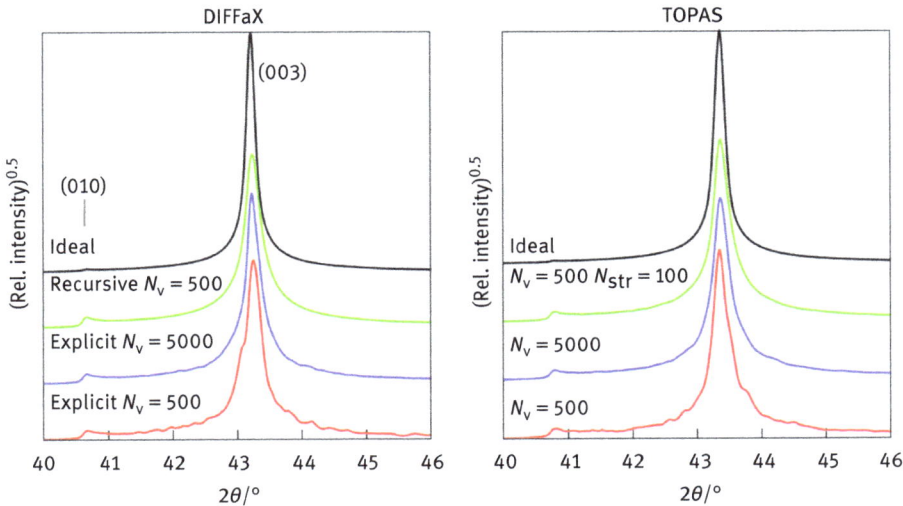

Figure 10.9: DIFFaX and TOPAS simulations of diffraction profiles for *ccp*-Cu. Top to bottom: unfaulted *ccp*-packing (black), 2% hexagonal intergrowth 500 layer recursive calculation (green), 5000 layers explicit (blue) and 500 layers explicit (red). Corresponding TOPAS simulations on the right. Hexagonal intergrowths cause the appearance of a peak close to the *ccp*-forbidden (010) reflection. All indices are given in the notation of the transformed hexagonal unit cell.

generated. If a series of these patterns is averaged the resulting pattern closely approximates the recursive or 5000 stack simulations.

10.5 TOPAS implementation

While DIFFaX provides a straightforward simulation tool for single-phase systems, it doesn't offer the power of a full Rietveld approach. There have been a number of Rietveld-compatible stacking-fault methods published, each with their own advantages and disadvantages (e.g., Leoni et al., 2004; Casas-Cabanas et al., 2016; Reynolds, 1985; Ufer et al., 2004; Wang et al., 2012). Other approaches to derive the layer constitution from vastly stacking faulted samples exist, for example, by using two-dimensional projections of the crystal structure and geometric considerations (Leineweber et al., 2012).

TOPAS v6 and onwards (Coelho et al., 2016) contains powerful tools for DIFFaX-like modeling of stacking-faulted patterns using a *super cell approach*. This allows stacking faults to be analyzed alongside all the other features of TOPAS. An extended unit cell is defined that contains a large (several 100 to several 1000) number, N_v, of layers that are stacked in the c'-direction and individual faults can be built into the supercell. With a sufficiently large c'-axis, the large number of *hkl* reflections of the supercell sum to give an excellent approximation of the broadened smooth peak shapes observed experimentally. As an example, diffraction profiles of polycrystalline copper with faultless *ccp*-stacking and hexagonal intergrowths have been simulated using TOPAS and are shown alongside the DIFFaX simulations in Figure 10.9. To achieve this the space group was reduced to *P*1 and the unit cell of copper was transformed as described in Table 10.2[2]: As copper is situated at the origin in both the original and the transformed unit cell, the fractional coordinates don't need to be transformed.

The simplest TOPAS process for simulating stacking-faulted XRPD patterns is comparable to the explicit mode of DIFFaX. First of all the number of generated layers in a stacked sequence, and therefore the extension of the unit cell, has to be defined:

```
number_of_stacks_per_sequence Nv 500 ' Generation of a supercell including 500 layers
c  = Get(generated_c);:0           ' Works out c-axis from number of layers and heights
```

The keyword *layer* identifies a site as belonging to a *layer* called *$layer_name*, which is A in our case. Here, this first layer consists of only one copper atom located at the

---

2 Note that the same transformation must be carried out for DIFFaX simulations.

Table 10.2: Transformation of the *ccp*-structure of copper to a pseudo-hexagonal supercell.

Copper	Original unit cell	Transformed unit cell
Space group	$Fm\bar{3}m$	$P1$
Unit cell	$a = b = c = 3.62\ \text{Å}$ $\alpha = \beta = \gamma = 90°$	$a' = b' = \frac{1}{\sqrt{2}}a = 2.56\ \text{Å};$ $c' = \frac{\sqrt{3}}{3}a = 2.09\ N_v\ \text{Å}$ $\alpha = \beta = 90;\ \gamma = 120$
Volume	$47.6\ \text{Å}^3$	$\frac{1}{3}\cdot N_v \cdot 47.6\ \text{Å}^3$
Stacking direction	[111]	[001]
Number of layers	3	N_v

origin. In general, the z-coordinate of the atomic sites belonging to a layer must be divided by the total number of layers, N_v:

```
site Cu1 x = 0; y = 0; z = 0.0/Nv; occ Cu 1 beq 0.2 layer A
```

TOPAS has certain rules that govern the behavior of sites marked with the *layer* keyword: a site marked with *layer* cannot take part in restraints and it is not displayed in the *view_structure* window. These can be overcome using the keyword *generate_these*.

Next the stacking vector (eq. (10. 1)) is defined as a transition from one layer to the subsequent one. For creating a pure *ccp*-type stacking only one layer type and one stacking vector (Figure 10.1, left) is necessary:

```
prm !h 2.0925      'this is the interlayer spacing, see Tab. 10.2
Transition(A, h)
   to A = 1; ' Translate from layer type A to layer type A with a probability of 1
   a_add = 1/3; b_add = -1/3;' For A → A stacking vector: sx =⅓, sy = -⅓,sz = h/(h·Nv)
```

In order to introduce a hexagonal intergrowth fault, we define a second type of layer, *layer B*, which is chemically identical to *layer A*. A second stacking vector and associated transition probabilities must be defined:

```
prm !PAA 0.98       ' PAB needs not to be defined as PAB = 1 - PAA (eq. 10. 8)
prm !PBA 0.98       ' PBB needs not to be defined as PBB = 1 - PBA (eq. 10. 8)
Transition(A, h)
   to A = PAA;     ' Translate from layer type A to layer A with a probability of 0.98
   a_add =  1/3;
   b_add = -1/3;   ' For A → A apply a stacking vector with sx = ⅓, sy = -⅓, sz = h/(h·Nv)
```

```
    to B = 1-PAA;      ' Translate from layer type A to layer type B with a probability of 0.02
    a_add = -1/3;
    b_add =  1/3;      ' For A → B apply a stacking vector with sx = -⅓, sy = ⅓, sz = h/(h·Nv)
Transition(B, h)
    to A = PBA;        ' Translate from layer type B to layer type A with a prob. of 0.98
    a_add =  1/3;
    b_add = -1/3;      ' For B → A apply a stacking vector with sx = ⅓, sy = -⅓, sz = h/(h·Nv)
    to B = 1-PBA;      ' Translate from layer type B to layer type B with a prob. of 0.02
    a_add = -1/3;
    b_add =  1/3;      ' For B → B apply a stacking vector with sx = -⅓, sy = ⅓, sz = h/(h·Nv)
```

The pattern simulated with this approach is shown in red on the right hand side of Figure 10.9. The use of only 500 layers causes ripples just as observed in the DIFFaX simulations. These can be reduced by using 5000 layers (blue line) to improve the statistics. While the use of 5000 layers improves the simulation, the significantly larger c axis leads to many more hkl reflections in the simulation. In the current case the number of reflections over a 5–150° range increases from 24946 to 249431 and the simulation time (on a standard PC) increases significantly from 0.29 to 2.59 s.

It is, however, possible to mimic the DIFFaX recursive mode using the TOPAS *number_of_sequences* N_{str} keyword. By setting N_{str} to 100, TOPAS will average the patterns of 100 individual sequences. Internally TOPAS uses the same set of hkl reflections for each of the sequences and the speed penalty for calculating the pattern of 100 sequences compared to 1 is minimal (0.29 to 0.63 s in this example). This gives the green pattern on the right of Figure 10.9 that closely matches the 5000 layer simulations or the DIFFaX recursive calculation. The specific language needed is included in the INP file below, which actually performs a Rietveld fit to the data simulated in DIFFaX in recursive mode (the green pattern in Figure 10.9, left).

```
seed               ' Reinitiates random number generator each time
xdd "Cu-rec500.xy"   ' Simulated in DIFFaX, green in Fig. 10.9 left
    start_X 20
    LP_Factor(0)
    bkg @  0.0333011786`  0.0289287388`  0.0405216319`
    Zero_Error(@,-0.00878`)
    lam ymin_on_ymax 0.00001 la 1 lo  1.5418 lh  1e-5
    str
        prm s  3484.09010` min 1e-15
        scale = s 1e-6/ (Nv Nstr);  ' Scale factor proportional to layers used
        phase_name "Cu-2%-faults"
        space_group P1
        a  !lpa  2.563066
        b  !lpa  2.563066
```

```
c   = Get(generated_c);:1046.230000 ' c-axis derived from total number of layers
ga 120

prm !PAA 0.980
prm !PBA 0.980
prm !h  2.09246
generate_stack_sequences {
    number_of_sequences Nstr 100        ' 100 stacking sequences
    number_of_stacks_per_sequence Nv 500 ' Each sequence has 500 layers
    Transition(A, h)
        to A = PAA;    a_add =  1/3; b_add = -1/3;
        to B = 1-PAA; a_add = -1/3; b_add =  1/3;
    Transition(B, h)
        to A = PBA;    a_add =  1/3; b_add = -1/3;
        to B = 1-PBA; a_add = -1/3; b_add =  1/3;
}
site Cu1 x = 0; y = 0;  z  = 0/Nv; occ Cu  1 beq bval 0.0 layer A
site Cu1 x = 0; y = 0;  z  = 0/Nv; occ Cu  1 beq bval 0.0 layer B
peak_buffer_based_on = Xo;
peak_buffer_based_on_tol 0.1            ' Limit number of hkl's used
SF_smooth(@, 5.28134`, 1)              ' Smooth peaks if Nv small, section 10.6
TCHZ_Peak_Type(,  -0.07754,,   0.13069,,  -0.03604,,  0,,  0.15771,,  0.00010) ' IRF
```

In this file we also make use of the *peak_buffer_based_on_tol* command to speed up calculations. This reduces the number of individual *hkl* peaks and different peak shapes that need to be summed to calculate the powder pattern. In this example the use of the peak buffer reduces the number of *hkl* reflections used from 24946 to 2430 and 249431 to 2872 for $N_v = 500$ and $N_v = 5000$, respectively, with negligible impact on the calculated pattern. More details of how this is done are given by Ainsworth et al. (2016) and Coelho et al. (2016). The final Rietveld fit to the DIFFaX-simulated data is shown in Figure 10.10.

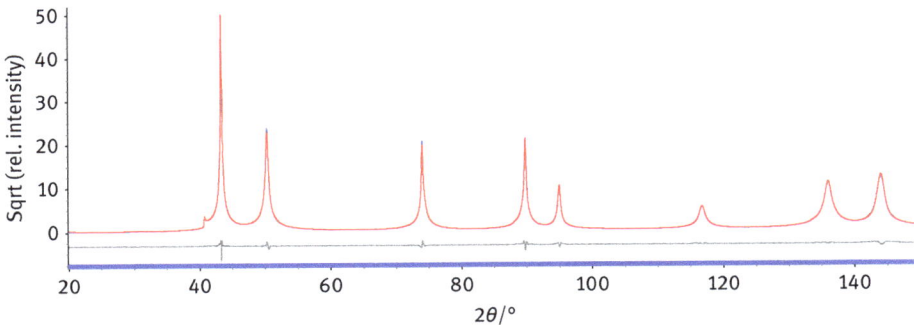

Figure 10.10: Full Rietveld fit of DIFFaX-simulated data for 2% faulted Cu. Note the effectively continuous blue bar of *hkl* tick marks.

10.6 Determining fault probabilities by Rietveld refinement

The speed of stacking fault calculations in TOPAS means that it is possible to determine stacking fault probabilities from experimental data, despite the stacking fault probabilities not being directly refinable parameters. One way of doing this is to make use of the *#list* command (discussed in detail in Chapter 12) to test a series of different fault probabilities (Ainsworth et al., 2016). Examples for dealing with more sophisticated faulting models have been given by, for example, Bette et al., (2015) and Kudielka et al., (2017).

For example, the INP file below performs a Rietveld refinement on a simulated data set of an intergrowth of 70% cubic diamond and 30% hexagonal lonsdaelite layers (discussed in the DIFFaX manual or Ainsworth, 2016). The script scrolls through 27 different values of the stacking probability *pa* and does a Rietveld fit for each. The lowest R_{wp} is found for *pa* = 0.7, as shown in Figure 10.11.

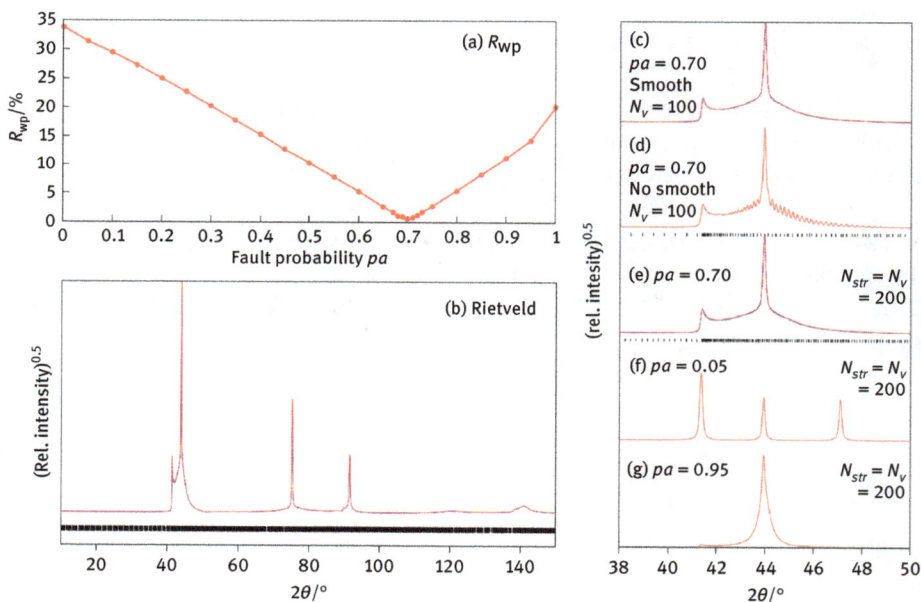

Figure 10.11: (a) R_{wp} for a series of Rietveld fits to a simulated data set of an intergrowth of diamond and lonsdaelite. The minimum R_{wp} is achieved with *pa* = 0.7 leading to the Rietveld fit in panel (b). The right hand plots show simulated patterns for different stacking fault probabilities. Plots (e) to (g) show simulations with N_{str} = 200 N_v = 200 and smoothing applied. Panel (d) shows the "ringing" observed in the calculated pattern when a small number of layers (N_v = 100) is simulated; panel (c) shows how smoothing allows a small value of N_v to simulate a much larger supercell. In each panel red lines are TOPAS calculated patterns and (b, c, e) blue DIFFaX-simulated "observed" data.

```
num_runs 27
#list paval {
   0.9999 0.9500 0.9000 0.8500 0.8000 0.7500 0.7300 0.7200 0.7100
   0.7000 0.6900 0.6800 0.6700 0.6500 0.6000 0.5500 0.5000 0.4500
   0.4000 0.3500 0.3000 0.2500 0.2000 0.1500 0.1000 0.0500 0.0010
}
seed
xdd diffax_dia.xye
   weighting 1
   LP_Factor(0) bkg 1
   start_X 10 finish_X 149.6
   rebin_with_dx_of 0.02
   lam ymin_on_ymax 0.0001 la 1 lo  1.5405754 lh  1e-5
   Zero_Error( , 0.00238)
   str
      space_group P1
      a 2.518156 b 2.518156 c = Get(generated_c); ga 120
      prm !pa =paval(Run_Number);:0.70000 prm !h 2.05870
      prm s  4983.64843 min 1e-15 scale = s 1e-6 / (Nv Nstr);
      generate_stack_sequences {
        number_of_sequences Nstr 200
        number_of_stacks_per_sequence Nv 200
        Transition(A, h)
           to A = pa;    a_add = 2/3; b_add = 1/3;
           to B = 1-pa; a_add = 0;    b_add = 0;
        Transition(B, h)
           to A = 1-pa; a_add = 0;    b_add = 0;
           to B = pa;    a_add =-2/3; b_add =-1/3;
      site C1 x =-1/3; y =-1/6; z = -0.125/Nv; occ C 1 beq 1 layer A
      site C2 x = 1/3; y = 1/6; z =  0.125/Nv; occ C 1 beq 1 layer A
      site C3 x = 1/3; y = 1/6; z = -0.125/Nv; occ C 1 beq 1 layer B
      site C4 x =-1/3; y =-1/6; z =  0.125/Nv; occ C 1 beq 1 layer B
      peak_buffer_based_on = Xo; peak_buffer_based_on_tol 0.1
      TCHZ_Peak_Type( ,  0.09408, , -0.07795, ,  0.02058, ,0, ,  0.11131, , 0.02457) 'IRF
      SF_smooth( ,  5.7, 1)
      }
```

This INP file contains one more "trick" to help speed up refinements by allowing the use of a small number of layers N_v through the *SF_smooth* macro. As shown in panel (d) of Figure 10.11, if one uses $N_v = 100$ the finite number of *hkl* reflections generated leads to a "ringing" in the calculated pattern caused by the peak shapes of the individual reflections. The *SF_smooth* macro applies a convolution based on the calculated intensity of adjacent *hkl* reflections to smooth out the profile of some reflections, while leaving others sharp. In this way a small number of layers can effectively mimic a much larger number, which speeds refinements up enormously.

References

Ainsworth, C.M., Lewis, J.W., Wang, C.H., Coelho, A.A., Johnston, H.E., Brand, H.E.A., Evans, J.S.O. (2016): *3D transition metal ordering and Rietveld stacking fault quantification in the new oxychalcogenides La$_2$O$_2$Cu$_{2-4x}$Cd$_{2x}$Se$_2$*. Chem. Mat. 28, 3184–3195.

Bette, S., Dinnebier, R.E., Freyer, D. (2015): *Structure solution and refinement of stacking-faulted NiCl (OH)*. J. Appl. Cryst. 48, 1706–1718.

Biscoe, J., Warren, B.E. (1942): *An X- ray study of carbon black*. J. Appl. Phys. 13, 364–371.

Casas-Cabanas, M., Reynaud, M., Rikarte, J., Horbach, P., Rodríguez-Carvajal, J. (2016): *FAULTS: a program for refinement of structures with extended defects*. J. Appl. Cryst. 49, 2259–2269.

Coelho, A.A., Evans, J.S.O., Lewis, J.W. (2016): *Averaging the intensity of many-layered structures for accurate stacking-fault analysis using Rietveld refinement*. J. Appl. Cryst. 49, 1740–1749.

Hulliger, F. (1976): *Structural chemistry of layer-type phases*, Springer Netherlands, Dordrecht.

Kudielka, A., Bette, S., Dinnebier, R.E., Abeykoon, M., Pietzonka, C., Harbrecht, B. (2017): *Variability of composition and structural disorder of nanocrystalline CoOOH materials*. J. Mater. Chem. C 5, 2899–2909.

Leineweber, A., Kreiner, G., Grüner, D., Dinnebier, R., Stein, F. (2012): *Crystal structure, layer defects, and the origin of plastic deformability of Nb$_2$Co$_7$*. Intermetallics 25, 34–41.

Leoni, M., Gualtieri, A.F., Roveri, N. (2004): *Simultaneous refinement of structure and microstructure of layered materials*. J. Appl. Cryst. 37, 166–173.

Martin, S., Ullrich, C., Šimek, D., Martin, U., Rafaja, D. (2011): *Stacking fault model of ε-martensite and its DIFFaX implementation*. J. Appl. Cryst. 44, 779–787.

Moore, D.M., Reynolds, R.C. (1997); *X-ray diffraction and the identification and analysis of clay minerals*, 2nd edition, Oxford University Press, Oxford; New York, 378 pages.

Radha, A.V., Vishnu Kamath, P., Subbanna, G.N. (2003): *Disorder in layered hydroxides: synthesis and DIFFaX simulation studies of Mg(OH)$_2$*. Mater. Res. Bull. 38, 731–740.

Ramesh, T.N., Jayashree, R.S., Kamath, P.V. (2003): *Disorder in layered hydroxides: diffax simulation of the X-ray powder diffraction patterns of nickel hydroxide*. Clay Clay Miner. 51, 570–576.

Ramesh, T.N., Kamath, P.V. (2008): *Planar defects in layered hydroxides: simulation and structure refinement of β-nickel hydroxide*. Mater. Res. Bull. 43, 3227–3233.

Reynolds, R.C., Jr. (1985): *NEWMOD: A computer program for the calculation of one-dimensional patterns of mixed-layered clays*, Hannover, NH (US).

Treacy, M.M.J., Newsam, J.M., Deem, M.W. (1991): *A general recursion method for calculating diffracted intensities from crystals containing planar faults*. Proc. R. Soc. London, Ser. A 433, 499–520.

Ufer, K., Kleeberg, R., Bergmann, J., Dohrmann, R. (2012): *Rietveld refinement of disordered Illite-Smectite mixed-layer structures by a recursive algorithm. II: Powder-pattern refinement and quantitative phase analysis*. Clay Clay Miner. 60, 535–552.

Ufer, K., Roth, G., Kleeberg, R., Stanjek, H., Dohrmann, R., Bergmann, J. (2004): *Description of X-ray powder pattern of turbostratically disordered layer structures with a Rietveld compatible approach*. Z. Kristallog. – Crystalline Materials 219, 519–527.

Wang, X., Hart, R.D., Li, J., McDonald, R.G., van Riessen, A. (2012): *Quantitative analysis of turbostratically disordered nontronite with a supercell model calibrated by the PONKCS method*. J. Appl. Cryst. 45, 1295–1302.

Warren, B.E. (1941): *X-Ray diffraction in random layer lattices*. Phys. Rev. 59, 693–698.

11 Total scattering methods

11.1 Introduction

A powder pattern can be regarded as the projection of three-dimensional reciprocal space onto a one-dimensional axis ($1/d$, 2θ, $sin\theta/\lambda$ or $Q = 4\pi \, sin\theta/\lambda$), as discussed in Chapter 1. Geometrically this means that all intensity in a spherical shell with inner radius $1/(d+\Delta d)$ and outer radius $1/d$ will be summed up to the corresponding step-scan intensity of width $1/(\Delta d)$ in the powder pattern. Everything that is present in the incident and diffracted beam paths will potentially contribute to the total intensity measured. The powder pattern therefore contains contributions both from the environment (instrument) and from the sample: scattering from the sample holder, from air, from inelastic Compton scattering and fluorescence all combine to give what is commonly called "background."

The coherent elastic scattering from the sample, which is what we are usually interested in, consists of two parts: Bragg scattering at the reciprocal lattice points and diffuse scattering due to deviations from the average periodic structure (local structure) in between the reciprocal lattice points. Both types of elastic scattering contain a wealth of structural information.

The Rietveld method retrieves information solely about the time- and space-averaged periodic crystal structure by fitting the Bragg intensities. A powder pattern for Rietveld analysis should, therefore, be measured with a focus on high resolution in reciprocal space and only up to the angle beyond which no Bragg scattering is visible above the background.

The total scattering method is an alternative approach for analyzing the elastic scattering by Fourier transforming the entire powder-diffraction data (Bragg peaks and diffuse features between them) to the so-called atomic pair distribution function (PDF). The real-space PDF can give information about local structure, and will include contributions from amorphous components, disorder and from lattice dynamics. The nature of Fourier transforms means that data is needed over a wide range of reciprocal space to give rise to high resolution in real-space. Powder-diffraction patterns therefore need to be measured so that the coherent scattering (Bragg and diffuse) is collected over as much of reciprocal space as possible.

A combination of Rietveld refinement and PDF analysis will reveal the maximum amount of structural information from a powder pattern, but one must be aware that the measurement strategies for the two techniques are somewhat different, often requiring that two powder patterns are measured with different setups or instruments.

In the following sections we will discuss the origin and form of PDF functions and how they are analyzed in TOPAS. At the time of writing, PDF analysis and its TOPAS implementation are developing rapidly. Methods and macros are being updated regularly. We give an introduction to what's possible, and the reader should check,

https://doi.org/10.1515/9783110461381-011

for example, the TOPAS wiki for links to the latest features (http://topas.dur.ac.uk/topaswiki/doku.php?id=pdf_fitting). There is also a gitHub site containing a TOPAS pdf.inc file, which contains various macros that are useful in PDF fitting (https://github.com/pachater/topas).

11.2 Total scattering and atomic pair distribution function (PDF) analysis

Powder-diffraction data for total scattering studies can be measured in much the same way as for regular studies, but it's usual to collect data to a Q value[1] of 30–50 Å^{-1} and with small statistical uncertainties. This makes synchrotron or neutron radiation from spallation sources the methods of choice. A typical data set recorded using a two-dimensional area detector at the NSLS synchrotron is shown in Figure 11.1. In the laboratory, Mo-K$_\alpha$ or Ag-K$_\alpha$ radiations are the best choice. Ag-K$_\alpha$ radiation offers the highest Q_{max} (approximately 22 Å^{-1} at 160° 2θ), while Mo-K$_\alpha$ radiation provides a compromise between Q_{max} and flux.

Corrections must be made for contributions to the scattered intensity from effects like Compton scattering, fluorescence, scattering from the sample holder and so on (Egami & Billinge, 2012). Instrument effects are usually removed by measuring an empty sample holder – typically a glass, amorphous silica or Kapton capillary (Figure 11.1). This background signal is then subtracted from the sample signal. Some scaling may be required if absorption by the sample significantly influences the instrument contribution that reaches the detector.

The resulting coherent scattering function $I(Q)$ is a continuous function of Q (Figure 11.2) with sharp features where there are Bragg peaks, and broad features in between. $I(Q)$ is normalized by dividing by the total scattering cross-section of the sample to give $S(Q)$, the *total-scattering structure function*. In the case of X-ray scattering, the sample scattering cross-section is the square of the average atomic form-factor, $\langle f(Q) \rangle^2$, which becomes very small at high-Q. Thus, during the normalization process the experimental data at high-Q are amplified. This has the effect that even weak signals at high-Q, which are usually insignificant in a conventional Rietveld analysis, can become important in a total-scattering experiment. Because the signal at high-Q is weak, it is important to collect the data in that region with good statistics. One way of doing this with one-dimensional detectors is to increase the counting time with increasing Q, ideally in a way inversely proportional to the global decrease in intensity.

The final dimensionless total-scattering structure function $S(Q)$ has an average value of unity (Egami & Billinge, 2012) and is given by:

---

[1] $Q = 2\pi/d = 2\pi s = 4\pi \sin\theta/\lambda$.

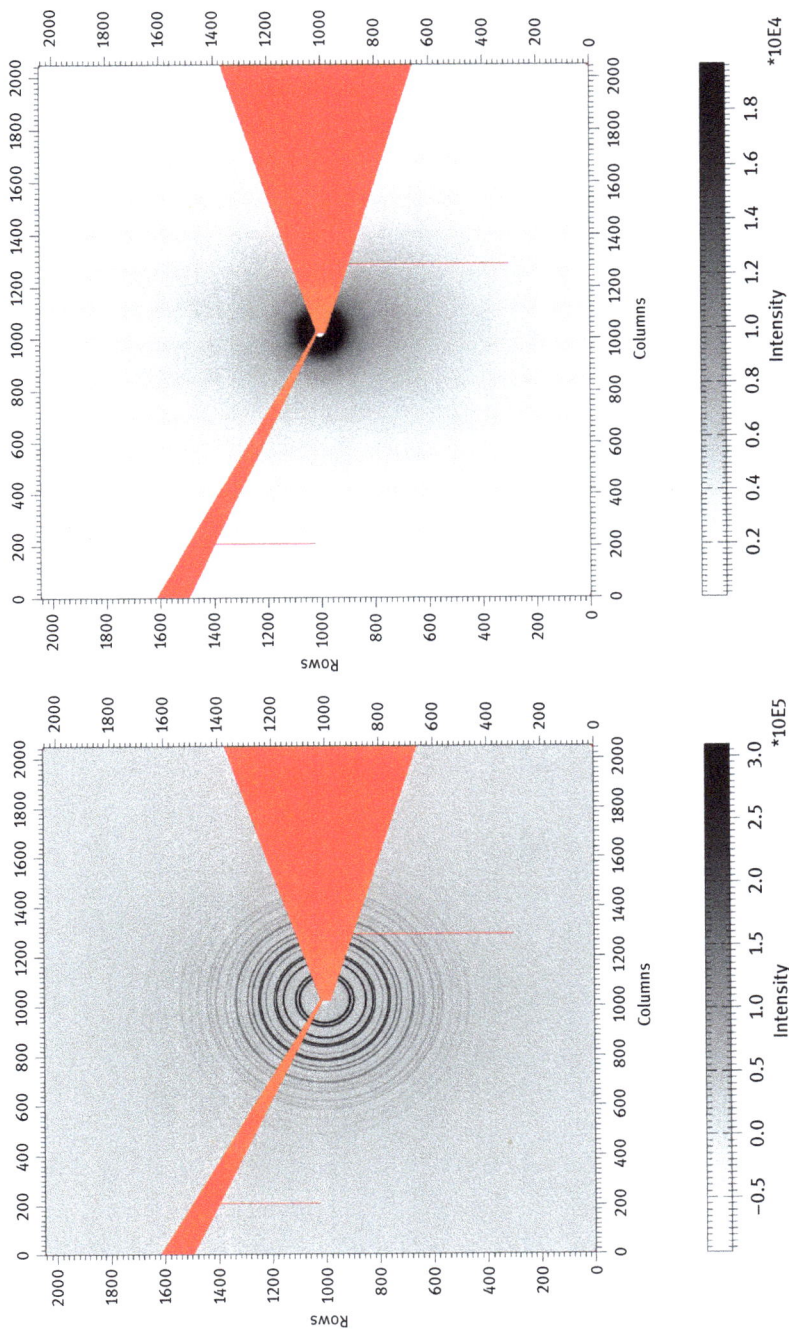

Figure 11.1: Screenshots of the Fit2D program (Hammersley et al., 1996) showing the two-dimensional powder pattern of nickel (left) and the background pattern of an empty capillary (right) measured at a wavelength of $\lambda = 0.1839$ Å at beamline X17A (NSLS) using a PerkinElmer amorphous silicon detector. The red areas denote masked regions omitted from the integration (due to primary beam stop, dead pixels, etc.).

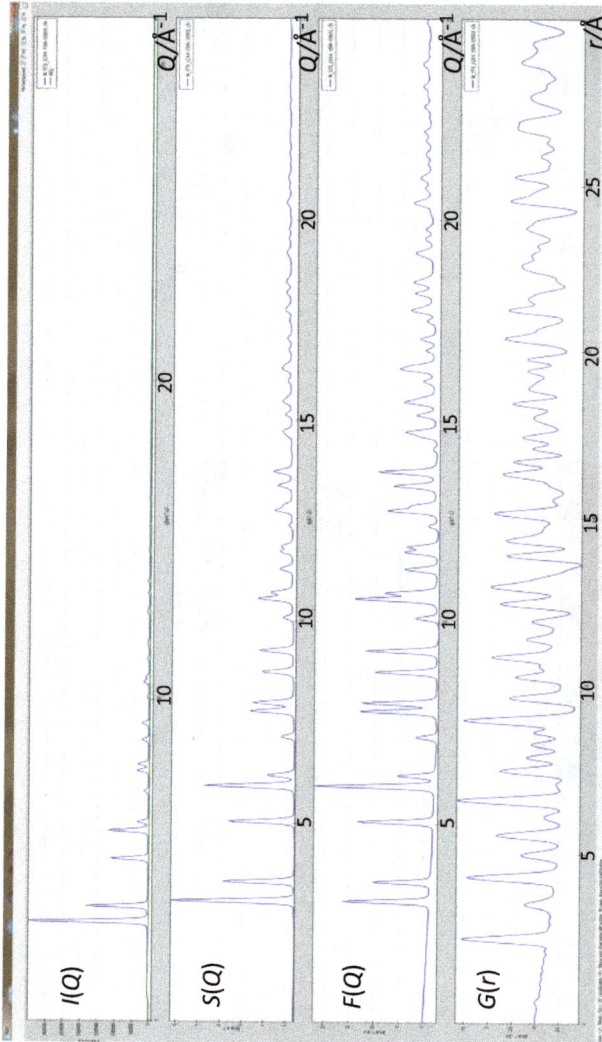

Figure 11.2: Screenshot of the PdfGetX3 program that is part of the xPDFsuite (Yang et al., 2014). From top to bottom: (1) raw integrated powder pattern $I(Q)$ from Figure 11.1; (2) total-scattering structure function $S(Q)$; (3) normalized reduced total scattering structure function $F(Q)$ and (4) atomic pair distribution function (PDF) $G(r)$. $F(Q) = Q[S(Q) - 1]$ is called the *reduced structure function* and tends to zero at high Q.

$$S(Q) = \frac{I(Q)}{\langle f(Q) \rangle^2} - \frac{\langle f^2(Q) \rangle}{\langle f(Q) \rangle^2} + 1, \quad \langle S(Q) \rangle = 1, \tag{11.1}$$

with $\langle f^2(Q) \rangle$ the mean of the squared form factors and $\langle f(Q) \rangle^2$ the square of the average form factor. For monoatomic materials this simplifies to:

$$S(Q) = \frac{I(Q)}{\langle f(Q) \rangle^2}. \tag{11.2}$$

In addition to the form factor fall off, the coherent intensities (the features) in $I(Q)$ and hence $S(Q)$ decrease with increasing Q due to the Debye-Waller factor, which comes from thermal and quantum zero-point motion of the atoms, and any static disorder in the material. By a Q value of 30–50 Å^{-1} (depending on temperature and the stiffness of the sample) there are usually no significant features in $I(Q)$, and there is no need to measure data to higher-Q.

From the total scattering structure function $S(Q)$ one can calculate the *reduced pair distribution function*, $G(r)$.[2] $G(r)$ is related to $S(Q)$ through a sine Fourier transform according to:

$$G(r) = \frac{2}{\pi} \int_{Q_{min}}^{Q_{max}} Q[S(Q) - 1] \sin(Qr) dQ. \tag{11.3}$$

It has slope $-4\pi\rho_0 r$ at low-r where ρ_0 is the atomic number density in atoms/Å^3, and oscillates around zero at high-r (Figure 11.2, panel 4). The values for the integration limits Q_{min} and Q_{max} are limited by the experimental setup. Q_{max} is usually lowered below the experimentally measured maximum to prevent noise in high-Q data from adversely affecting the Fourier transform. $G(r)$ is one of several functions called the PDF (see footnote 3).

Formally, $G(r)$ is the autocorrelation (Section 13.7) of the atomic density. This is obtained by taking the atomic density (the atoms at their respective positions) of the molecule, cluster or crystal and convoluting it with a replica of itself. This object is then averaged over all orientations to obtain the PDF. The PDF therefore contains peaks at positions, r, that separate pairs of atoms in the solid with high probability (Figure 11.3). For a completely random distribution of atoms, the PDF would be flat (if atoms are allowed to overlap). For nonrandom distributions the PDF contains valuable information on both short-range nearest-neighbor interactions and on pair-correlations extending to much higher values of r. In fact, with high Q-space resolution data, PDFs can be measured out to hundreds of nanometers (thousands of

---

2 There are a number of different naming conventions in use for different "flavors" of radial distribution functions. Unfortunately different communities (and often continents!) use the same $X(r)$ label for different functions. Keen (2001) has provided a useful collation of the different formalisms.

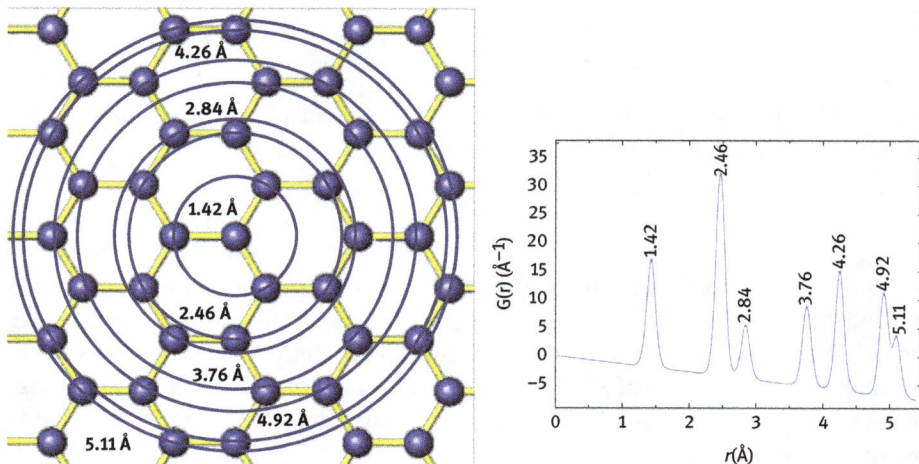

Figure 11.3: Two-dimensional graphene layer and the corresponding PDF, with peaks corresponding to pairs of atoms separated by distance r. The circles highlight radial distances from the central atom where other atoms are found.

Angstroms) with the structural information remaining quantitatively reliable (Levashov, Billinge & Thorpe, 2005).

There is a large number of other correlation functions related to $G(r)$ that are commonly used.[3] $G(r)$ has a number of advantages over other functions. First, it arises directly from the experimental $S(Q)$ with no assumption made about the atomic number density ρ_0 that appears in other expressions. Indeed, ρ_0 is contained experimentally in the low-r slope of $G(r)$. The uncertainties in the data are also constant as a function of r meaning that difference plots between observed and calculated $G(r)$ functions should show constant scatter with r. Finally, oscillations in $G(r)$ continue with equal amplitude up to $r = \infty$ for crystalline materials, meaning that departures from this behavior (Section 11.3.2) give direct visual information on various sample or experimental factors.

One useful and intuitive correlation function is the *radial distribution function* (RDF), $R(r)$, which is related to $G(r)$ by:

$$G(r) = \frac{R(r)}{r} - 4\pi\rho_0 r. \tag{11.4}$$

---

3 For example, $g(r)$ is the *atomic pair distribution function* and gives the probability of finding 2 atoms at distance r. It is defined via $G(r) = 4\pi\rho_0 r(g(r) - 1)$ such that $g(r) \to 0$ as $r \to 0$ and $g(r) \to 1$ as $r \to \infty$. $\rho(r) = \rho_0 g(r)$ is the *atomic pair density function*. $\rho(r) \to 0$ as $r \to 0$ and $\rho(r) \to \rho_0$ as $r \to \infty$. $G(r)$, $g(r)$, and $\rho(r)$ are all commonly abbreviated as PDF.

The RDF is important because it is most directly related to the physical structure: $R(r)dr$ gives the number of atoms in an annulus of thickness dr at distance r from another atom. This means that the coordination number or the number of neighbors, N_C, around a chosen atom is given by:

$$N_C = \int_{r_1}^{r_2} R(r)dr, \tag{11.5}$$

where r_1 and r_2 define the beginning and ending positions of the RDF peak corresponding to the coordination shell in question. One disadvantage of the RDF is that it depends on r^2 at high r making it inconvenient for plotting and visualization.

The RDF suggests a simple way that PDFs can be calculated from atomic models. We can set up a model consisting of a large number of atoms situated in an arbitrary box at positions $\mathbf{r}$ with respect to the origin. The relative atomic positions will give rise to a set of atomic distances r_{jk}, where $r_{jk} = |\mathbf{r}_j - \mathbf{r}_k|$, for each pair of atoms in the structure. We can therefore express the RDF as a sum of delta functions:

$$R(r) = \frac{1}{n}\sum_{j=1}^{n}\sum_{k=1}^{n}\delta(r - r_{jk}), \tag{11.6}$$

where the double sum runs twice over all atoms in the sample (Figure 11.3). If several types of atoms are present, the expression for $R(r)$ is:

$$R(r) = \frac{1}{n}\sum_{j=1}^{n}\sum_{k=1}^{n}\frac{f_j f_k}{\langle f \rangle^2}\delta(r - r_{jk}), \tag{11.7}$$

where the fs are the form factors, evaluated at $Q = 0$, for the jth and kth atoms and $\langle f \rangle$ is the sample average form factor.[4]

When eqs. (11.4) and (11.6) are used to calculate the PDF from a structural model, the δ-functions are expressed as Gaussians with a width dependent on various sample-related factors. The equation is also modified to account for various experimental effects (Olds et al., 2018) as outlined in the following sections.

Since we can calculate $R(r)$ and therefore $G(r)$ from any atomic configuration, an experimental $G(r)$ can easily be fitted using a Rietveld-like approach, or analyzed using methods such as simulated annealing. This is often called *real space Rietveld refinement*. The size of box used for modeling depends on the type of system being studied. If one is looking for small local deviations from an ordered crystal structure, it's common to use a box that corresponds to a single unit cell, often reduced to lower symmetry. To model longer-range effects (or amorphous samples) a large-box approach might be used.

---

4 Or scattering lengths for neutrons.

11.3 Parameters influencing the PDF

The appearance of an experimental PDF will depend strongly on a number of parameters related to the structure and microstructure of the sample, and also on the experimental conditions used. Each PDF peak is characterized by position, width and area. All of these can be affected by the instrument resolution function and by PDF processing.

TOPAS calculates PDFs as a summation of Gaussian functions (Coelho, 2015). The minimum/maximum extent of the Gaussians along the X-axis (r) defaults to the value where y falls to 10^{-3} of its maximum height, but can be modified by changing the expression:

```
pdf_ymin_on_ymax  0.001
```

11.3.1 Peak position in a PDF

The peak positions in a calculated PDF are determined by the bond lengths and angles in the structural model. If fractional coordinates are used, these will depend on the size of the simulation box containing the atoms (often a single unit cell in small-box modeling of crystalline materials). It should be noted that the range of convergence for coordinates is much lower for a PDF than for regular Rietveld refinement, since coordinate shifts change peak positions in the PDF rather than intensities.

Any zero point can normally be satisfactorily described by a simple linear function. Fixed offsets are sometimes needed to correct for different conventions used by different packages to produce the PDF. Note that the offset is applied in r-space so that the correction needs to be at the TOPAS *str* level:

```
str
    prm z0  0.00216 min -.1 max .1 val_on_continue = Rand(-0.05, 0.05);
    prm z1 -0.03542 min  -1 max  1 val_on_continue = Rand(-0.05, 0.05);
    pdf_zero = z0 + z1 0.01 X;
```

11.3.2 Peak area of a PDF

The areas of PDF peaks are determined by coordination number, occupancies and the scattering power of the atom pair. For materials in which the coherent domains have a finite size, peak areas will decrease as a function of r due to a decreasing number of pair–pair distances. For example, for nanoparticles there can clearly be no peaks at r values greater than the particle size. Functions to describe this so-called peak damping have been calculated for various sample shapes (e.g., Kodama et al., 2006; Korsunskiy et al., 2007). An example showing how to describe the damping function of 5 nm spherical particles in TOPAS is given below:

```
prm !r 5 `radius of spherical nanoparticle in nm
scale_phase_X = IF X > 2 r
                THEN 0
                ELSE (Pi X^2 ((0.25 (X/r)^3)-(3 X/r)+4)) / (4 Pi X^2 1)
                ENDIF ;
```

A phenomenological peak damping can be achieved using a unit area Gaussian centered at $r = 0$ with a *fwhm* that can be refined to give some kind of size measure:

```
prm damp_fwhm  51.08423 min 1e-6 max = 10 X2;
prm damp = Gauss(0, damp_fwhm);
scale_phase_X = damp;
```

The resolution dQ in Q-space in which the data are collected also influences the intensity of peaks in the PDF and is another source of peak damping. This can be modeled by a normalized Gaussian function (Toby & Egami, 1992) with dQ as an adjustable or fixed parameter (Figure 11.4). In TOPAS this is realized by:

```
prm !dQ 0.01 min 0.001 max 0.5          ' Instrumental resolution of S(Q) data
scale_phase_X = Exp(-0.5 X^2 (dQ/2.35482)^2);
```

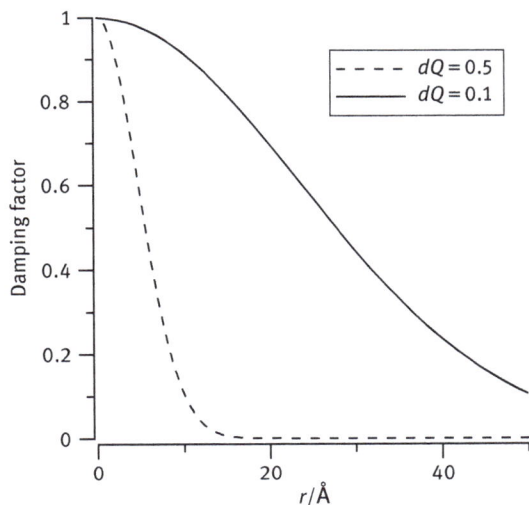

Figure 11.4: Gaussian damping factor with different resolutions dQ.

Since the low-r region is relatively unaffected by the Q-space resolution, PDF collections are often optimized for Q_{max} at the expense of dQ. The resulting damping causes peak decay that is similar to that in a nanoparticle. Qualitative estimates of particle size from visual inspection of PDFs should therefore be treated with caution.

For multiphase samples one will see peaks in a PDF from each component. It is therefore possible to perform quantitative phase analysis in a similar manner to that described in Chapter 5. The scale factor of each phase is proportional to its mole fraction.[5]

11.3.3 *r*-space resolution of a PDF

Since the density of interatomic distances in a PDF increases rapidly with r (Figure 11.3), it becomes difficult to associate a PDF peak with a specific atom–atom pair past the first few Å, even in simple structures. The ability to distinguish different distances is related to the PDF resolution, Δr, which is dominated by the Q_{max} used in the Fourier transform. According to Farrow et al. (2011) this can be approximated by:

$$\Delta r \approx \frac{\pi}{Q_{max}} \text{ unless } Q_{max} > \frac{\pi}{\sqrt{\langle u^2 \rangle}} \tag{11.8}$$

where $\sqrt{\langle u^2 \rangle}$ is the root-mean-squared atomic displacement. The Q_{max} limitation relates to the fact that local thermal vibrations will cause an inherent smearing out of interatomic distances and hence a broadening of peaks. High Q values may therefore not help resolution for materials with large values of $\sqrt{\langle u^2 \rangle}$. Q_{max} values that can be reached using laboratory instruments are 8 Å^{-1} for Cu-K$_{\alpha1}$, 17 Å^{-1} for Mo-K$_{\alpha1}$ and 22 Å^{-1} for Ag-K$_{\alpha1}$, while at a high-energy synchrotron or neutron TOF beamline Q_{max} values > 50 Å^{-1} might be measurable.

As we've just discussed, the sample-related peak width of a PDF is determined by vibrations (phonons), and disorder. In TOPAS this can be modeled in several ways. The simplest is to use an isotropic temperature factor, *beq*. This approach neglects the fact that atomic motions of directly–bonded atoms are likely to be highly correlated, which will lead to an overestimation of peak widths at low-r compared to high-r. A better approximation might be to allow *beq* to vary as a function of r (X in the TOPAS equation) using the error function (see Chapter 13):

```
prm erf_a  0.3  min 0
prm b1     1.0  min 0
...
site Ni1  x 0  y 0  z 0  occ Ni 1  beq = b1 Erf_Approx( erf_a X);
```

---

[5] In TOPAS v6, the weight fractions reported for PDF fits in the gui and default *MVW* macros that are defined for Rietveld refinement should therefore not be used.

In many samples, we might expect different broadenings for different types of inter-actions. For example, in relatively rigid molecules long range intramolecular distances might be significantly stiffer than much shorter intermolecular (VDW controlled) distances. To account for this, the keyword *pdf_gauss_fwhm* can be used to write specific width equation for pairs specified by *pdf_for_pairs*. If all of the pairs possible are described by *pdf_for_pairs* (e.g., *pdf_for_pairs* * *) then the associated *beqs* are not used and they become redundant. An example is given below:

```
pdf_for_pairs C1 "C2 C3 C4"    ' C2 to C4 in a different molecule to C1
pdf_gauss_fwhm = 0.1;
```

Note that the keyword *pdf_for_pairs* simply overwrites *fwhms* of the PDF peaks from *beq* values, and values set by *pdf_gauss_fwhm* are used instead.

11.3.4 Processing artefacts

The fact that $S(Q)$ can only be Fourier transformed over a finite Q range gives rise to termination ripples in the PDF (Chung and Thorpe, 1997). As outlined in Chapter 13, the Fourier transform of a box function is a sinc function with termination ripples. It is important that these termination ripples aren't misinterpreted as pair–pair distances.

Termination ripples can be reduced using, for example, a Lorch (or super-Lorch) function (Soper & Barney, 2012). The function is applied at the Fourier-transform step and effectively smears the density in $G(r)$ space over a finite volume removing ripples. There is, however, a price to pay in that the real-space resolution is degraded making it harder to distinguish closely separated pair distances. If this type of function is used when producing the experimental PDF, an appropriate correction (e.g., *convolute_SoperLorch(!d_zero, 0.08)* in TOPAS) must be applied to the calculated PDF.

Alternatively, it is possible to convolute additional functions like the *sinc* func-tion directly into the PDF peak profile to describe the ripples. This can be done using the keyword *pdf_convolute*. The following example (courtesy of Phil Chater) will convolute a sinc function into the PDF profile:

```
prm !Step_Size = 0.02;
prm !q2 0.25 min= Step_Size / 10; max = Step_Size / 2;
prm !q3 0       min= -Pi; max = Pi;
pdf_for_pairs * *
   pdf_gauss_fwhm @ 0.1
   local conv_max = (5 Qmax - Mod(5 Qmax,1))/5 2 Pi / Qmax;
   pdf_convolute = If(Abs(X) > Yobs_dx_at(Xo),(Sin(Qmax) X)/(X)),Qmax);
      min_X = Min(-Xo,-conv_max) ;
      max_X = Max( Xo, conv_max) ;
```

11.4 Example: Ni standard

A Ni standard sample was measured at ambient conditions at beamline 28-ID-2 of the National Synchrotron Light Source II using the rapid acquisition PDF mode (Chupas et al., 2003), with an X-ray wavelength of 0.1827 Å using a PerkinElmer two-dimensional detector (2048 × 2048 pixels and 200 × 200 μm pixel size) mounted orthogonal to the incident beam path with a sample-to-detector distance of 219.431 mm. Calibration and integration of the two-dimensional images were performed using the program Fit2D (Hammersley et al., 1996), and Fourier transformation to the PDF was performed using PDFgetX3 in xPDFsuite (Juhas et al., 2013; Yang et al., 2014).

The INP file below shows the TOPAS commands needed to perform a PDF fit to the experimental $G(r)$. The only significant difference to a standard Rietveld INP file is the inclusion of the line *pdf_data*. The data are assumed to be in a two (.xy) or three (.xye) column format file in PDFgetX3.gr format. If other formats are used, an appropriate scaling should be applied.[6] The INP file uses the r-dependent displacement parameter discussed in Section 11.3.3 and a linear correction for peak positions. In general, the absolute scaling of the $G(r)$ data coming out of PDFgetX3 is arbitrary and will later be accounted for by the scale factor of the model. The fit is shown in Figure 11.5.

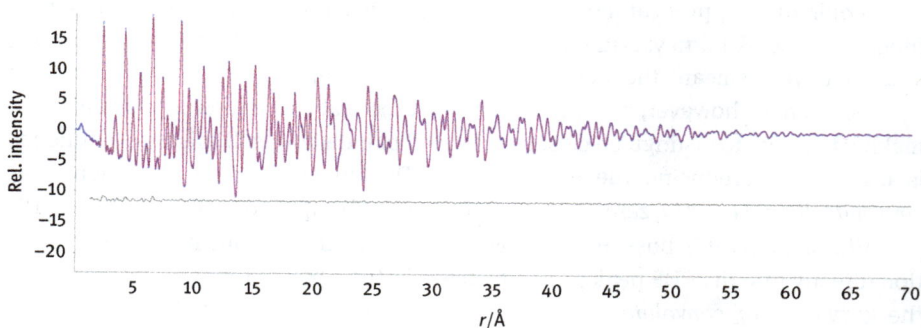

Figure 11.5: Fit of the PDF of a nickel reference material. Note the termination ripples apparent at low r.

---

6 Different software packages produce PDFs with different normalizations. TOPAS expects the PDF to vary from $-4\pi\rho_0 r$ at low-r to 0 at high-r. This is the .gr format of PDFgetX2 and PDFgetX3 (with a scale factor applied), the .mdor01 format produced by GudrunN when the sample is set to normalize to <b>², the .dofr format produced by GudrunX when "Divide by <F>²" is selected in the normalization or the dofr.xy format of DAWN. The equivalent function in the large-box RMCPROFILE package is $D_{norm}(r)$. Alternative $X(r)$ formats can be fitted using TOPAS *scale_phase_X* and *fit_obj* commands. For example, if the data are in the $G'(r)$ format of Keen (2001), the following commands can be used: *prm number_-density = num_atoms / Get(cell_volume); scale_phase_X = If(X>0.01,1/(4 Pi number_ density X),0); fit_obj = If(X>0.01,1,0);*

```
#include "pdf.inc"    ' Read pdf.inc, by Phil Chater if not already in your local.inc

xdd "Ni_standard_real-space.xy"
   start_X 1.4
   finish_X 70.0
   pdf_data

' Refine Qresolution damping and set the window for the Fourier truncation of the data
   dQ_damping(dQ, 0.09817)
   convolute_Qmax_Sinc(!Qmax, 22)
str
   phase_name "Ni"
   space_group Fm-3m
   Cubic( @  3.523588 )
   site Ni1  x 0  y 0 z 0     occ Ni 1
   scale @  0.99676674  ' 1.0 for a fully normalised G(r)
   ' Define beq by PDFfit2 method, using Uiso, delta1 for correlated motion of nearby atoms,
   ' and qbroad for Q-dependent broadening
   beq_PDFfit2(uiso, 0.00632, !rcut, 0.0, !sratio, 1.0, delta1, 0.82068, !delta2, 0.0,
   qbroad, 0.01422)
```

11.5 Calculation of a PDF

A PDF can be simulated in a similar way to a regular powder diffraction pattern (see Chapter 12). The following TOPAS script calculates a PDF for nickel from $r = 0$ to 100 Å with a step width of $dr = 0.01$ Å without any damping (Figure 11.6):

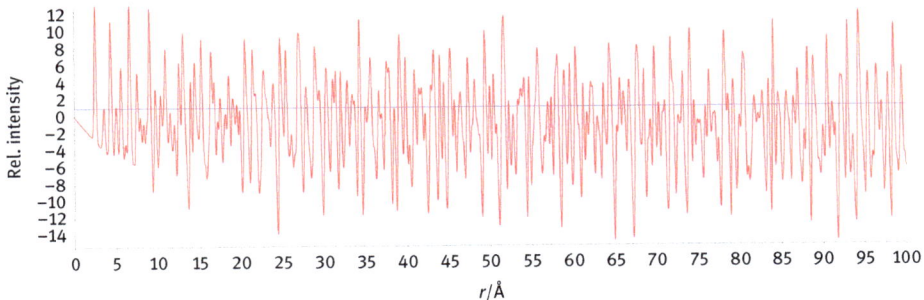

Figure 11.6: Simulated PDF of nickel without damping.

```
iters 0
yobs_eqn !Ni_PDF.xy = 1;  min 0  max 100  del 0.01
   pdf_data
   Out_X_Ycalc(Ni_PDF.xy)
```

```
str
    phase_name "Ni"
    space_group Fm-3m
    Cubic(3.524)
    scale 1
    prm  z0  0.05000  prm z1  0.50000  pdf_zero = z0 + z1 0.01 X;
    site Ni1  x 0  y 0  z 0  occ Ni 1 beq = 0.75 Erf_Approx( 0.3 X);
```

11.6 Joint refinements | Bragg + PDF

One useful feature of TOPAS is the ability to combine PDF fitting with Rietveld refinement and/or the global optimization method of simulated annealing. With disordered materials this approach can, for example, help constrain bond lengths in the Rietveld fit to their better-defined PDF values. The PDF dataset can either be derived from the powder pattern used for Rietveld refinement, or from one measured with PDF-optimized parameters (good statistics at high Q_{max}). Both patterns are then evaluated by refining common parameters derived from both. Data sets are weighted according to the summed intensities, but this can be modified by the user.

A simple constrained PDF+Rietveld refinement of the Ni standard is shown in Figure 11.7, and the corresponding TOPAS script is given below. The cubic lattice parameter and the isotropic displacement parameter are jointly refined parameters:

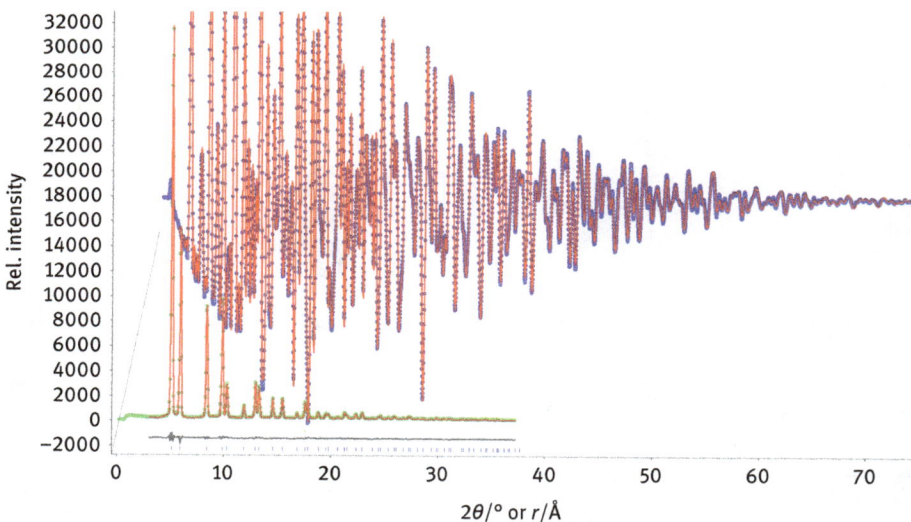

Figure 11.7: Rietveld (lower) and PDF plots (upper) from a combined Rietveld and PDF refinement of Ni.

```
#include "pdf.inc"    ' Read pdf.inc, by Phil Chater if not already in your local.inc

'------[ corefined parameters ]------
prm aNi  3.52427                  ' lattice parameter of Ni
prm bNi  0.42202 min 0.0          ' displacement parameter of Ni

'-----------[ PDF data ]-----------
xdd "Ni_standard_real-space.xy"
   xdd_sum !sum1 = Abs(Yobs);
   weighting = 1/sum1;
   start_X 1.4
   finish_X 70.0
   dQ_damping(dQ, 0.09279)
   convolute_Qmax_Sinc(!Qmax, 22)
   pdf_data

   str
     phase_name "Ni PDF"
     space_group Fm-3m
     Cubic(=aNi;)
     scale @  0.997
     site Ni num_posns  4 x  0 y  0 z  0 occ Ni  1
     beq_PDFfit2(uiso, =bNi/8/Pi^2;, !rcut, 0.0, !sratio, 1.0, delta1, 0.63, !delta2, 0.0,
qbroad, 0.02)

'-----------[ XRD data ]-----------
xdd "Ni_standard_2th.xy"
   xdd_sum !sum2 = Abs(Yobs);
   weighting = 1/sum2;
   x_calculation_step 0.02
   convolution_step 5
   bkg @  210.26 -64.69 -22.25 3.36 -1.68 -0.35 0.62 -0.15 -2.88 2.51 -3.18
   start_X  3
   LP_Factor( 90)
   lam
     ymin_on_ymax  0.0001
     la  1 lo  0.1827 lh  1e-005
   str
     phase_name "Ni XRD"
     LVol_FWHM_CS_G_L( 1, 6.16204, 0.89, 5.97526,@, 6.89316,@, 138.30578)
     e0_from_Strain( 0.00110,@, 0.46394,@, 0.07777)
     scale @  0.000216628033
     space_group Fm-3m
     Cubic(=aNi;)
     site Ni num_posns  4 x  0 y  0 z  0 occ Ni  1 beq=bNi;
```

References

Coelho, A.A., Chater, P.A., Kern, A. (2015): *Fast synthesis and refinement of the atomic pair distribution function*. J. Appl. Cryst. 48, 869–875.

Chung, J.S., Thorpe, M.F. (1997): *Local atomic structure of semiconductor alloys using pair distribution functions*. Phys. Rev. B 55, 1545–1153.

Chupas, P.J., Qiu, X., Hanson, J.C., Lee, P.L., Grey, C.P., Billinge, S.J.L. (2003): *Rapid-acquisition pair distribution function (RA-PDF) analysis*. J. Appl. Cryst. 36, 1342–1347.

Egami, T. Billinge, S.J.L. (2012): *Underneath the Bragg peaks, structural analysis of complex materials*, Volume 16, 2nd edition, Pergamon, Oxford (UK), 422 pages.

Farrow, C.L., Shaw, M., Kim, H., Juhás, P., Billinge, S.J.L. (2011): *Nyquist-Shannon sampling theorem applied to refinements of the atomic pair distribution function*. Phys. Rev. B 84, 134105.

Hammersley, A.P., Svensson, S.O., Hanfland, M., Fitch, A.N., Hausermann, D. (1996): *Two-dimensional detector software: from real detector to idealised image or two-theta scan*. High Pressure Research 14, 235–248.

Juhas, P., Davis, T., Farrow, C.L., Billinge, S.J.L. (2013): *PDFgetX3: a rapid and highly automatable program for processing powder diffraction data into total scattering pair distribution functions*. J. Appl. Cryst. 46, 560–566.

Keen, D.A. (2001): *A comparison of various commonly used correlation functions for describing total scattering*. J. Appl. Cryst. 34, 172–177.

Kodama, K., Tikubo, S., Taguchi, T., Shamoto, S. (2006): *Finite size effects of nanoparticles on the atomic pair distribution functions*. Acta. Cryst. A 62, 444–453.

Korsunskiy, V.I., Neder, R.B., Hofmann, A., Dembski, S., Graf, C., Rühl, E. (2007): *Aspects of the modelling of the radial distribution function for small nanoparticles*. J. Appl. Cryst. 40, 975–985.

Levashov, V.A., Billinge, S.J.L., Thorpe, M.F. (2005): *Density fluctuations and the pair distribution function*. Phys. Rev. B 72, 024111.

Olds, D., Saunders, C.N., Peters, M., Proffen, T., Neuefeind, J., Page, K. (2018): *Precise implications for real-space pair distribution function modeling of effects intrinsic to modern time-of-flight neutron diffractometers*. Acta Cryst. A 74, 293–307.

Soper, A.K., Barney, E.R. (2012): *On the use of modification functions when Fourier transforming total scattering data*. J. Appl. Cryst. 45, 1314–1317.

Toby, B.H., Egami, T. (1992): *Accuracy of pair distribution function analysis applied to crystalline and non-crystalline materials*. Acta Cryst. A 48, 336–346.

Yang, X., Juhas, P., Farrow, C.L., Billinge, S.J.L. (2014): *xPDFsuite: an end-to-end software solution for high throughput pair distribution function transformation, visualization and analysis*, arXiv 1402.3163v3.

12 Multiple data sets

A major strength of powder diffraction lies in its ability to measure diffraction patterns or PDFs under the influence of external variables such as temperature, pressure, time, changing chemical environment, exposure to light and so on. Recent progress in detectors in the laboratory (silicon strip position sensitive detectors) and at the synchrotron (large two-dimensional area detectors) allows the collection of huge numbers of powder patterns during *in situ* or *operando* powder diffraction experiments.

There are, in principle, two ways to analyze such a series of powder patterns: sequential and parametric. In sequential analysis, Rietveld analysis of each powder pattern is done separately based on the results of the previous refinement, with the entire set of all relevant parameters refined separately for each pattern. Further analysis of the values of these parameters, for example, fitting to empirical or physically based functions, is then performed *after* the Rietveld refinements.

In contrast, in parametric refinement (sometimes called surface refinement), all powder patterns of a series are analyzed simultaneously with selected parameters refined across all the powder patterns using an appropriate functional form. The parametric approach can greatly reduce the number of parameters and can allow the refinement of nonstructural parameters like transition temperatures or critical exponents. It also allows simple physical models to be imposed on the refinement, which helps constrain it to "sensible" solutions when there is more than one equivalent or similar model. The caveat is, of course, that the validity of the applied functions must be carefully checked.

In addition, if structural data are available, powder patterns or PDFs can easily be simulated as a function of internal variables like a particular torsion angle, crystallite size and so on. This can be a powerful method of assessing where key information is contained in the data.

In this chapter, we will show how powder pattern simulation, sequential and parametric refinement can be realized using the scripting language of TOPAS.

12.1 Simulation of angle dispersive powder patterns

Simulating powder patterns is a powerful way to visualize how changes in the instrumental configuration, the application of specific correction functions or small changes of the crystal structure will change a data set. Simulation can be very useful in designing experiments and data analysis strategies.

12.1.1 Simulation of a single powder pattern

To simulate a pattern in TOPAS, the keyword *yobs_eqn* is used in place of the *xdd* keyword. It replaces the observed data by an equation. The name given to the equation

https://doi.org/10.1515/9783110461381-012

!name is used for identifying the plot of the equation in the GUI. The simulated powder pattern(s) can be saved, for example, as an .xy file using the *Out_X_Ycalc macro*. In the following example, an angle dispersive, equal step powder pattern of the LaB_6 line profile standard (NIST SRM 660a LaB_6) is created from 12° to 120° 2θ with a step of 0.01° 2θ and saved to the file "D8_Mo_LaB6_capillary-simulated.xy" (Figure 12.1).

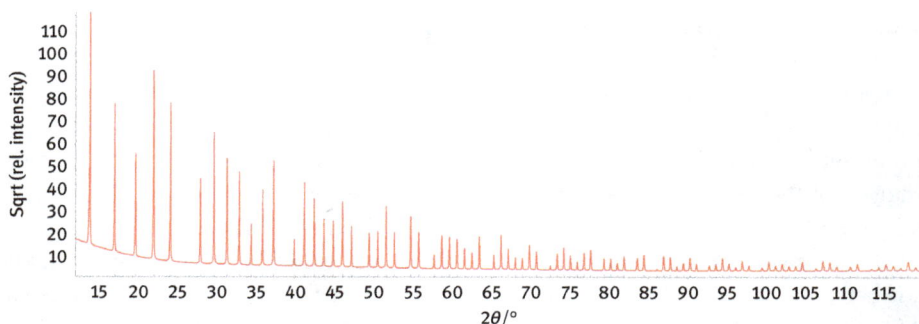

Figure 12.1: Simulated powder pattern of the LaB_6 line profile standard in Debye–Scherrer geometry using Mo-$K_{\alpha1}$ radiation.

The instrumental parameters describe a Bruker D8 advance powder diffractometer in Debye–Scherrer geometry with Mo-$K_{\alpha1}$ radiation (λ = 0.7093 Å) from a Ge(220) primary beam monochromator (angle for Lorentz polarization factor of 20.5°), and a silicon strip position sensitive detector. Parameters for the peak shape function (*TCHZ_Peak_Type*) and for the asymmetry due to axial divergence (*Simple_Axial_Model*) are taken from a previously determined instrumental resolution function of this diffractometer. The background is simulated by a Chebyshev polynomial of 10th order. The number of iterations (*iters*) is set to zero as no parameter is refined. The entire INP file for simulation of the powder pattern is given here:

```
' Create a simulated angle dispersive powder pattern of LaB6
iters 0
yobs_eqn !calc.xy = X;  min 12  max 120 del .01
Out_X_Ycalc( D8_Mo_LaB6_capillary-simulated.xy )

bkg  67.72 -85.24 64.89 -44.30 31.82 -21.40 12.48 -6.41 2.38 -0.58 -0.41
LP_Factor( 20.5)
Zero_Error(, 0.0)
Rs 217.5
Simple_Axial_Model( 9.5)
lam
```

```
   ymin_on_ymax  1e-005
    la  1 lo  0.7093 lh  0.2695
str
   TCHZ_Peak_Type(, 0.0106885,, -0.0011298,, 0.000521,, 0,, 0.041852,, 0)
   phase_name "LaB6 simulated"
   space_group Pm-3m
   scale  0.0003
   cubic( 4.154782417)
   site La num_posns  1 x =0;: 0.0  y =0;   : 0.0 z =0;   : 0.0 occ La+3  1 beq  0.39
   site B  num_posns  6 x  0.19986  y =1/2; : 0.5 z =1/2; : 0.5 occ B      1 beq  0.24
```

For a more realistic pattern, random noise can be added. In the following script, Poisson statistics are assumed (typical for single counter detectors), and noise proportional to the square root of the intensity is added to the calculated profile:

```
xdd_out sim_noise.xy load out_record out_fmt out_eqn
    {
        " %11.6f " = X;
        " %11.6f\n" = Rand_Normal(Ycalc, Sqrt(Ycalc));
    }
```

12.1.2 Simulation of a series of powder patterns

It is often useful to simulate a series of powder patterns as a function of one or more variables. Instead of manually changing and rerunning the INP file, TOPAS batch commands can perform all the simulations in a single step. TOPAS calls each simulation a "run".

In the following example, an increasing Lorentzian strain broadening is added to the previous LaB$_6$ simulation, creating a series of 10 simulated powder patterns (Figure 12.2). The number of simulations (runs) is defined by the variable *num_runs* and the current simulation number is defined by *Run_Number*. These variables control the way TOPAS executes the INP file:

```
'Create a series of 10 simulated patterns with increasing Lorentzian strain broadening
num_runs 10
iters 0
' Save INP file as an OUT
out_file = Concat(String(INP_File), ".out");
system_after_save_OUT { copy INP_File##.out INP_File##_##Run_Number##.out }

yobs_eqn !calc##Run_Number##.xy = X;    min 12    max 120    del .01
' Save each simulated powder pattern in the XY file format
   Out_X_Ycalc( D8_Mo_LaB6_capillary-simulated_##Run_Number##.xy )
```

```
bkg  67.72 -85.24 64.89 -44.30 31.82 -21.40 12.48 -6.41 2.38 -0.58 -0.41
LP_Factor( 20.5)
Zero_Error(, 0.0)
Rs 217.5
Simple_Axial_Model( 9.5)
lam
   ymin_on_ymax  1e-005
   la  1 lo  0.7093 lh  0.2695
str
   phase_name "LaB6 series simulated"
   TCHZ_Peak_Type(, 0.0106885,, -0.0011298,, 0.0005207,, 0,, 0.041852,, 0)
   lor_fwhm = 0.1 Run_Number Tan(Th);
   space_group Pm-3m
   scale  0.0002823909984
   Cubic( 4.154782417)
   site La num_posns 1 x =0;: 0.0  y =0;    : 0.0 z =0;  : 0.0 occ La+3   1 beq 0.39
   site B  num_posns 6 x  0.19986  y =1/2;  : 0.5 z =1/2;: 0.5 occ B       1 beq  0.24
```

The line *out_file* creates an output (OUT) file for each simulation in case the simulated data are needed again. The OUT file comprises the INP file but with parameter values updated. In this example, the OUT file is created using the name of the INP file plus the extension ".out" then copied to a unique name. More complex examples can be performed using the *#list* command discussed in Section 12.2.

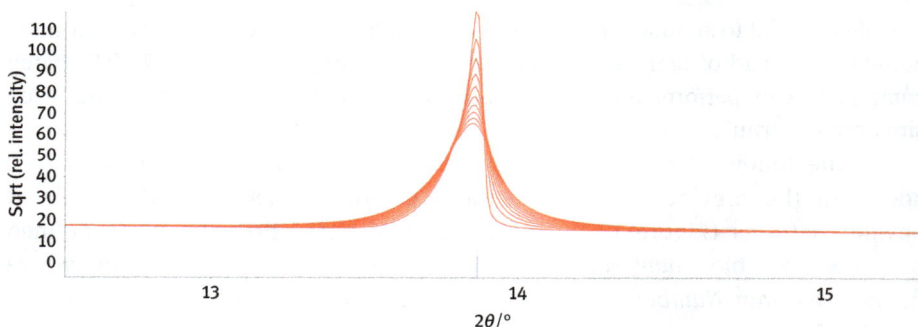

Figure 12.2: Blow up of a series of simulated powder patterns of LaB$_6$ in Debye–Scherrer geometry using Mo-K$_{\alpha 1}$ radiation with increasing Lorentzian strain broadening.

12.1.3 Scripting and the command line mode

For more complicated examples needing loops or parameter input from other sources or software, more elaborate scripting languages like Perl or Python might be used. In this situation, TOPAS can be called stand-alone from the windows command line or from a powershell. Specific parameters and instructions can then be automatically passed to TOPAS using macros and #define commands.

An example of the command line format is as follows:

```
C:\topas_v5> tc C:\directory\file.inp "macro filename {C:\directory\d9_02592} macro rangeuse {22}
macro information {prm !te 300 } #define CMD_LINE"
```

The corresponding INP file, file.inp, can contain any of the usual TOPAS scripting language, but would need the additional sections:

```
#ifdef !CMD_LINE 'use these lines when in gui mode
   macro filename {d8_022592}
   macro rangeuse { 12}
   macro information {prm !te 300 }
#endif

information

' Instead of giving the raw filename and range number in the INP file use the words
' "filename" and "rangeuse" which get replaced by information in the macros above or
' from command line.

RAW(filename)
range rangeuse
```

To execute this command line instruction, one needs to be in the main TOPAS directory. tc calls the program tc.exe, which is the command line version of TOPAS. It reads instructions from the INP file file.inp, which is best stored in a different directory. Any text placed in the quotation marks on the command line is passed to TOPAS. Here the first two macros define the name of the Bruker RAW data file to be analyzed and the data set range number within that file. The next macro passes information to the refinement as parameters – here the temperature at which the data were recorded. The final section "*#define CMD_LINE*" is a convenient trick that allows the same INP file to be run by TOPAS in either command line mode (tc.exe) or gui mode (topas.exe or ta.exe). This turns out to be very useful when testing INP files. The INP file contains a small section "*#ifdef !CMD_LINE ... #endif*". This syntax means that TOPAS will read this part of the INP file if *CMD_LINE* has *not* (!) been defined and will ignore it otherwise. When ta.exe or topas.exe is run macros *filename, rangeuse* and *information* are therefore defined via the INP file. When tc.exe is run, the command line flag *#define CMD_LINE* means that they are instead defined from the command prompt instruction.

12.2 Sequential Rietveld refinement

12.2.1 Introduction

Evaluation of a large numbers of powder patterns measured as a function of an external variable is potentially very time consuming. Fortunately, TOPAS can automate the data handling in most cases. For instance, if the step change of the external variable (e.g., temperature) between data sets is small, the refined parameters of one diffraction pattern can be used as starting values for the following diffraction pattern in an automated manner, provided that the important parameters don't change significantly. This can happen if, for example, a phase transition occurs. This approach is called sequential Rietveld refinement. If there are major step changes, TOPAS offers some flexibility through conditional statements to change the type and number of *strs* used as a function of the external variable(s).

12.2.2 Sequential refinement

From TOPAS v6 onward sequential refinements can be realized within a single INP file using a few keywords and macros.[1] In order to tell the program which data sets (powder patterns) should be analyzed, a one-dimensional array (*list*) of strings (*#list list_name*) is used. Each item in the list denotes a filename of a powder pattern, and the position in the list is given an index (starting at zero) which allows access to the corresponding powder pattern (e.g., *list_name(2)*)[2]:

```
#list File_Name
{
file1.xy
file2.xy
file3.xy
}
```

The number of runs that should be performed (usually equivalent to the number of data files) is given by the constant *num_runs*. Once started, the running variable *Run_Number* contains the running index from 0 until *num_runs–1*. This variable allows access to the powder pattern to be analyzed from the *#list* by *list_name (Run_Number)* (e.g., *xdd File_Name(Run_Number)*):

---

1 The same process can be readily achieved using scripting and the command line mode in v5 and earlier.

2 Hint: The contents of a directory can be conveniently copied into a text file by typing e.g. *dir *.raw > filelist.txt* at the command prompt of the windows command interpreter cmd.exe.

```
num_runs 3
...
xdd File_Name(Run_Number)
```

Parameters like temperature or time, which might be needed for identification or calculation purpose, may also be defined as a function of *Run_Number*. In the following example, the starting temperature is 25 °C which is increased by 2.6667 °C for each 10 s run until it reaches 450 °C after which it stays constant:

```
prm !Nr= Run_Number;
prm !Seconds= 10 Nr;
prm !DegCelsius = If( (Nr * 2.6667 + 25) > 450, 450, Nr * 2.6667 + 25);
```

If temperature values do not have a simple functional dependency on *Run_number*, and cannot be calculated, they can be included in the #list as a second column of real numbers. Any additional metadata and/or parameters can be included in the same way:

```
#list File_Name Temperature
{
file1.xy 25.0
file2.xy 27.8
file3.xy 39.5
}
```

Most frequently, the output of the current run should serve as input for the following run. Therefore, an OUT file must be created after each convergence and then copied to the INP filename. It sometimes helps to keep a copy of each OUT file for quality control, which can be achieved, for example, by converting the *Run_number* to a string to be included in the filename. In TOPAS syntax this can all be achieved with the following:

```
out_file = Concat(String(INP_File), ".out");           ' create an output file
system_after_save_OUT
{
 copy INP_File##.out INP_File##Run_Number##.out 'save an individ. copy of the output file
 copy INP_File##.out INP_File##.inp            ' create the next input file
}
```

Two different keywords can be used, depending on whether the system commands should be executed before (*system_before_save_OUT*) or after (*system_after_save_OUT*) the OUT file is created.

There are various ways to report refined values of parameters of interest for each *Run_Number* to an external file. The least controlled is the command *out_prm_vals_ on_convergence*, which writes each parameter in the model to a file. If only a small number of parameters are of interest, then the *out* command can be used either directly or via macros. For example, the macro *Append_0(file)* creates a new file at *Run_Number* = 0 then appends the data of the following runs. The following macro can be called from different phases with a different filename for each phase:[3]

```
macro Parameters(file, time, temperature)
{
  Append_0(file)
  Out_String("\n")      ' puts new line in file
  Out(time,             " %V")
  Out(temperature,      " %V")
  . . .
  Out(Get(cell_volume),  " %V")
  Out(Get(r_bragg),      " %V")
}
```

Additional parameters of interest can always be added to a previously used results file. For example, the following script first calculates the integral breath and then writes it to the results file. If a parameter doesn't exist, TOPAS will stop writing the results file. This can be circumvented with the command *Prm_There()*, which tests if the parameter exists, before it is used for calculation or report:

```
LVol_FWHM_CS_G_L( 1, 14.22392`, 0.89, 15.58663`,prm_CL, 22.07950`,prm_CG, 49.73158`)
prm prm_IB = 1 / IB_from_CS(CV(, prm_CL), CV(, prm_CG));
if Prm_There(prm_IB)  { Out( prm_IB,   " %V") }
```

Once all refinements are done, the results files can be read by a scientific plotting program like ™Origin, ™Grapher, ™Excel or gnuplot to analyze the data posteriori. Section 12.3.3 provides some simple gnuplot tricks for automating this.

For multiphase samples, phases that disappear during an experiment often refine to give unrealistically broad Bragg reflections, which can correlate with the background leading to incorrect phase fractions. This makes the definition of a lower size limit useful:

---

[3] The %V in the output command is a c++ like format descriptor. The various options are described in the Technical Reference.

```
macro lower_size_limit { 11 }

…

LVol_FWHM_CS_G_L( 1, 7.202120694, 0.89, 10.06647809,,,@, 11.30909565 min lower_size_limit)
```

Alternatively, all phases with fractions lower than, for example, 0.7 weight% can be automatically removed from the refinement using the following:

```
for strs { Remove_Phase 0.7 }
```

A more elegant way is to define the occurrence of phases using precompiler conditional statements. In the following example, phase 1 exists from 158 K < t < 198 K and phase 2 from 198 K < t < 226 K. The phases are then automatically turned on and off during the refinement:

```
' Phase boundaries
#prm t = 100 + (Run_Number 4);    'Temperature = 100 + Run_Number * 4
#if Or(t < 158, t > 226);
   #define Phase1
#elseif And(t > 158, t < 198);
   #define Phase1
   #define Phase2
#elseif And(t > 198, t < 226);
   #define Phase2
#endif
…
#ifdef Phase1
   str
   …
#endif
```

We'll see in Section 12.3 that parametric refinement can remove the need for such tricks, allowing a consistent model for all data sets.

12.2.3 Example of a sequential Rietveld refinement

The phase composition and particle size of carbonyl iron powder (CIP) was investigated on heating in a 2 bar stream of nitrogen gas (König, 2017). CIP contains 0.7–1 weight% carbon, and, depending on process conditions, nitrogen and oxygen below 1 weight%. A series of synchrotron X-ray powder diffraction patterns was collected as the CIP sample was heated from room temperature (RT) to 450 °C at a heating rate of 10 °C/min. The measurement time was fixed to 10 s/scan. Two hundred and fifty-five patterns were recorded as the sample was heated, then another 93 powder patterns while the final temperature was held constant. α-Fe, Fe_3C, Fe_4N and Fe_3O_4 were found in varying amounts as crystalline phases. Full quantitative sequential Rietveld refinements as a function of time and temperature were performed. A typical Rietveld plot is shown in Figure 12.3 and selected refined quantities in Figure 12.4.

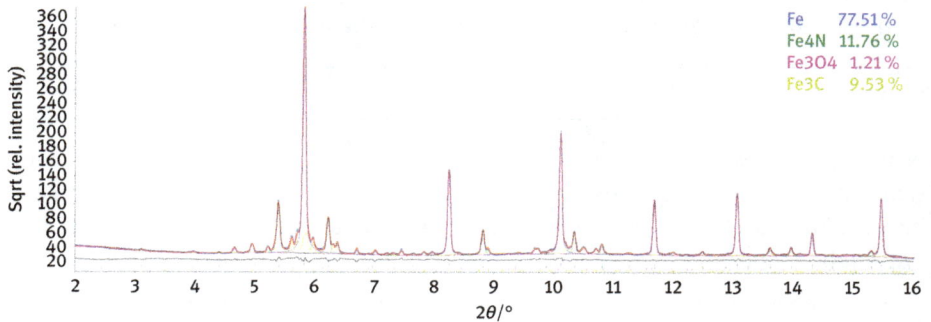

Figure 12.3: Typical Rietveld plot showing a full quantitative analysis of a single powder pattern of CIP at high temperature in a hot nitrogen gas stream. The square root of the intensity is displayed for better visualization of the minor phases.

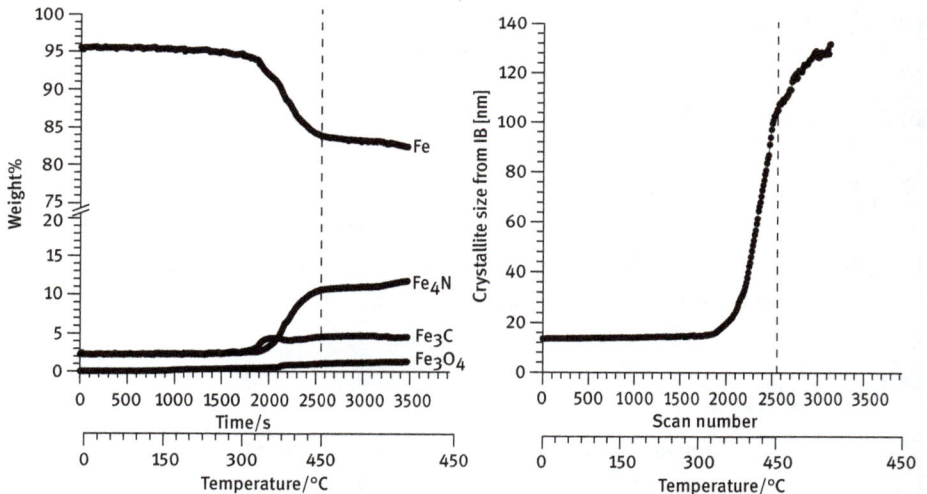

Figure 12.4: Results of the sequential Rietveld refinements of CIP in a hot nitrogen gas stream as a function of time and temperature. Left: weight fractions of all phases. Right: crystallite sizes of iron particles from integral breath data.

The TOPAS script for the sequential quantitative Rietveld refinements is given below. Only two phases (α-Fe and Fe_4N) are shown. In addition to the parameters listed in the macro *Parameters()*, the integral breath and two anisotropic microstrain parameters

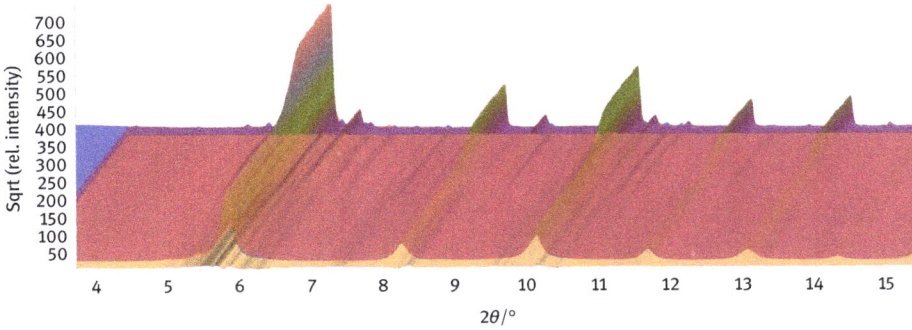

Figure 12.5: A 3D Rietveld plot for a sequential Rietveld refinement series of CIP in a hot nitrogen gas stream depending on scattering angle and time/temperature.

for α-Fe are reported for each run. The 3D Rietveld plot for the sequential Rietveld refinement series is shown in Figure 12.5.

```
 #list File_Name
{
CIP_3_SDD986_10s-00266.chi.xy
CIP_3_SDD986_10s-00267.chi.xy
...
CIP_3_SDD986_10s-00613.chi.xy
}
num_runs 348
out_file = Concat(String(INP_File), ".out");
 system_after_save_OUT
    {
       copy INP_File##.out INP_File##Run_Number##.out
       copy INP_File##.out INP_File##.inp
    }
prm !Nr= Run_Number;
prm !Seconds= 10 Nr;
prm !DegCelsius = If( (Nr * 2.6667 + 25) > 450, 450, Nr * 2.6667 + 25);

' lower crystallite domain size limit of the crystalline (!) phases in nm
macro lower_size_limit{ 11 }

macro Parameters(file, time, temperature)
{
            Append_0(file)
               Out_String("\n")
               Out(time,                  " %V")
               Out(temperature,           " %V")
               Out(Get(weight_percent),   " %V")
               Out(Get(a),                " %V")
```

```
                    Out(Get(b),                 " %V")
                    Out(Get(c),                 " %V")
                    Out(Get(al),                " %V")
                    Out(Get(be),                " %V")
                    Out(Get(ga),                " %V")
                    Out(Get(cell_volume),       " %V")
                    Out(Get(r_bragg),           " %V")
}
xdd File_Name(Run_Number)
   x_calculation_step 0.02
   bkg @  338.64 -66.14 32.72 -53.26 17.48 -14.59 -5.33 -2.01 6.04 7.99 -3.68
   start_X    3
   finish_X  16.3
   LP_Factor( 90)
   Zero_Error(@, -0.01674` min -0.02 max 0.02)
   exclude  4  4.1
   lam
      ymin_on_ymax  0.001
      la  1 lo  0.20717 lh  1e-005
   str                          ' α -Fe
      Stephens_cubic(, 1 min =0;,prm_S400, 67953.87 min =0;,prm_S220, 0.0 min =0;)
      TCHZ_Peak_Type( 0,,  -0.00124,, 0.00123,, 0,, 0.0106,, 0.00426)
      LVol_FWHM_CS_G_L( 1, 14.22392`, 0.89,15.58663`,prm_CL, 22.07950`,prm_CG, 49.73158`)
      phase_name "Fe"
      space_group Im-3m
      scale @  0.000441464724`
      Cubic(@  2.859845`)
      site Fe1 num_posns  2 x  0 y  0 z  0 occ Fe  1 beq B1  0.46191` min =0; max =1;
      prm prm_IB = 1 / IB_from_CS(CV(, prm_CL), CV(, prm_CG));

      Parameters("Fe_CIP3.dat", Seconds, DegCelsius)
      if Prm_There(prm_S400) { Out( prm_S400, " %V") }
      if Prm_There(prm_S220) { Out( prm_S220, " %V") }
      if Prm_There(prm_IB)   { Out( prm_IB,   " %V") }
   str                              ' Fe₄N
      LVol_FWHM_CS_G_L( 1, 7.00282`, 0.89, 9.79000`,,,@, 11.00 min lower_size_limit)
      TCHZ_Peak_Type(, 0,,  -0.00124,, 0.00123,, 0,, 0.0106,, 0.00426)
      phase_name "Fe4N"
      space_group Pm-3m
      scale @  4.39949789e-006`
      Cubic(@  3.774593`)
      site Fe1  num_posns  3 x  0.5 y  0     z  0     occ Fe  1 beq  B2  1.0
      site Fe2  num_posns  1 x  0.5 y  0.5 z  0.5 occ Fe  1 beq =B2;
      site N1   num_posns  1 x  0     y  0     z  0     occ N   1 beq =B2;

      Parameters("Fe4N_CIP3.dat", Seconds, DegCelsius)
...
```

12.3 Parametric Rietveld refinement

In the previous sections, we've discussed sequential refinements where individual data sets are analyzed independently. There are, of course, many situations where it's necessary to analyze an ensemble of data collected under evolving conditions (a "surface" of diffraction data) in a more sophisticated way. For example, you might want to derive some parameters from all the data sets rather than from a single (noisy) pattern. A trivial example might be the diffractometer zero point during a series of time-resolved experiments. The instrument misalignment leading to the zero point is unlikely to change during the course of the experiment, and it is therefore best determined from all data sets simultaneously. Similarly, when analyzing a variable temperature experiment in Bragg–Brentano geometry, one might need to refine a height correction to allow for thermal expansion of the sample mount. The correction is likely to change smoothly with temperature, and is probably best described using a simple empirical function derived from all the data. The fitting of a smooth function to describe parameters from a surface of diffraction data is called parametric Rietveld refinement. Since zero point and sample height both correlate with cell parameters, describing these parameters parametrically would lead to reduced uncertainties in cell parameters, that is, we have a method to extract "good information" from "bad data."

The flexibility of the TOPAS scripting language means that it's possible to parametrize a wide range of different parameters. Some of these and typical functions used to describe them are given in Table 12.1.

The parametric approach is most powerful in cases where there is a significant correlation between different refined parameters, or multiple models that could fit an individual data set equally well. In this situation, the parametric equations can guide the refinement toward a unique or physically sensible minimum. The second major strength comes from the ability to refine "non-crystallographic" parameters, which wouldn't be defined by a single data set. Examples include kinetic parameters, critical exponents of phase transitions or temperature calibration curves for *in situ* experiments.

When looking for subtle changes in a parameter of interest (e.g., a specific site occupancy or atomic coordinate), the best way to use the parametric approach is probably to parameterize the variables you're *not* interested in and to freely refine the parameter of most interest.

The danger of the parametric approach is that a physical model is imposed on the refinement as a mathematical constraint and it is important to ensure that it is valid. One way to assess this is to compare the R-factor for each data set from a parametric fit with that from sequential refinement. If the parametric fit is significantly worse, then the model may not be describing the experiment well. That said, identifying an inappropriate model is actually a major strength of the parametric approach: In fitting an individual data set, refined model parameters can adopt physically non-sensical values and incorrectly fit features of the data. This is prevented in the more demanding parametric approach. Discrepancies in plots of R_{wp} over the refined

Table 12.1: Examples of functions used in parametric refinements.

Quantity	Expression	Comment
Cell parameters or adps	$a(T) = a_0 + \dfrac{c_1\theta_1}{e^{(\theta_1/T)} - 1}$	Simple Einstein-like model with a_0, c_1 and θ_1 as refinable parameters. Ensures a physically sensible zero gradient at $T = 0$ K.
Fractional coordinates	$x(T) = x_0(1 + c_1 T + c_2 T^2 + c_3 T^3)$	Simple polynomial form.
Critical behavior	$v(T) = c_1\left(1 - \dfrac{T}{T_c}\right)^\beta$	Site occupancy, magnetic moment or other quantity related to an order parameter approaching a phase transition.
Kinetic parameters	$frac(t) = c_1\left(1 - e^{-k_{frac}t}\right) + c_2$ $cell(t) = c_1\left(1 - e^{-k_{cell}t}\right) + c_2$	Simple rate expression. $k(t - t_0)$ or $k(t - t_0)^n$ to describe more complex evolution.
Zero point	$zero(t) = const$	Zero point correction unchanging over data surface.
Sample height	$height(T) = h_0(1 + c_1 T + c_2 T^2)$	Sample height as a smooth function of temperature.
Temperature error	$\Delta T = c_0\left(1 + c_1 T_{set} + c_2 T_{set}^2\right)$	Offset between furnace set point and sample temperature as a smooth function.

surface can therefore reveal inadequacies in the model that wouldn't otherwise be identified (Magdysyuk et al., 2014).

12.3.1 Parametric quantitative refinement

As an example of parametric refinement we'll consider analysis of a set of laboratory powder data recorded on cooling a sample of WO_3 from 300 to 90 K. Over this temperature range WO_3 undergoes two transitions from a RT $P2_1/n$ monoclinic structure to triclinic $P\bar{1}$ at ~230 K then to monoclinic Pc at ~190 K (Figure 12.6). These are displacive transitions that give rise to relatively minor structural changes and therefore relatively small changes in the powder diffraction pattern (Figure 12.6). In fact, if we consider a small portion of the diffraction pattern around 24° 2θ, all three phases of WO_3 have peaks in this 2θ region, with similar intensities (Figure 12.7). This makes quantitative analysis of a single data set difficult as it's possible for the "incorrect" phase to distort to fit the diffraction data. For example, even though the material is predominantly in the monoclinic $P2_1/n$ form at RT, the $P\bar{1}$ structural model, which has more degrees of freedom, can distort to fit the experimental data as well or better than a $P2_1/n$ model.

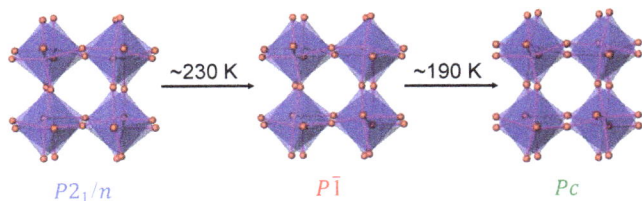

Figure 12.6: Illustrations of the crystal structures of the different polymorphs of WO_3 at temperatures below 300 K. Tungsten is located in the center of the blue octahedra.

Figure 12.8(a) and (c) show the phase fractions (as weight%) and cell volumes of each phase when the data are analyzed sequentially. Fifty-six parameters are used at each temperature meaning 5600 parameters for all data sets. Each refinement was started from what we know (from the later parametric analysis) to be an ideal model and simulated annealing was performed at each temperature to ensure the lowest R_{wp} solution was found. As such this is the "best possible" sequential fitting. It is clear that the results don't make physical sense: abrupt changes and reversals in the phase evolution and cell parameters are observed. We note that less careful ("normal") sequential fitting models give more marked discrepancies.

Figure 12.8(b) and (d) show equivalent plots from a parametric approach in which all patterns were evaluated simultaneously using three simple assumptions: (1) The cell parameters of each individual phase show a smooth variation with temperature with the functional form given in Tab. 12.1, with coefficients a_0, c_1 and θ refined from the diffraction data; for non-90° angles a second-order polynomial was used. (2) The peak shape description was set up such that each individual phase had an identical peak shape at all temperatures. (3) Fractional atomic coordinates were refined from all data sets simultaneously. In this way a single Rietveld refinement was performed with 1167 parameters being refined simultaneously from all 100 data sets. This led to the phase fractions and cell volumes shown in Figure 12.8(b) and (d). The extracted phase factions make much more chemical sense and vary in a smooth manner with temperature, even though they were not constrained in any way. Treating quantities of minor interest (here the cell) parametrically has given better information on the quantities of interest (phase fractions). The parametrically fitted cell parameters also make physical sense in that they give rise to comparable volume coefficients of expansion for each phase as expected.

The success of this approach has two basic origins. The most important is the fact that the three-phase model, with each phase constrained to have cell parameters that evolve with temperature in a physically sensible way, must simultaneously fit every experimental data set. This prevents, for example, the high-temperature $P2_1/n$ phase distorting to fit minor discrepancies in the low-temperature data as happens in the sequential analysis. If it did, the distortions would prevent accurate fitting of

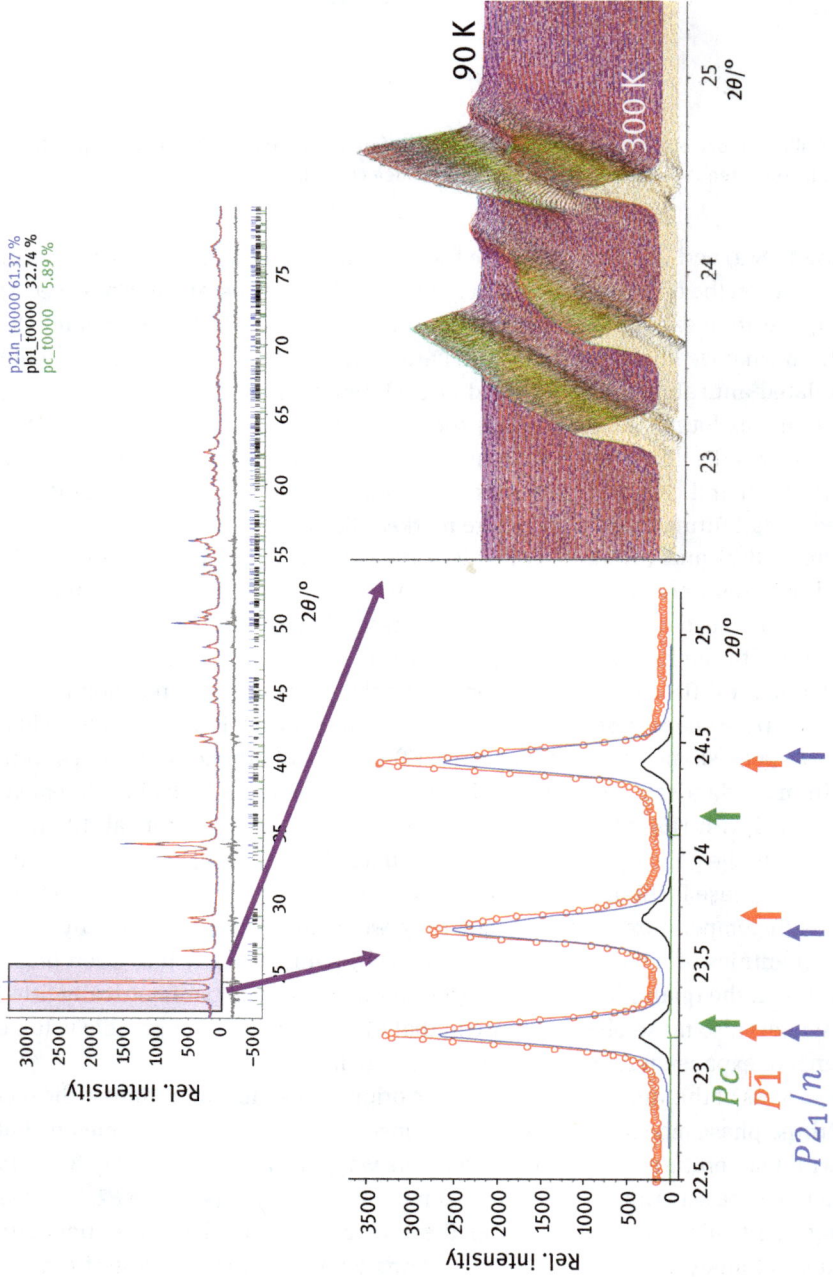

Figure 12.7: Powder diffraction data of WO₃. Top: 300 K Rietveld fit. Bottom left: reflections contributing to peaks at ~24° 2θ. Bottom right: parametric Rietveld fit on cooling 300–90 K showing phase evolution.

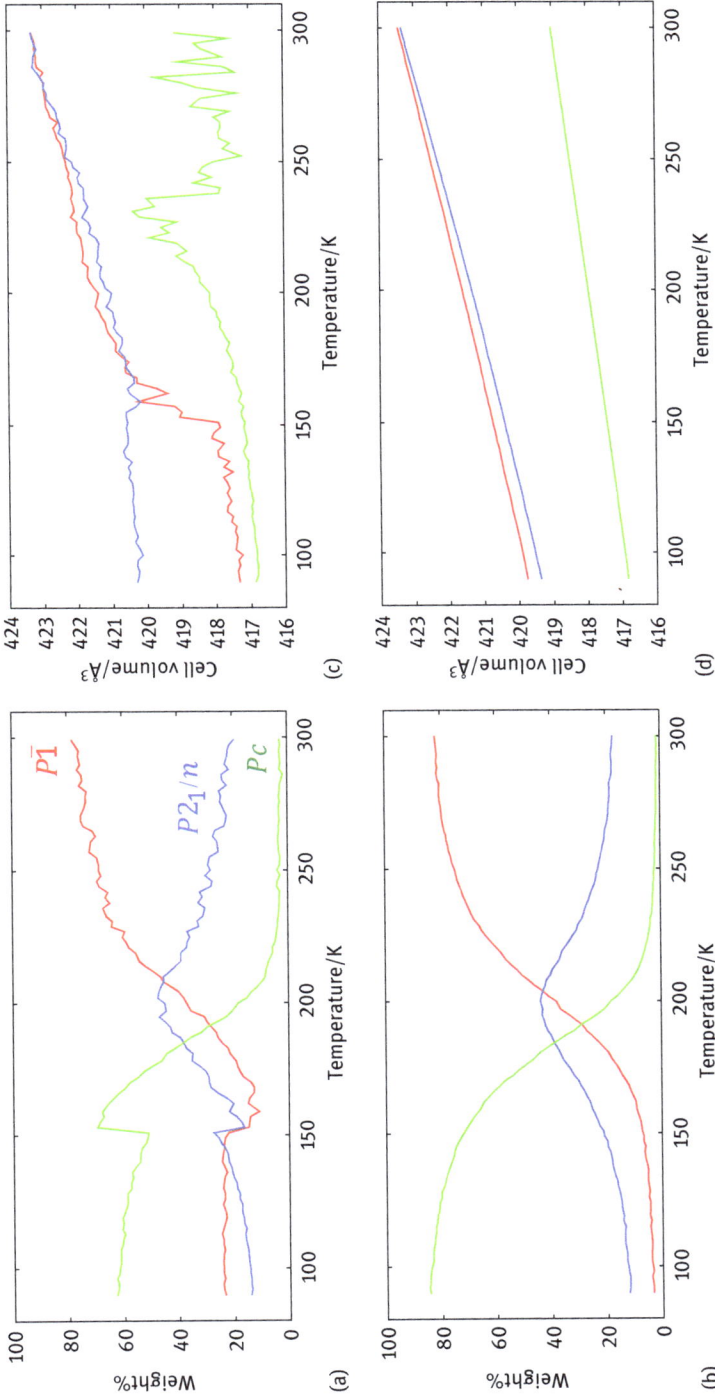

high-temperature data, where more of the $P2_1/n$ phase is present. Secondly, the parametric approach to modeling peak shapes means that peak shape parameters for an individual phase are controlled by the temperature regions of the diffraction surface where it is actually present. This prevents peak shapes for a phase becoming excessively broad in regions of the data where the phase is not present, which would allow it to "mop up" minor errors in fitting the experimental background or peak tails. In essence the parameters describing an individual phase are determined by regions of the data where the phase is present rather than regions where it is not.

The INP file needed to perform this refinement makes use of many of the scripting ideas we've met before. In particular, it makes extensive use of *#define* and *#ifdef ... #endif* statements to control refinement; it uses the *local* keyword instead of *prm* to allow reuse of parameter names in different sections of the file; it uses *"for xdds"* and *"for strs"* sections to feed information to individual data sets and it uses macros to insert blocks of text into multiple sections of the file. We can break the file down into the six separate sections shown below, where each section has been folded in jEdit so that the ~12000 line input file looks remarkably simple:

```
'------------------------------------------------------------
' Parametric fit of WO3 data collected on cooling 300 K to 90 K
' See http://community.dur.ac.uk/john.evans/topas_workshop/tutorial_surface_new.htm
'------------------------------------------------------------

'{{{ 1. R-factors, refinement choices, iters, etc [16 lines]
'{{{ 2. Select data sets to analyse: define filenames, temperatures, metadata [119 lines]
'{{{ 3. Refinement instructions for each dataset [11701 lines]
'{{{ 4. Overall refined parameters for every data set [20 lines]
'{{{ 5. for xdds section with for strs inside it [75 lines]
'{{{ 6. Some overall simplifying macros [34 lines]
```

If we expand each of the six sections in turn what we see is as follows:

Section 1 contains overall instructions for the refinement and *#define* flags telling TOPAS specific instructions. For example, *#param_cell* is a flag telling TOPAS to read the section of the INP file describing a smooth cell parameter across the data; if it's not defined, then cell parameters are refined independently. *#define write_out* flags that parameters should be written to an output file on convergence. To make later plotting of results easier, there is also a command that deletes any old parameter output file "results.txt", then writes a column header line to a new file:

```
'{{{ 1. R-factors, refinement choices, iters, etc
r_wp 12.7080712 r_exp 8.94672233 r_p 9.82045561 r_wp_dash 18.7386884 r_p_dash 16.1765072
r_exp_dash 13.1923908 weighted_Durbin_Watson 102.986766 gof 1.42041641
chi2_convergence_criteria 0.001
```

```
continue_after_convergence
conserve_memory
iters 10000
'#define riet_out
#define write_out
#define param_cell
Backup_INP    ' Writes a .bck backup of the INP in case of divergence
system_before_save_OUT { del results.txt }
out "results.txt" append
Out_String(" Temp(K) Rwp(%) Perc_P21n Esd Scale_P21n Esd a_P21n Esd b_P21n Esd c_P21n Esd
alpha_P21n ESD beta_P21n Esd gamma_P21n Esd Vol_P21n Esd Perc_Pb1 Esd Scale_Pb1 Esd a_Pb1 Esd
b_Pb1 Esd c_Pb1 Esd alpha_Pb1 ESD beta_Pb1 Esd gamma_Pb1 Esd Vol_Pb1 Esd Perc_Pc Esd Scale_Pc Esd
a_Pc Esd b_Pc Esd c_Pc Esd alpha_Pc ESD beta_Pc Esd gamma_Pc Esd Vol_Pc Esd")
'}}}
```

Section 2 defines the filename, temperature, range number and any other metadata for each data set using macros. A flag *#define use_t0001* on each line tells TOPAS to analyze this specific data set, where *t0001* is a simple label. Individual data sets can then be included/excluded rapidly by commenting these lines in/out of the INP file. The philosophy is similar to the *#list* command:

```
'{{{ 2. Select data sets to analyse: define filenames, temperatures, metadata
'#define use_some 'flag this to just fit 10 ranges
#define use_t0001 macro filename_t0001 {d9_02592} prm !t0001  300  macro r_t0001 { 1 }
#define use_t0002 macro filename_t0002 {d9_02592} prm !t0002  297  macro r_t0002 { 2 }
#define use_t0003 macro filename_t0003 {d9_02592} prm !t0003  295  macro r_t0003 { 3 }
.
.
.
#define use_t0100 macro filename_t0100 {d9_02592} prm !t0100  90  macro r_t0100 {100}
#endif

'}}}
```

Section 3 contains the "guts" of the INP file – 11701 lines in this example![4] Within an *#ifdef use_t0001 ... #endif* section instructions are given for Rietveld refinement in the usual TOPAS format. The data set is defined using the macro*"filename_t0001"* and temperature is defined as a *local* parameter; both use information specified in Section 2. These are followed by *local* parameters defining an overall temperature factor and sample height. Finally, there are lines that give the option of outputting the *R*-factor to a file and/or writing ASCII files of observed calculated and difference patterns to separate directories for later plotting:

---

4 The whole INP file contains 51731 "words" – almost double Shakespeare's longest play!

```
'{{{ 3. Refinement instructions for each dataset
/* {{{ (Foldable) Information for dataset number -> 0001 */
#ifdef use_t0001
    RAW(filename_t0001)
    local te = t0001;:300.00000 range r_t0001
    bkg  @  62.1081942`  -58.3869804`  33.8200838`  -23.7210244`  14.0400682`  -5.0985472`
    local  individual_b -1.01942` min -4 max 10 'single bvalue for all phases
    local height -0.08002` min = -0.5; max = 0.5; del 0.001  th2_offset = -2
57.2957795130823 (height) Cos(Th) / Rs;

    '{{{ write out xdd specific results here; phase-specific info done in the for_strs section
    #ifdef write_out
        out "results.txt" append
        Out(te, "\n %11.5f")
        Out(Get (r_wp), " %11.5f")
    #endif
    'write out rietveld fits here
    #ifdef riet_out
        Rietveld_Plot(riet_plots\rie_t0001.xyd)
        xdd_out  obs_plots\obs_t0001.xy load out_record out_fmt out_eqn { "%11.5f "=X; " %11.5f
\n" = Yobs; }
        xdd_out calc_plots\cal_t0001.xy load out_record out_fmt out_eqn { "%11.5f "=X; " %11.5f
\n" = Ycalc; }
        xdd_out diff_plots\dif_t0001.xy load out_record out_fmt out_eqn { "%11.5f "=X; " %11.5f
\n" = Yobs-Ycalc; }
    #endif
    '}}}

    '{{{ P21/n structure information
    '{{{ Pb1 structure information
    '{{{ Pc structure information

#endif
```

Section 3 also contains information on each of the phases in a series of *str* sections. These are folded away in the script above, but are shown in unfolded form below. In this example there are two options set up for describing the unit cell: either with a parametric approach (*#ifdef param_cell*) using the cell parameter expression given in Tab. 12.1 or (*#else*) free refinement at each temperature. Fractional atomic coordinates aren't given within the *str* section, but are read in from the later Section 5 using "*for strs*":

```
'{{{ P21/n structure information
str
phase_name p21n
#ifdef param_cell
```

```
      local mlpa    = zero_ma + (grad_ma/(Exp(theta_ma/t0001)-1));:7.30115`
      local mlpb    = zero_mb + (grad_mb/(Exp(theta_mb/t0001)-1));:7.53897`
      local mlpc    = zero_mc + (grad_mc/(Exp(theta_mc/t0001)-1));:7.69259`
      local mlpbeta = beta_m0 * (1+beta_m1*1e-6*t0001+beta_m2*1e-9*t0001^2);:90.84889`
      a = mlpa;:7.301151`
      b = mlpb;:7.538973`
      c = mlpc;:7.692591`
      al   90.
      be = mlpbeta;:90.84889`
      ga   90
   #else
      a  @   7.302503 mlpaminmax
      b  @   7.537414 mlpbminmax
      c  @   7.690997 mlpcminmax
      al     90.
      be @  90.86670 mbetminmax
      ga     90.
   #endif
   space_group "P121/n1"
   scale @  1.38975664e-05`
   weight_percent  82.160`
   Phase_Density_g_on_cm3( 7.27438`)
   r_bragg  100
   output_lines
   '}}}}
```

Section 4 of the file is used to define overall parameters that are simultaneously refined from all data sets. In this case, the parameters are coefficients
of the functions used to describe smooth unit cell parameters via the equations
of Tab. 12.1:

```
'{{{ 4. Overall parameters for every data set
#ifdef param_cell
'overall variables controlling cell dimensions here for P21/n (m), P-1 (t) and Pc (p) phases
'cell edges
    prm zero_ma   7.29410` prm   grad_ma 0.00447` prm  theta_ma  147.30008`
    prm zero_mb   7.50500` prm   grad_mb 0.02154` prm !theta_mb = theta_ma;:147.30008`
    prm zero_mc   7.65447` prm   grad_mc 0.02417` prm !theta_mc = theta_ma;:147.30008`
    prm zero_ta   7.30852` prm   grad_ta 0.00275` prm !theta_ta = theta_ma;:147.30008`
    prm zero_tb   7.48849` prm   grad_tb 0.02081` prm !theta_tb = theta_ma;:147.30008`
    prm zero_tc   7.65227` prm   grad_tc 0.02814` prm !theta_tc = theta_ma;:147.30008`
    prm zero_pa   5.27198` prm   grad_pa 0.00669` prm !theta_pa = theta_ma;:147.30008`
    prm zero_pb   5.15781` prm   grad_pb 0.00490` prm !theta_pb = theta_ma;:147.30008`
    prm zero_pc   7.65858` prm   grad_pc 0.01523` prm !theta_pc = theta_ma;:147.30008`
'angles
    prm  beta_m0  91.08706` prm    beta_m1 -4.74099`   prm   beta_m2 -13.25062`
    prm  alph_t0  88.92748` prm    alph_t1 -15.83299`  prm   alph_t2  63.55158`
    prm  beta_t0  89.71326` prm    beta_t1 136.08812`  prm   beta_t2 -365.55447`
```

```
    prm  gamm_t0  89.96058` prm   gamm_t1  64.99380` prm   gamm_t2 -94.22736`
    prm  beta_p0  91.78854` prm   beta_p1 -19.65621` prm   beta_p2  66.08487`
#endif
'}}}
```

Section 5 contains information that applies to all the data sets within a *"for xdds {...}"* block, information that applies to all structures *"for strs {...}"* and information for each individual structure *"for strs n to n {...}"*. Using this format means that a change on a single line in the INP file (e.g., *finish_X 80*) will be applied to all the data sets. Defining atomic sites with parameter names also means that a single set of coordinates is fitted across the entire data surface. Alternatively coordinates could be refined using a smooth function with a language analogous to that for the unit cells in Sections 3 and 4. Or, in the other extreme, an @ flag could be used so they refine independently for each data set. The coordinates in this example are contained within a macro that prevents them moving too far from literature values and allows them to be reset in a *continue_after_convegence* process:

```
'{{{ 5. for xdds section with for strs inside it
for xdds {
    x_calculation_step = Yobs_dx_at(Xo); convolution_step 4
    start_X 20
    finish_X 80.0
    LP_Factor(!th2_monochromator, 0)
    CuKa2(0.0001)
    Zero_Error(!zero, 0.0) 'refining height for each phase so not needed
    Simple_Axial_Model(!axial, 5.74444)
    scale_pks = (Exp(-2*individual_b/(2 D_spacing)^2)); 'bvalue for all phases in each
dataset

    for strs {
    TCHZ_Peak_Type(pkuc,-0.78214`,pkvc, 0.53092`,pkwc,-0.07771`,!pkxc,0.0001,pkyc,
0.16135`,!pkzc, 0.0001)
    }
    macro A1(param,val,val2) {x param val min = val2 - 0.05; max = val2 + 0.05;
val_on_continue = val2;}
    macro A2(param,val,val2) {y param val min = val2 - 0.05; max = val2 + 0.05;
val_on_continue = val2;}
    macro A3(param,val,val2) {z param val min = val2 - 0.05; max = val2 + 0.05;
val_on_continue = val2;}
    for strs 1 to 1 {
        'ideal val2 coordinates from Woodward P21/n
        site W1 A1( xW1, 0.25258`, 0.25130) A2( yW1, 0.03122`, 0.02770) A3( zW1, 0.27932`,
0.28650) occ W 1.0  beq !bval 0
        site W2 A1( xW2, 0.24584`, 0.24810) A2( yW2, 0.03034`, 0.03420) A3( zW2, 0.78571`,
0.78150) occ W 1.0  beq !bval 0
        site O3 A1(!xO3, 0.00143,  0.00143) A2(!yO3, 0.03550,  0.03550) A3(!zO3, 0.21526,
```

```
0.21526) occ O 1.0  beq !bval 0
     site O4 A1(!xO4, 0.99858,  0.99858) A2(!yO4, 0.46301,  0.46301) A3(!zO4, 0.21782,
0.21782) occ O 1.0  beq !bval 0
     site O5 A1(!xO5, 0.28165,  0.28165) A2(!yO5, 0.26251,  0.26251) A3(!zO5, 0.28384,
0.28384) occ O 1.0  beq !bval 0
     site O6 A1(!xO6, 0.20905,  0.20905) A2(!yO6, 0.25985,  0.25985) A3(!zO6, 0.73375,
0.73375) occ O 1.0  beq !bval 0
     site O7 A1(!xO7, 0.27950,  0.27950) A2(!yO7, 0.04128,  0.04128) A3(!zO7, 0.00461,
0.00461) occ O 1.0  beq !bval 0
     site O8 A1(!xO8, 0.28259,  0.28259) A2(!yO8, 0.48854,  0.48854) A3(!zO8, 0.99449,
0.99449) occ O 1.0  beq !bval 0
     Strain_G(strainp21n, 0.12499`)
    }
  for strs 2 to 2 {
.
.
  }
  for strs 3 to 3 {
.
.
  } 'end of for xdds
'}}}
```

Section 6 contains macros that simplify the input file. For example, the macro *output_line* is included in each of the *strs* of Section 4. By changing the macro, one can instantly change the information written to results.txt at the end of each refinement. There are also macros to automatically control non-parameterized cell parameters in the case of divergence:

```
'{{{ 6. Some overall simplifying macros
  ' These set min and max values for cells if allowed to refine individually
  ' Useful for resetting things to "sensible" values if refinement diverges
  ' val_on_continue values set at typical values for each phase from parametric fit
  prm !tol 1.1
  macro mlpaminmax  { min =7.301/tol; max = 7.301*tol; val_on_continue = 7.301;}
  macro mlpbminmax  { min =7.539/tol; max = 7.539*tol; val_on_continue = 7.539;}
  macro mlpcminmax  { min =7.692/tol; max = 7.692*tol; val_on_continue = 7.692;}
  macro tlpaminmax  { min =7.310/tol; max = 7.310*tol; val_on_continue = 7.310;}
  macro tlpbminmax  { min =7.508/tol; max = 7.508*tol; val_on_continue = 7.508;}
  macro tlpcminmax  { min =7.678/tol; max = 7.678*tol; val_on_continue = 7.678;}
  macro plpaminmax  { min =5.274/tol; max = 5.274*tol; val_on_continue = 5.274;}
  macro plpbminmax  { min =5.159/tol; max = 5.159*tol; val_on_continue = 5.159;}
  macro plpcminmax  { min =7.662/tol; max = 7.662*tol; val_on_continue = 7.662;}
  macro pbetminmax  { min 91.6 max 91.8 val_on_continue = 91.68;}
  macro mbetminmax  { min 90.0 max 92.0 val_on_continue = 90.86;}
  macro tlpalminmax { min 87.0 max 90.0 val_on_continue = 88.87;}
  macro tlpbeminmax { min 90.0 max 92.0 val_on_continue = 90.85;}
  macro tlpgaminmax { min 90.0 max 92.0 val_on_continue = 90.79;}
```

```
' Output key information on each phase to a file through this macro
macro output_lines {
#ifdef write_out
   Out(Get(weight_percent), " %11.6f", " %11.6f")
   Out(Get(scale),          " %11.9f", " %11.9f")
   Out(Get(a),              " %11.6f", " %11.6f")
   Out(Get(b),              " %11.6f", " %11.6f")
   Out(Get(c),              " %11.6f", " %11.6f")
   Out(Get(al),             " %11.6f", " %11.6f")
   Out(Get(be),             " %11.6f", " %11.6f")
   Out(Get(ga),             " %11.6f", " %11.6f")
   Out(Get(cell_volume),    " %11.6f", " %11.6f")
 #endif
 }
'}}}
```

One of the challenges of producing a file like this is, of course, writing the 11701 lines needed in Section 4 for the 100 data sets. However, this is much less daunting than it might seem! In essence, the information within each of the *#ifdef use_t0001 … #endif* folds is identical to that needed for any single Rietveld refinement. The only difference is that it contains various local parameters and equations containing the temperature (or any other external variable) as *t0001*. Once a single Rietveld has been set up one just needs to reproduce this section of the file once for each data set with the *t0001* string swapped to *t0002, t0003, … , t0100*. This can be easily done with a simple computer program or PYTHON script. It can also be done in seconds on a linux system (e.g., the ubuntu shell that is included in Windows 10) using the built in function *sed*, which performs simple text editing functions from a shell prompt.

In this case, all that's needed is a text file (single.riet) that contains the Section 3 information for a single refinement. Then create a file (e.g., expand.sh) containing the following lines:

```
sed '1,$s/t0000/t0001/g' single.riet
sed '1,$s/t0000/t0002/g' single.riet
.

.
sed '1,$s/t0000/t0100/g' single.riet
```

These commands can be executed with "*sh expand.sh >all_riet.txt.*" This will create a single text file all_riet.txt containing the 11701 lines of TOPAS script needed. expand.sh and Section 2 information on filenames can be generated in seconds using column editing in jEdit from any list of experimental details (e.g., *dir *.xye > files.list*).

12.3.2 Non-crystallographic parameters

The parametric approach also allows the refinement of noncrystallographic para-meters. An example is the extraction of smooth temperature calibration curves for nonambient experiments as described by Stinton et al. (2007). In this approach internal standards such as Si or Al_2O_3, which have known thermal expansion, are mixed with the sample of interest. During parametric analysis of variable temperature data the cell parameters of the standards aren't refined freely, but are instead calculated using expressions of the type given in Tab. 12.1 with coefficients known from the literature. If there is any discrepancy between the furnace set temperature and the actual set temperature ($T_{sample} = T_{set} + \Delta T$) then calculated and observed peak positions of the standards will not match. This can be corrected by refining ΔT either individually from each data set or (better) using a function that varies smoothly with temperature. In a published example of analyzing 871 data sets on ZrP_2O_7 to understand its unusual thermal expansion, using this latter approach both the true phase transition tempera-ture and cell parameters could be extracted more reliably with the parametric approach than sequential refinements. Figure 12.9 (left) shows cell parameters extracted sequen-tially as scattered open points and those from a parametric refinement as closed points.

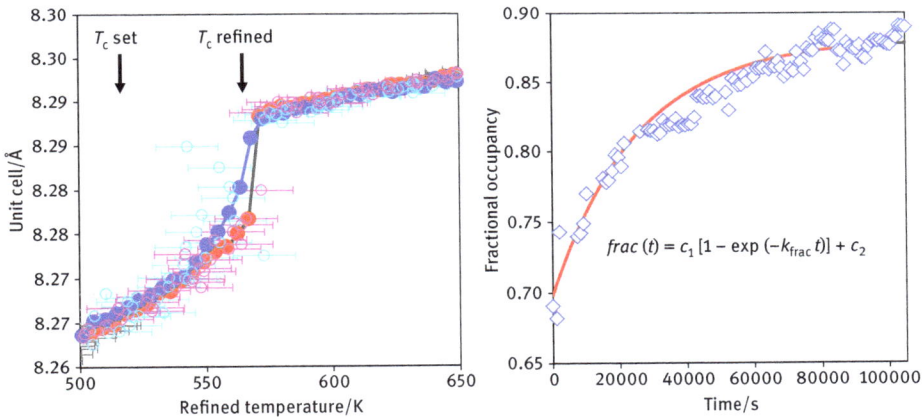

Figure 12.9: Left: unit cell parameters of ZrP_2O_7 close to a displacive phase transition derived either sequentially (open points) or parametrically (closed points) on warming (red) and cooling (blue). A parametric temperature calibration gives significantly lower uncertainty in cell parameters. The uncorrected phase transition temperature, T_c, was ~520 K, whereas the refined T_c of 567–571 K agreed perfectly with DSC data. Right: a comparison between parametrically (red solid line) and sequentially (blue points) refined site occupancies for an order–disorder phase transition in $ZrWMoO_8$. The parametric approach allows direct Rietveld refinement of the rate constant, k_{frac}.

A second published example followed the kinetics of an oxygen migration process in $ZrWMoO_8$ using time-resolved X-ray data. Here either site occupancies or accompanying cell parameter changes could be expressed using simple rate laws, allowing rate constants to be refined directly from a surface of data. In fact, in unpublished work, it was possible to parametrically analyze >1000 data sets collected as a function of time and temperature (4D data) using a simple Arrhenius expression to describe the temperature dependence of rate constants. This allowed the activation energy of the process to be determined directly by Rietveld refinement.

12.3.3 Plotting tricks

Thanks to the speed of TOPAS, one of the slowest parts of either sequential or parametric work is often visualizing the refined parameters to see if they make physical sense. This is particularly time consuming if you need to compare multiple refinement models. The text files of refined parameters produced by the methods in this chapter can, of course, be plotted in any standard program (e.g., Origin, Excel, etc.). For speed we recommend the plotting program gnuplot (http://www.gnuplot.info/), which is freely available for most operating systems. One good feature is that it can be driven by a simple scripting language just like TOPAS, and can produce multiple plots very quickly. For example, the simple script below will automatically produce plots both on the screen and saved as individual .gif files for the 28 columns of the file "results.txt" in the WO_3 example. It can be adapted to autoplot any results file by editing just five lines. The file can be run either by double clicking in windows or using "*load 'filename.gnu'*" at the gnuplot command:

```
# Script to plot results file from multiple refinements
# Plot columns with no error first then columns with errors
# File has a dummy header line labelling the columns
# Temperature in col 1 Then data are in col 2 to 14; col 2 has no esd
# Change the column numbers to your example

reset

file = 'results.txt'
xcol     = 1        # x in column 1 is temperature
startcol = 2        # first column to plot is column 2
numnoesd = 1        # 1 column without an esd (Rwp in column 2)
numesd   = 27       # 27 columns with an esd

set style line 1 lt rgb "red" linewidth 2 pt 7

set key autotitle columnheader font "Arial,12" Left noenhanced
set term win font "Arial,24"
set ytics mirror
set xlabel 'Temperature/K'
```

```
#plot each column to the screen then to a gif file for columns with no esd
do for [col=startcol:startcol+(numnoesd-1):1] {
  plot file using xcol:col w linesp ls 1
  set term gif size 640,480 font "arial,12"; set output 'pic_col_'.col.'.gif'; replot; set
term win
  pause -1 "<cr> to see the next plot"
}

#plot each column to screen then to a gif file for columns with associated esd
do for [col=startcol+numnoesd:startcol+numnoesd+2*(numesd-1):2] {
    plot file using xcol:col w linesp ls 1
  replot file using xcol:col:col+1 w errorbars ls 1 notitle
  set term gif ; set output 'pic_col_'.col.'.gif'; replot; set term win
  pause -1 "<cr> to see the next plot"
}
```

References

König, R., Müller, S., Dinnebier, R.E., Hinrichsen, B., Müller, P., Ribbens, A., Hwang, J., Liebscher, R., Etter, M., Pistidda, C. (2017): *The crystal structures of carbonyl iron powder – revised using in situ synchrotron XRPD*. Z. Kristallogr. Cryst. Mater. 232, 835–842.

Magdysyuk,O.V., Müller, M., Dinnebier, R.E., Lipp, C., Schleid, T. (2014): *Parameterization of the coupling between strain and order parameter for LuF[SeO₃]*. J. Appl. Cryst. 47, 701–711.

Stinton, G.W., Evans, J.S.O. (2007): *Parametric Rietveld refinement*. J. Appl. Cryst. 40, 87–95.

13 Appendix: Mathematical basics

The aim of this chapter is to remind the reader of some basic mathematical tools which are commonly used in crystallography and, in particular, in Rietveld analysis. This not intended as a substitute for a math (e.g., Papula, 2014) or general crystallography textbook (e.g., Giacovazzo, 2011; Müller, 2013). For brevity we give a collection of formulas and mathematical concepts rather than provide full definitions or proofs. We assume the reader is familiar with basic vector algebra, analysis and analytical geometry.

Conventions: Tensors of rank one or greater are always in bold. Transposed matrices are denoted by an additional bar. Lowercase letters are used for one-dimensional tensors (vectors) and uppercase letters for multidimensional tensors (matrices). Positive rotations are always counterclockwise when looking toward the origin of the rotation axis/vector.

13.1 Complex numbers

A complex number z codes a length and an angle, like a vector in a two-dimensional Cartesian coordinate system spanned by the two axes of the real and imaginary numbers (Figure 13.1).

$$z = x + \mathrm{i}y = |z|(\cos\varphi + \mathrm{i}\sin\varphi) \tag{13.1}$$

with the complex number $\mathrm{i} = \sqrt{-1}$. The complex axis has a phase shift of $\pi/2$ (or 90°) with respect to the real axis. The conjugate complex number z^* is defined as:

$$z^* = x - \mathrm{i}y = |z|(\cos\varphi - \mathrm{i}\sin\varphi). \tag{13.2}$$

The length of the complex vector can be calculated as:

$$|z| = \sqrt{zz^*} = \sqrt{x^2 + y^2}. \tag{13.3}$$

The angle is determined by:

$$\varphi = \arctan\left(\frac{y}{x}\right) \text{ if } x > 0,$$

$$\varphi = \arctan\left(\frac{y}{x}\right) + \pi \text{ if } x < 0 \text{ and } y \geq 0,$$

$$\varphi = \arctan\left(\frac{y}{x}\right) - \pi \text{ if } x < 0 \text{ and } y < 0$$

https://doi.org/10.1515/9783110461381-013

Imaginary axis

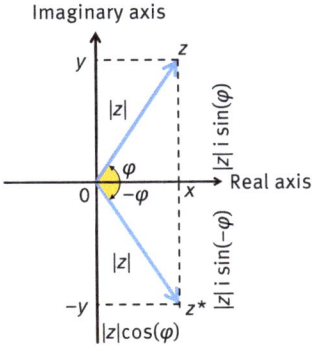

$\varphi = \dfrac{\pi}{2}$ if $x = 0$ and $y > 0$,

$$\phi = -\frac{\pi}{2} \text{ if } x = 0 \text{ and } y < 0 \tag{13.4}$$

For $x = y = 0$, φ is undetermined.

13.2 Power series

Derivatives of nonlinear and transcendental functions are important in the Rietveld method. Such functions can be approximated by a special power series (Taylor series) around the current values of their parameters. Such a series is usually terminated after the first or second term. When used in least squares this truncation means that shifts applied to a model won't lead to a fully minimized solution, but to a hopefully better approximation.

A general representation of a power series with the center of the series x_0 is given by:

$$P(x) = \sum_{n=0}^{\infty} a_n (x - x_0)^n \tag{13.5}$$

with real numbers $a_0, a_1, \ldots$ as coefficients. Under several boundary conditions, it is in principle possible to approximate a function $f(x)$ by a power series. The most common power series of this kind is the Taylor series:

$$f(x) = \sum_{n=0}^{\infty} \frac{f^{(n)}(x_0)}{n!} (x - x_0)^n, \tag{13.6}$$

Figure 13.1: The vector (pointer) z and its conjugate complex z^* in the complex plane. Dashed lines represent the Cartesian coordinates.

where the coefficient a_n is determined by the n-th derivative $f^{(n)}(x_0)$ of the approximated function around the point x_0 with $f^{(0)}(x_0) = f(x_0)$ (Figure 13.2). If the derivatives of the original function cannot be determined analytically, numerical approximations can be used. If the development point x_0 is at the origin of the function, the so-called MacLaurin series is obtained:

$$f(x) = \sum_{n=0}^{\infty} \frac{f^{(n)}(0)}{n!} x^n. \tag{13.7}$$

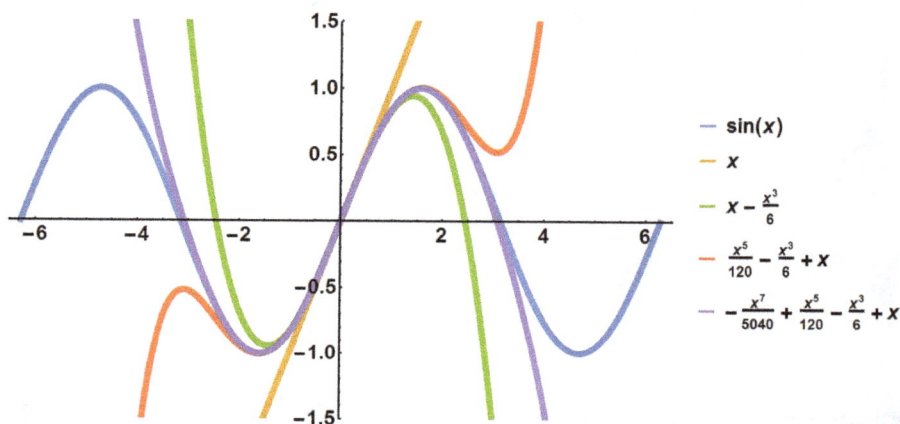

Figure 13.2: Taylor expansion of a sine-function around the origin as development point x_0 using derivatives of second, fourth, sixth and eighth order.

MacLaurin series for the exponential, sine and cosine functions are given by:

$$e^x = \sum_{n=0}^{\infty} \frac{x^n}{n!}, \quad e = \lim_{n \to \infty} \left(1 + \frac{1}{n}\right)^n = 2.718281\ldots$$

$$\sin x = \sum_{n=0}^{\infty} (-1)^n \frac{x^{2n+1}}{(2n+1)!},$$

$$\cos x = \sum_{n=0}^{\infty} (-1)^n \frac{x^{2n}}{(2n)!}. \tag{13.8}$$

A common estimate for the error between the original and approximation function using a Taylor expansion up to the nth term is given by the remainder term after Lagrange:

$$R_n(x) = \frac{f^{(n+1)}(\delta x)}{(n+1)!} x^{n+1} \text{ with } (0 < \delta < 1). \tag{13.9}$$

13.3 The Euler formula

Using a MacLaurin series, the famous Euler formula (Figure 13.3) can be calculated by:

$$e^{i\varphi} = \sum_{n=0}^{\infty} \frac{(i\varphi)^n}{n!} = 1 + \frac{(i\varphi)^1}{1!} + \frac{(i\varphi)^2}{2!} + \frac{(i\varphi)^3}{3!} + \frac{(i\varphi)^4}{4!} + \frac{(i\varphi)^5}{5!} + \frac{(i\varphi)^6}{6!} + \frac{(i\varphi)^7}{7!} +$$

$$1 + i\varphi - \frac{\varphi^2}{2!} - i\frac{\varphi^3}{3!} + \frac{\varphi^4}{4!} + i\frac{\varphi^5}{5!} - \frac{\varphi^6}{6!} - i\frac{\varphi^7}{7!} + \ldots$$

$$= \left(1 - \frac{\varphi^2}{2!} + \frac{\varphi^4}{4!} - \frac{\varphi^6}{6!} + \ldots\right) + i\left(\varphi - \frac{\varphi^3}{3!} + \frac{\varphi^5}{5!} - \frac{\varphi^7}{7!} + \ldots\right),$$

$$(13.10)$$

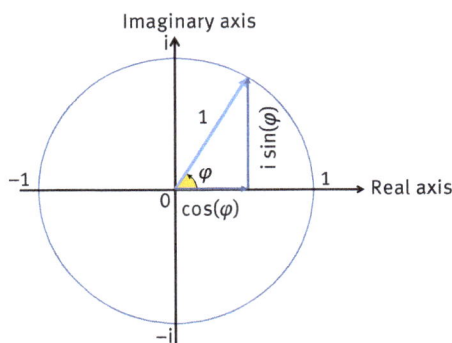

Figure 13.3: The unity vector (pointer) in the complex number plane.

which is the same as:

$$e^{i\varphi} = \cos\varphi + i\sin\varphi \tag{13.11}$$

from which it follows:

$$\cos\varphi = \frac{1}{2}\left(e^{i\varphi} + e^{-i\varphi}\right) \quad \text{and} \quad i\sin\varphi = \frac{1}{2}\left(e^{i\varphi} - e^{-i\varphi}\right). \tag{13.12}$$

13.4 Summing pointers

A pointer (e.g., amplitude) consists of a magnitude (length) and a phase. Summing pointers can be done by vector-like addition in two-dimensional space using the complex number plane (Figure 13.4).

Imaginary axis

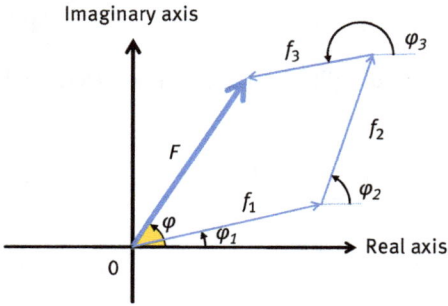

Real axis

Figure 13.4: Summation of complex pointers f_1, f_2 and f_3 to yield complex F.

The resulting complex amplitude F:

$$F = |F|\ e^{i\varphi} = |F|(\cos\varphi + i\sin\varphi) \tag{13.13}$$

can be calculated by complex summation over all complex pointers f_j :

$$F = \sum_{j=1}^{n} |f_j| e^{i\varphi_j} = \sum_{j=1}^{n} |f_j| (\cos\varphi_j + i\sin\varphi_j). \tag{13.14}$$

The sum above basically represents a Fourier series.

13.5 Fourier transformation

The Fourier integral transforms a function $f(x)$ of the variable x into an integral over cosine und sine functions $F(s)$ of the variable s:

$$F(s) = \int_{-\infty}^{\infty} f(x)\ e^{2\pi i s x} dx = \int_{-\infty}^{\infty} f(x)\ (\cos(2\pi s x) + i\ \sin(2\pi s x)) dx \tag{13.15}$$

while the back transformation is given by:

$$f(x) = \int_{-\infty}^{\infty} F(s)\ e^{-2\pi i s x} dx$$
$$= \int_{-\infty}^{\infty} F(s)(\cos(2\pi s x) - i\ \sin(2\pi s x)) ds. \tag{13.16}$$

Since the exponent of the exponential function is dimensionless, s is reciprocal to x. If x is in position space, s is in "Fourier space" or "reciprocal space", if x is time, s is a frequency.

The function $f(x)$ must satisfy the conditions of Dirichlet:

- The interval of definition can be divided into a finite number of intervals in which $f(x)$ is continuous and monotonous.
- The limits of any point of discontinuity are defined on both sides $\lim_{\varepsilon \to \infty} f(x_v - \varepsilon) = f(x_v - 0)$ and $\lim_{\varepsilon \to \infty} f(x_v + \varepsilon) = f(x_v + 0)$.
- The integral $\int_{-\infty}^{+\infty} |f(x_v)| dx$ must converge.

The Fourier transformation (denoted by $\leftrightarrow$) is a linear operation:

$$\alpha f(x) + \beta g(x) \leftrightarrow \alpha F(s) + \beta G(s), \tag{13.17}$$

with the complex numbers α, β and $f(x) \leftrightarrow F(s)$ and $g(x) \leftrightarrow G(s)$.

Shifting of a function $f(x)$ on the x-axis by a constant u to the right simply leads to an additional phase factor before the Fourier transform. Using the substitution $w = x - u$ with the same derivative $dx = dw$ leads to:

$$\int_{-\infty}^{\infty} f(x - u) e^{2\pi i s x} dx = \int_{-\infty}^{\infty} f(w) e^{2\pi i s (w + u)} dw = e^{2\pi i s u} F(s). \tag{13.18}$$

The Fourier transform of a differentiated function leads to an additional pre-factor:

$$\frac{\partial^n}{\partial x^n} f(x) \leftrightarrow (-2\pi i s)^n F(s). \tag{13.19}$$

13.6 Convolution

Convolution or folding is a basic concept in crystallography and of particular important in powder diffraction analysis. The process of convolution is one in which the product of two functions $f(x)$ and $g(x)$ is integrated over all space:

$$h(x') = \int_{-\infty}^{\infty} f(x)\, g(x' - x) dx = f \circ g, \tag{13.20}$$

where $h(x')$ is the convolution product, x' is the variable of integration in the same domain as x and $\circ$ denotes the convolution process. Convolution can be understood as "blending" one function with another, producing a kind of very general "moving average." One can illustrate the formation of the convoluted function by setting down the origin of the first function in every possible position of the second, multiplying the values of both functions in each position and taking the sum of all these operations. The convolution operation is commutative and associative.

An illustrative example for a typical convolution function is given in Figure 13.5, where a triangular and a rectangular intensity function are convoluted leading to a function approximating a Gaussian (see eq. (13.83)). Most functions cannot be convoluted analytically and the convolution integral needs to be calculated numerically.

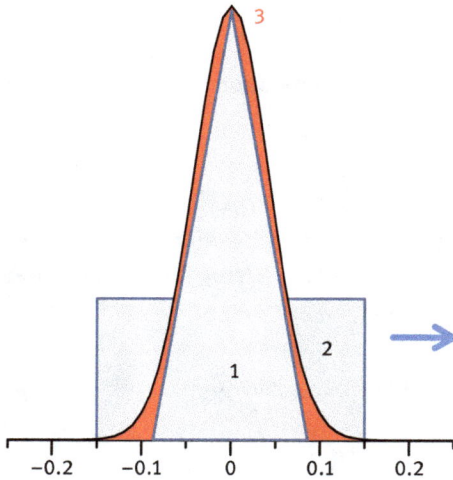

Figure 13.5: Convolution of a triangular (1) with a rectangular (2) function leading to a function approximating a Gaussian (3).

An alternative way of calculation follows from the Fourier transform of the convolution function. The Fourier transform of the convolution function can be calculated by:

$$\int_{-\infty}^{\infty} h(x')e^{2\pi i s x'}dx' = \int_{-\infty}^{\infty}\int_{-\infty}^{\infty} f(x)\ g(x'-x)dx\ e^{2\pi i s x'}dx'$$

$$= \int_{-\infty}^{\infty}\int_{-\infty}^{\infty} f(x)\ g(x'-x)dx\ e^{2\pi i s(x'-x+x)}dx'. \qquad (13.21)$$

This can be rewritten by using the substitution $u = x' - x$ and therefore $du = dx'$ as:

$$= \int_{-\infty}^{\infty}\int_{-\infty}^{\infty} f(x)\ g(u)e^{2\pi i s x}dx\ e^{2\pi i s u}du$$

$$= \int_{-\infty}^{\infty}\int_{-\infty}^{\infty} f(x)e^{2\pi i s x}dx\ g(u)e^{2\pi i s u}du = F(s)\ G(s). \qquad (13.22)$$

From this it follows directly that the Fourier transform of the convolution integral is the product of the Fourier transforms of all functions participating in the convolution:

$$f \circ g \leftrightarrow F(s)\ G(s), \qquad (13.23)$$

while the back-transformation of a convolution is the product of the back-transformed functions that participate in the convolution:

$$f(x)\, g(x) \leftrightarrow F \circ G \tag{13.24}$$

As a practical example, a crystal structure can be regarded as a convolution of the structural motif (atom, group of atoms or molecule) and a three-dimensional lattice. The scattered amplitude of a crystal is the product of the Fourier transforms of the motif (density distribution of the unit cell) and the lattice function (three-dimensional sum of equally spaced δ-functions).

In TOPAS, the operation of convolution can be performed by means of direct convolution or by Fast Fourier Transform (FFT) convolution.

13.7 Correlation function

If $f(x)$ and $g(x)$ are functions with $F(s)$ and $G(s)$ being their Fourier transforms, then the correlation function is defined as:

$$h_C(x') = \int_{-\infty}^{\infty} f(x)\, g(x'+x)\mathrm{d}x \equiv f_- \circ g$$
$$f_- \circ g \leftrightarrow F(-s)\, G(s) \tag{13.25}$$

For real functions, $F(-s)$ is the same as the conjugate complex function $F^*(s)$. A special form of the correlation function is the autocorrelation (Patterson) function, where a function is correlated with itself:

$$h_K(x') = \int_{-\infty}^{\infty} f(x)\, f(x'+x)\mathrm{d}x = f_- \circ f$$
$$f_- \circ f \leftrightarrow F(-s)\, F(s) \tag{13.26}$$

A typical example is the diffracted intensity, which can be considered as the autocorrelation of the structure factor amplitude.

13.8 The delta *(δ)* function

The δ-function is defined by its integration properties:

$$F(s) = \int_A^B \delta(x - x_0)\mathrm{d}x \text{ is 1 if } A < x_0 < B \text{ otherwise } 0$$
$$F(s) = \int_A^B f(x)\, \delta(x - x_0)\mathrm{d}x \text{ is } f(x_0) \text{ if } A < x_0 < B \text{ otherwise } 0. \tag{13.27}$$

In practice, a δ-function can be approximated by a normalized Gaussian function with a width approaching 0: $\lim(\sigma \to 0)$ where σ is related to the width via eq. (13.85). The Fourier transform of a δ-function can be easily calculated:

$$F(s) = \int_{-\infty}^{\infty} \delta(x-x_0) \, e^{2\pi i s x} dx = e^{2\pi i s x_0}, \tag{13.28}$$

which in case of $x_0 = 0$ equals to 1. The back transformation then is:

$$\delta(x-x_0) = \int_{-\infty}^{\infty} e^{2\pi i s x_0} e^{-2\pi i s x} ds = \int_{-\infty}^{\infty} e^{-2\pi i s(x-x_0)} ds. \tag{13.29}$$

A series of N equally spaced δ-functions separated by distance a represents a one-dimensional lattice:

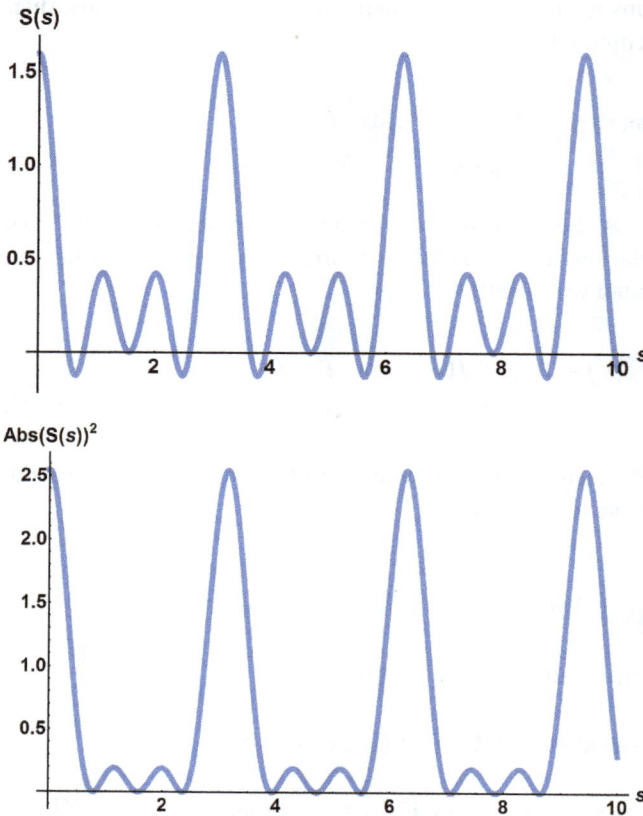

Figure 13.6: Amplitude (top) and squared amplitude (bottom) of the Laue function with $N = 3$ and $a = 2$.

$$f_\nu(x) = \sum_{\nu=0}^{N} \delta(x - a\nu). \tag{13.30}$$

The Fourier transform of this one-dimensional lattice can be directly calculated with help of eq. (13.12)[1] as (Figure 13.6):

$$S(s) = \int \sum_{\nu=0}^{N} \delta(x - a\nu) e^{2\pi i s x} dx = \sum_{\nu=0}^{N} e^{2\pi i a\nu s}$$

$$= \frac{e^{2\pi i(N+1)as} - 1}{e^{2\pi ias} - 1} = \frac{e^{\pi i(N+1)as}}{e^{\pi ias}} \frac{e^{\pi i(N+1)as} - e^{-\pi i(N+1)as}}{e^{\pi ias} - e^{-\pi ias}}$$

$$= e^{\pi iNas} \frac{\sin(\pi(N+1)as)}{\sin(\pi as)}, \tag{13.31}$$

which can easily be expanded to higher-dimensional lattices.

13.9 Matrix algebra

A rectangular array of real numbers in m rows and n columns is called a real $(m \times n)$ matrix **A**:

$$\mathbf{A} = \begin{pmatrix} A_{11} & A_{12} & \cdots & A_{1n} \\ A_{21} & A_{22} & \cdots & A_{2n} \\ \vdots & \vdots & \ddots & \vdots \\ A_{m1} & A_{m2} & \cdots & A_{mn} \end{pmatrix} = (A_{ik}) \tag{13.32}$$

The left index i, running from 1 to m, is called the row index, the right index k, running from 1 to n, is the column index of the matrix. An $n \times n$ matrix is called a square matrix, an $(m \times 1)$ matrix a column matrix (or column vector) and a $(1 \times n)$ matrix a row matrix (or row vector). A transposed matrix is obtained by exchanging rows and columns:

$$\overline{\mathbf{A}} = (A_{ki}). \tag{13.33}$$

A symmetric matrix is called a diagonal matrix if $A_{ik} = 0$ for $i \neq k$. A diagonal matrix with all elements $A_{ik} = 1$ is called the unit matrix **I**. A matrix consisting of zeroes only,

---

1 $P_N = \sum_{\nu=0}^{N-1} e^{x\,\nu} = 1 + e^x + e^{2x} + \ldots + e^{(N-1)x}$ $P_N\,e^x\,P_N = e^x + e^{2x} + \ldots + e^{Nx}$ $P_N - e^x\,P_N = 1 - e^{Nx} \rightarrow P_N(1 - e^x) =$
$(1 - e^{Nx}) \rightarrow P_N = \frac{1-e^{Nx}}{1-e^x}$ or $P_N = \frac{e^{Nx}-1}{e^x-1}$

that is, $A_{ik} = 0$ is called the **0** matrix. The sum of the diagonal terms of a square matrix is called the trace of the matrix:

$$\mathrm{tr}(\mathbf{A}) = \sum_{j=1}^{n} A_{jj}. \tag{13.34}$$

An $(m \times n)$ matrix $\mathbf{A}$ is multiplied by a (real or complex) number λ by multiplying each element with λ:

$$\lambda\mathbf{A} = (\lambda A_{ik}). \tag{13.35}$$

Two $(m \times n)$ matrices $\mathbf{A}$ and $\mathbf{B}$ can be summed (subtracted) by summing (subtracting) each pair:

$$\mathbf{C} = (A_{ik} \pm B_{ik}). \tag{13.36}$$

Two matrices $\mathbf{A}$ and $\mathbf{B}$ can be multiplied if the matrix $\mathbf{A}$ has the same number r of columns as $\mathbf{B}$ has rows:

$$\mathbf{C} = \mathbf{AB} \text{ with } C_{ik} = \sum_{j=1}^{r} A_{ij}B_{jk}, \ i = 1, ..., \mathrm{m}; \ k = 1, ..., n. \tag{13.37}$$

The determinant is a number, which can be calculated for any square matrix. The formula for a (2×2) matrix is:

$$\det(\mathbf{A}) = \begin{vmatrix} A_{11} & A_{12} \\ A_{21} & A_{22} \end{vmatrix} = A_{11}A_{22} - A_{12}A_{21} \tag{13.38}$$

and for a (3×3) matrix:

$$\det(\mathbf{A}) = \begin{vmatrix} A_{11} & A_{12} & A_{13} \\ A_{21} & A_{22} & A_{23} \\ A_{31} & A_{32} & A_{32} \end{vmatrix}$$

$$= A_{11}A_{22}A_{33} + A_{12}A_{23}A_{31} + A_{13}A_{21}A_{32}$$

$$- A_{11}A_{23}A_{32} - A_{12}A_{21}A_{33} - A_{12}A_{22}A_{31}. \tag{13.39}$$

If $\det(\mathbf{A}) \neq 0$ the matrix $\mathbf{A}$ is called regular, otherwise singular. The inverse of a matrix is defined by:

$$\mathbf{A}^{-1}\mathbf{A} = \mathbf{I} \tag{13.40}$$

with:

$$\left(A^{-1}\right)_{ik} = (\det(\mathbf{A}))^{-1}(-1)^{i+k}B_{ki}, \tag{13.41}$$

where B_{ki} is the determinant obtained from $\det(\mathbf{A})$ by cancelling the kth row and ith column.

If $\mathbf{A} = \bar{\mathbf{A}}$ ($A_{ik} = A_{ki}$), the matrix is called symmetric. If $\mathbf{A}^{-1} = \bar{\mathbf{A}}$, the matrix is called orthogonal.

13.10 Crystallographic computing

A crystallographic coordinate system D consists of a set of three nonlinear base vectors spanning in a right handed manner $D = \{\mathbf{a}, \mathbf{b}, \mathbf{c}\}$ (Figure 13.7).

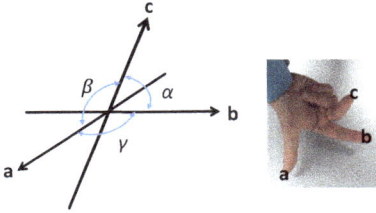

Figure 13.7: Right-handed base $D = \{\mathbf{a}, \mathbf{b}, \mathbf{c}\}$ in three-dimensional space.

An arbitrary vector within such a coordinate system can be written as:

$$\mathbf{r} = x\mathbf{a} + y\mathbf{b} + z\mathbf{c} = (\mathbf{abc})\begin{pmatrix} x \\ y \\ z \end{pmatrix} = \bar{\mathbf{A}}\mathbf{x}. \tag{13.42}$$

The scalar product of two vectors $\mathbf{r}_1$ and $\mathbf{r}_2$ enclosing the angle ϕ is defined as:

$$\mathbf{r}_1\mathbf{r}_2 = r_1 r_2 \cos\phi = (x_1\mathbf{a} + y_1\mathbf{b} + z_1\mathbf{c}) \cdot (x_2\mathbf{a} + y_2\mathbf{b} + z_2\mathbf{c})$$

$$= (x_1\ y_1\ z_1)\begin{pmatrix} \mathbf{a}\cdot\mathbf{a} & \mathbf{a}\cdot\mathbf{b} & \mathbf{a}\cdot\mathbf{c} \\ \mathbf{b}\cdot\mathbf{a} & \mathbf{b}\cdot\mathbf{b} & \mathbf{b}\cdot\mathbf{c} \\ \mathbf{c}\cdot\mathbf{a} & \mathbf{c}\cdot\mathbf{b} & \mathbf{c}\cdot\mathbf{c} \end{pmatrix}\begin{pmatrix} x_2 \\ y_2 \\ z_2 \end{pmatrix} = \bar{\mathbf{x}}_1\mathbf{G}\mathbf{x}_2 \tag{13.43}$$

with the metric tensor:

$$\mathbf{G} = \begin{pmatrix} \mathbf{a}\cdot\mathbf{a} & \mathbf{a}\cdot\mathbf{b} & \mathbf{a}\cdot\mathbf{c} \\ \mathbf{b}\cdot\mathbf{a} & \mathbf{b}\cdot\mathbf{b} & \mathbf{b}\cdot\mathbf{c} \\ \mathbf{c}\cdot\mathbf{a} & \mathbf{c}\cdot\mathbf{b} & \mathbf{c}\cdot\mathbf{c} \end{pmatrix} = \begin{pmatrix} a^2 & a\,b\,\cos\gamma & a\,c\,\cos\beta \\ a\,b\,\cos\gamma & b^2 & b\,c\,\cos\alpha \\ a\,c\,\cos\beta & b\,c\,\cos\alpha & c^2 \end{pmatrix}. \tag{13.44}$$

In the case of a Cartesian coordinate system, the metric tensor simplifies to $\mathbf{G} = \mathbf{I}$.

In general, the lengths of a vector $\mathbf{r}$ (e.g., the distance between two points) can be calculated as:

$$r = \sqrt{\overline{\mathbf{x}}\mathbf{G}\mathbf{x}} \tag{13.45}$$

and, therefore, the angle ϕ between two vectors $\mathbf{r}_1$ and $\mathbf{r}_2$ is:

$$\cos\phi = \frac{\overline{\mathbf{x}}_1\mathbf{G}\mathbf{x}_2}{\sqrt{\overline{\mathbf{x}}_1\mathbf{G}\mathbf{x}_1}\sqrt{\overline{\mathbf{x}}_2\mathbf{G}\mathbf{x}_2}}. \tag{13.46}$$

A unit vector $\hat{\mathbf{r}}$ is defined by dividing the vector by its length:

$$\hat{\mathbf{r}} = \frac{\mathbf{r}}{r} = \frac{\overline{\mathbf{A}}\mathbf{x}}{\sqrt{\overline{\mathbf{x}}\mathbf{G}\mathbf{x}}}. \tag{13.47}$$

The cross product of two vectors $\mathbf{r}_1$ and $\mathbf{r}_2$ leads to a vector, which is perpendicular to $\mathbf{r}_1$ and $\mathbf{r}_2$, and which completes a right-handed system:

$$\mathbf{r}_1 \times \mathbf{r}_2 = (x_1\mathbf{a} + y_1\mathbf{b} + z_1\mathbf{c}) \times (x_2\mathbf{a} + y_2\mathbf{b} + z_2\mathbf{c}) =$$
$$(y_1z_2 - y_2z_1)\mathbf{b} \times \mathbf{c} + (z_1x_2 - z_2x_1)\mathbf{c} \times \mathbf{a} + (x_1y_2 - x_2y_1)\mathbf{a} \times \mathbf{b}, \tag{13.48}$$

which in terms of reciprocal lattice parameters (see below) can be written as:

$$\mathbf{r}_1 \times \mathbf{r}_2 = \left(\mathbf{a}^*\mathbf{b}^*\mathbf{c}^*\right)\begin{pmatrix} y_1z_2 - y_2z_1 \\ z_1x_2 - z_2x_1 \\ x_1y_2 - x_2y_1 \end{pmatrix}, \tag{13.49}$$

and, if $\mathbf{a}$, $\mathbf{b}$ and $\mathbf{c}$ are base vectors of a Cartesian coordinate system, $\mathbf{a} \times \mathbf{b} = \mathbf{c}$, $\mathbf{b} \times \mathbf{c} = \mathbf{a}$ and $\mathbf{c} \times \mathbf{a} = \mathbf{b}$ simplify to:

$$\mathbf{r}_1 \times \mathbf{r}_2 = (\mathbf{abc})\begin{pmatrix} y_1z_2 - y_2z_1 \\ z_1x_2 - z_2x_1 \\ x_1y_2 - x_2y_1 \end{pmatrix}. \tag{13.50}$$

The length of the cross product is:

$$|\mathbf{r}_1 \times \mathbf{r}_2| = r_1 r_2 \sin\phi = \sqrt{\overline{\mathbf{x}}_1\mathbf{G}\mathbf{x}_1}\sqrt{\overline{\mathbf{x}}_2\mathbf{G}\mathbf{x}_2} \sin\phi. \tag{13.51}$$

The scalar triple product of three vectors $\mathbf{r}_1$, $\mathbf{r}_2$ and $\mathbf{r}_3$ is defined as:

$$\mathbf{r}_1 \cdot \mathbf{r}_2 \times \mathbf{r}_3 = V\det\begin{pmatrix} x_1 & y_1 & z_1 \\ x_2 & y_2 & z_2 \\ x_3 & y_3 & z_3 \end{pmatrix} \tag{13.52}$$

where V is the volume of the unit cell and can be calculated according to:

$$V = \sqrt{\det(\mathbf{G})}. \tag{13.53}$$

In addition to the "normal" crystal lattice with the lattice parameters $a, b, c, \alpha, \beta, \gamma$, and volume V of the unit cell, a second lattice with lattice parameters of $a^*, b^*, c^*, \alpha^*, \beta^*, \gamma^*$, and the volume V^* with the same origin can be defined such that:

$$\mathbf{a} \cdot \mathbf{b}^* = \mathbf{a} \cdot \mathbf{c}^* = \mathbf{b} \cdot \mathbf{c}^* = \mathbf{a}^* \cdot \mathbf{b} = \mathbf{a}^* \cdot \mathbf{c} = \mathbf{b}^* \cdot \mathbf{c} = 0$$

$$\mathbf{a} \cdot \mathbf{a}^* = \mathbf{b} \cdot \mathbf{b}^* = \mathbf{c} \cdot \mathbf{c}^* = 1. \tag{13.54}$$

This is known as the reciprocal lattice,[2] which exists in so-called reciprocal space. The length of the reciprocal base vectors can be defined according to:

$$\mathbf{a}^* = x(\mathbf{b} \times \mathbf{c}), \tag{13.55}$$

where the scale factor x can easily be deduced, using the equations above as:

$$\mathbf{a}^* \cdot \mathbf{a} = x(\mathbf{b} \times \mathbf{c} \cdot \mathbf{a}) = xV \Rightarrow x = \frac{1}{V} \tag{13.56}$$

leading to:

$$\mathbf{a}^* = \frac{1}{V}(\mathbf{b} \times \mathbf{c}), \ \ \mathbf{b}^* = \frac{1}{V}(\mathbf{c} \times \mathbf{a}), \ \ \mathbf{c}^* = \frac{1}{V}(\mathbf{a} \times \mathbf{b}) \tag{13.57}$$

and vice versa:

$$\mathbf{a} = \frac{1}{V^*}\left(\mathbf{b}^* \times \mathbf{c}^*\right), \ \ \mathbf{b} = \frac{1}{V^*}\left(\mathbf{c}^* \times \mathbf{a}^*\right), \ \ \mathbf{c} = \frac{1}{V^*}\left(\mathbf{a}^* \times \mathbf{b}^*\right). \tag{13.58}$$

The relationship between the reciprocal and the real lattice parameters expressed geometrically rather than in the vector formalism above is:

$$a^* = \frac{bc \sin \alpha}{V},$$

$$b^* = \frac{ac \sin \beta}{V},$$

$$c^* = \frac{ab \sin \gamma}{V},$$

---

2 The reciprocal lattice is a commonly used construct in solid state physics, but with a different normalization: $\mathbf{a} \cdot \mathbf{a}^* = 2\pi$.

$$\cos \alpha^* = \frac{\cos \beta \cos \gamma - \cos \alpha}{\sin \beta \sin \gamma},$$

$$\cos \beta^* = \frac{\cos \gamma \cos \alpha - \cos \beta}{\sin \gamma \sin \alpha},$$

$$\cos \gamma^* = \frac{\cos \alpha \cos \beta - \cos \gamma}{\sin \alpha \sin \beta},$$

$$V = abc\sqrt{1 + 2\cos \alpha \cos \beta \cos \gamma - \cos^2\alpha - \cos^2\beta - \cos^2\gamma,}$$

(13.59)

which is the most general expression for non-orthogonal lattices. The expressions simplify considerably for higher-symmetry crystal systems.

An arbitrary vector within such a coordinate system can be written as:

$$\mathbf{r}^* = x\mathbf{a}^* + y\mathbf{b}^* + z\mathbf{c}^* = \left(\mathbf{a}^*\mathbf{b}^*\mathbf{c}^*\right)\begin{pmatrix} x \\ y \\ z \end{pmatrix} = \overline{\mathbf{A}}^*\mathbf{x}.$$

(13.60)

Accordingly, the reciprocal metric tensor is:

$$\mathbf{G}^* = \begin{pmatrix} \mathbf{a}^* \cdot \mathbf{a}^* & \mathbf{a}^* \cdot \mathbf{b}^* & \mathbf{a}^* \cdot \mathbf{c}^* \\ \mathbf{b}^* \cdot \mathbf{a}^* & \mathbf{b}^* \cdot \mathbf{b}^* & \mathbf{b}^* \cdot \mathbf{c}^* \\ \mathbf{c}^* \cdot \mathbf{a}^* & \mathbf{c}^* \cdot \mathbf{b}^* & \mathbf{c}^* \cdot \mathbf{c}^* \end{pmatrix} = \mathbf{G}^{-1}$$

(13.61)

A vector and its reciprocal counterpart are connected via the metric tensor:

$$\mathbf{r}^* = \mathbf{Gr}.$$

(13.62)

13.11 Basis transformations

If a matrix $\mathbf{T}$ transforms the base vector $\mathbf{a}_{D_1}$ in base D_1 into the base vector $\mathbf{a}_{D_2}$ in base D_2 and the transformation leaves the origin invariant then:

$$\mathbf{a}_{D_2} = \mathbf{Ta}_{D_1} \quad \text{and} \quad \mathbf{a}_{D_1} = \mathbf{T}^{-1}\mathbf{a}_{D_2}.$$

(13.63)

Any vector can then be transformed from one base to the other and vice versa by

$$\mathbf{x}_{D_2} = \overline{\overline{\mathbf{T}}}^{-1}\mathbf{x}_D \quad \text{and} \quad \mathbf{x}_{D_1} = \overline{\overline{\mathbf{T}}}\mathbf{x}_D.$$

(13.64)

The metric tensors can be calculated as

$$\mathbf{G}_{D_1} = \overline{\overline{\mathbf{T}}}\mathbf{G}_{D_2}\mathbf{T} \quad \text{and} \quad \mathbf{G}_{D_2} = \overline{\overline{\mathbf{T}}}^{-1}\mathbf{G}_{D_1}\mathbf{T}^{-1}.$$

(13.65)

13.12 Rotations in the Cartesian system

A counterclockwise active rotation of an angle α around an axis of a Cartesian coordinate system, which transforms a point P to a point P′ can be written as a (3×3) rotation matrix $\mathbf{R}_{\text{axis}}(\alpha)$ (Figure 13.8).

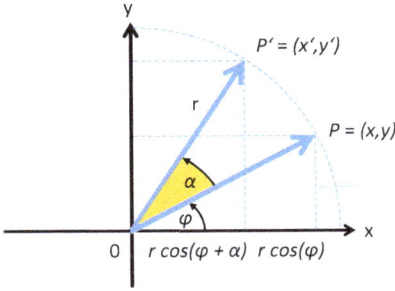

Figure 13.8: Counterclockwise rotation of a point P around z-axis in a Cartesian coordinate system.

A rotation α around the z-axis (axis perpendicular to the xy-plane) follows from Figure 13.8:

$$\begin{pmatrix} x \\ y \end{pmatrix} = \begin{pmatrix} r\cos\varphi \\ r\sin\varphi \end{pmatrix} \tag{13.66}$$

and:

$$\begin{pmatrix} x' \\ y' \end{pmatrix} = \begin{pmatrix} r\cos(\varphi+\alpha) \\ r\sin(\varphi+\alpha) \end{pmatrix} = \begin{pmatrix} r\cos\varphi\cos\alpha - r\sin\varphi\sin\alpha \\ r\sin\varphi\cos\alpha + r\cos\varphi\sin\alpha \end{pmatrix} = \begin{pmatrix} x\cos\alpha - y\sin\alpha \\ y\cos\alpha + x\sin\alpha \end{pmatrix}, \tag{13.67}$$

giving in three dimensions:

$$\mathbf{R}_z(\alpha) = \begin{pmatrix} \cos\alpha & -\sin\alpha & 0 \\ \sin\alpha & \cos\alpha & 0 \\ 0 & 0 & 1 \end{pmatrix} \tag{13.68}$$

Similarly:

$$\mathbf{R}_x(\alpha) = \begin{pmatrix} 1 & 0 & 0 \\ 0 & \cos\alpha & -\sin\alpha \\ 0 & \sin\alpha & \cos\alpha \end{pmatrix} \text{ and } \mathbf{R}_y(\alpha) = \begin{pmatrix} 1 & 0 & 0 \\ 0 & \cos\alpha & -\sin\alpha \\ 0 & \sin\alpha & \cos\alpha \end{pmatrix}. \tag{13.69}$$

For an active rotation, the coordinate system is not rotated but fixed.

13.13 Eulerian angles

If the coordinate system is not fixed but rotated, any rotation in three-dimensional space can be described by a rotation ϕ around z-axis, followed by a rotation around the new x'-axis θ, followed by a third rotation around the new z''-axis ψ (Figure 13.9). The three angles are called Eulerian angles.

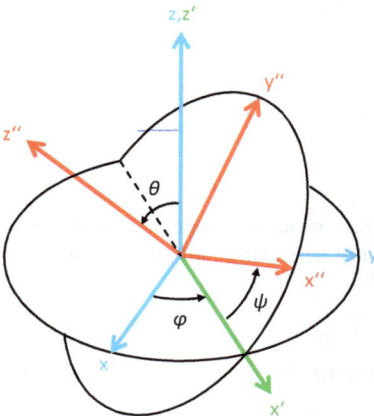

Figure 13.9: Proper Eulerian angles representing rotations about z, x' and z'' axes. The xyz (original) system is shown in blue, the rotated systems are shown in green and red.

The convention used here is only one of several possibilities:

$$\mathbf{R}_{\text{Euler}} = \mathbf{R}_z(\psi)\mathbf{R}_x(\theta)\mathbf{R}_z(\varphi)$$

$$= \begin{pmatrix} \cos\psi\cos\varphi - \sin\psi\cos\theta\sin\varphi & \sin\psi\cos\varphi + \cos\psi\cos\theta\sin\varphi & \sin\theta\sin\varphi \\ -\sin\psi\cos\theta\cos\varphi - \cos\psi\sin\varphi & -\sin\psi\sin\varphi + \cos\psi\cos\theta\cos\varphi & \sin\theta\cos\varphi \\ \sin\psi\sin\theta & -\cos\psi\sin\theta & \cos\theta \end{pmatrix}$$

$$(13.70)$$

13.14 Spherical coordinates

In many cases, it is more convenient to describe a point in three-dimensional space in spherical instead of Cartesian coordinates. Spherical coordinates consist of the radial distance r of that point from a fixed origin, its polar angle ϕ measured from a fixed zenith direction, and the azimuth angle θ of its orthogonal projection onto a reference plane that passes through the origin and is orthogonal to the zenith, measured from a fixed reference direction on that plane (Figure 13.10):

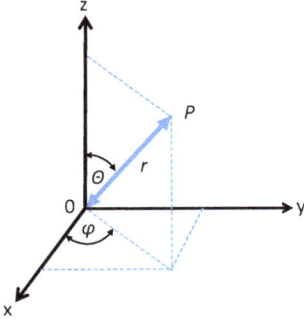

Figure 13.10: Relation between spherical and Cartesian coordinates at the point P.

$$r = \sqrt{x^2 + y^2 + z^2}$$
$$\theta = \arccos\left(\tfrac{z}{r}\right) \tag{13.71}$$
$$\varphi = \arctan\left(\tfrac{y}{x}\right)$$

The inverse tangent denoted in $\varphi = \arctan(y/x)$ must be suitably defined, taking into account the correct quadrant of (x, y). The ranges of the spherical coordinates are usually restricted to, for example, $r \in [0, \infty]$, $\theta \in [0, \pi]$, $\varphi \in [0, 2\pi]$. Then the Cartesian coordinates may be retrieved from the spherical coordinates by:

$$x = r \sin\theta \cos\varphi$$
$$y = r \sin\theta \sin\varphi \tag{13.72}$$
$$z = r \cos\theta.$$

13.15 Triaxial ellipsoid

A general orthogonal triaxial ellipsoid is a quadratic surface. In Cartesian space with the main axes aligned along the base vectors, such an ellipsoid is defined by:

$$\frac{x^2}{r_x^2} + \frac{y^2}{r_y^2} + \frac{z^2}{r_z^2} = 1 \tag{13.73}$$

with semi-axes of lengths r_x , r_y and r_z (Figure 13.11). In spherical coordinates as defined in Section 13.14 the Cartesian coordinates can be parameterized by

$$x = r_x \sin\theta \cos\varphi$$
$$y = r_y \sin\theta \sin\varphi \tag{13.74}$$
$$z = r_z \cos\theta.$$

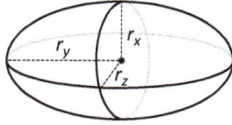

Figure 13.11: Definition of the main radii for an orthogonal triaxial ellipsoid.

The volume of a triaxial ellipsoid is:

$$V = \frac{4}{3}\pi r_x r_y r_z.$$ (13.75)

13.16 Symmetry operator

Crystallographic symmetry operations are special affine mappings leaving the crystal structure invariant. In three-dimensional space they are represented by a (3 × 3) rotation matrix and a (3 × 1) translation vector. The symmetry operation that maps a point:

$$\mathbf{x} = \begin{pmatrix} x \\ y \\ z \end{pmatrix}$$ (13.76)

onto a symmetry equivalent point:

$$\mathbf{x}' = \begin{pmatrix} x' \\ y' \\ z' \end{pmatrix}$$ (13.77)

is given by:

$$\mathbf{x}' = \begin{pmatrix} R_{11} & R_{12} & R_{13} \\ R_{21} & R_{22} & R_{23} \\ R_{31} & R_{32} & R_{33} \end{pmatrix} \mathbf{x} + \begin{pmatrix} t_1 \\ t_2 \\ t_3 \end{pmatrix} = \mathbf{R}\mathbf{x} + \mathbf{t} = (\mathbf{R}|\mathbf{t}).$$ (13.78)

$(\mathbf{R}|\mathbf{t})$ is called a symmetry operator in Seitz notation. Some important properties of the symmetry operator are given in the following:

If a symmetry operation $(\mathbf{U}|\mathbf{u})$ is followed by a second symmetry operation, then the combined symmetry operator is:

$$(\mathbf{V}|\mathbf{v})(\mathbf{U}|\mathbf{u}) = (\mathbf{VU}|\mathbf{Vu} + \mathbf{v}).$$ (13.79)

The symmetry operator for the identity operation is:

$$(\mathbf{I}|0) = (\mathbf{U}|\mathbf{u})^{-1}(\mathbf{U}|\mathbf{u}), \tag{13.80}$$

from which the operator for the inverse symmetry operation can be deduced as:

$$(\mathbf{U}|\mathbf{u})^{-1} = (\mathbf{U}^{-1}| - \mathbf{U}^{-1}\mathbf{u}). \tag{13.81}$$

If a (3×3) matrix $\mathbf{P}$ transforms the old to the new base vectors and a vector $\mathbf{p}$ translates the old to the new origin, then the symmetry operator changes according to:

$$(\mathbf{U}'|\mathbf{u}') = (\mathbf{P}^{-1}\mathbf{U}\mathbf{P}|\mathbf{u} + (\mathbf{U} - \mathbf{I})\mathbf{p}). \tag{13.82}$$

13.17 Probability density functions

13.17.1 Gaussian distribution

The Gaussian function is a very common continuous probability density function (normal distribution, Figure 13.12). Physical quantities that are expected to be the sum of many independent processes (such as measurement errors) often have distributions that are nearly normal. A Gaussian function normalized to an area of unity centered at a mean or expectation value x_0 is:

$$G(x - x_0) = \frac{1}{\sigma\sqrt{2\pi}} e^{-\frac{1}{2}\left(\frac{x - x_0}{\sigma}\right)^2} \tag{13.83}$$

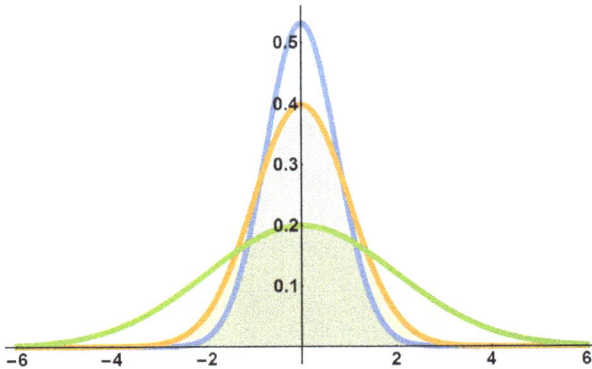

Figure 13.12: Series of normalized Gaussians with different $fwhm_G$.

where σ is the standard deviation and σ^2 the variance. The pre-factor follows from normalization:

$$\int\limits_{-\infty}^{+\infty} G(x - x_0)\mathrm{d}x = 1. \tag{13.84}$$

The full width at half-maximum (*fwhm*) is nowadays the most commonly used measurement in powder diffraction of peak widths. The *fwhm*$_G$ of a Gaussian follows the simple relation:

$$fwhm_G = 2\sqrt{2\ln 2}\,\sigma \approx 2.355\sigma. \tag{13.85}$$

Using this, the Gaussian function can be written as:

$$G(x - x_0) = \frac{2\sqrt{\frac{\ln 2}{\pi}}}{fwhm_G}\, e^{-4\ln(2)\left(\frac{x - x_0}{fwhm_G}\right)^2}. \tag{13.86}$$

13.17.2 Lorentz distribution

The Lorentzian (or Cauchy) function is another important continuous probability density distribution. The normalized Lorentz distribution with an *fwhm*$_L$ (Figure 13.13) is defined as:

$$L(x - x_0) = \frac{\frac{2}{\pi fwhm_L}}{1 + 4\left(\frac{x - x_0}{fwhm_L}\right)^2}. \tag{13.87}$$

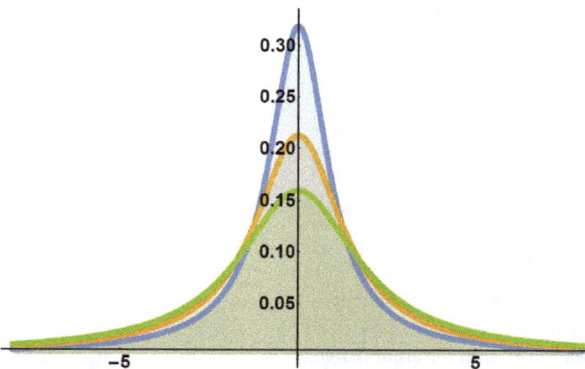

Figure 13.13: Series of normalized Lorentzians with different *fwhm*$_L$.

The Lorentz distribution is an example of a distribution that has no mean, variance or higher moments defined. Its mode and median are well defined and are both equal to x_0.

13.17.3 The Voigt distribution

The Voigt distribution is the convolution of a Gaussian and a Lorentzian. No analytical solution exists for the convolution integral but it can be expressed by the real part of the complex error function for which good approximations exist:

$$V(x) = (G \circ L)(x) = \frac{\text{Re}[w(z)]}{\sigma\sqrt{2\pi}} \quad \text{with} \quad z = \frac{x + \text{i}y}{\sigma\sqrt{2}} \qquad (13.88)$$

where $w(z)$ is called Faddeeva-function (also called Kramp-function or relativistic plasma-dispersion-function), and is a scalable complex conjugated error function (see Section 13.17.6).

Voigt functions are important profile functions, and typically normalized Gaussians and Lorentzian functions of different $fwhm$ values, $fwhm_\text{G}$ and $fwhm_\text{L}$ are convoluted to produce them. The total $fwhm$ of such a Voigt function can be approximated by a polynomial of fifth order (Thompson et al. 1987):

$$fwhm = \left(\begin{matrix} fwhm_\text{G}^5 + 2.69269\ fwhm_\text{G}^4 fwhm_\text{L} + 2.42843\ fwhm_\text{G}^3 fwhm_\text{L}^2 \\ +4.47163\ fwhm_\text{G}^2 fwhm_\text{L}^3 + 0.07842\ fwhm_\text{G} fwhm_\text{L}^4 + fwhm_\text{L}^5 \end{matrix} \right)^{\frac{1}{2}} \qquad (13.89)$$

An important property of both Gaussian and Lorentzian functions is that the convolution of a Gaussian with a Gaussian is a Gaussian and of a Lorentzian with a Lorentzian is a Lorentzian. Thereby, it holds:

$$fwhm_\text{G}^2 = fwhm_{\text{G},1}^2 + fwhm_{\text{G},2}^2 \qquad (13.90)$$

and:

$$fwhm_\text{L} = fwhm_{\text{L},1} + fwhm_{\text{L},2} \qquad (13.91)$$

where $fwhm_{\text{G/L},1/2}$ are the $fwhms$ of the two Gaussian or the two Lorentzian functions to be convoluted, where $fwhm_{\text{G/L}}$ are the $fwhms$ of the Gaussian and Lorentzian function resulting from the convolution.

In combination with eq. (13.88), the convolution properties of two Gaussian and Lorentzian functions illustrated above imply that the convolution of two Voigt functions V_1 and V_2 is also a Voigt:

$$V = \quad V_1 \circ V_2 \tag{13.92}$$

where the Gaussian and Lorentzian *fwhm* values *fwhm*$_G$ and *fwhm*$_L$ of V can be obtained from the corresponding values pertaining to V_1 and V_2 via eqs. (13.88) and (13.89). The functions can be centered by introducing the variable $X = x - x_0$ with the position x_0 at the average or maximum of the function.

13.17.4 The pseudo-Voigt distribution

The Voigt function can easily be convoluted, which is convenient when modeling the profile of a diffraction peak. The requirement for a numerical approximation of the Voigt function (due to lack of an analytical formulation), however, makes the use of the true Voigt function computationally expensive. A popular approximation to the Voigt function is the pseudo-Voigt function PV, which is defined as:

$$PV(x) = (1 - \eta) \ G(x) + \eta \, L(x), \tag{13.93}$$

with G and L usually being a Gaussian and Lorentzian function with equal *fwhm*s, that is, *fwhm*$_G$ = *fwhm*$_L$. These *fwhm* values are equal to the *fwhm* of the whole PV function. η is a weighting factor, where $\eta = 0$ implies a pure Gaussian (Figure 13.12) and $\eta = 1$ a pure Lorentzian function (Figure 13.13). Typically, the values of η are confined $0 \leq \eta \leq 1$. As illustrated in Figure 13.14, $\eta < 0$ implies negative values for some ranges of x, and η values somewhat above 1 give PV functions which still have unimodal shapes. These may occasionally agree with experimentally observed shapes, for example, in some line broadening cases. Such shapes are often referred to super-Lorentz. Even larger values of η result in multimodal distributions and or distributions with negative values in certain ranges of x (not shown in Figure 13.14).

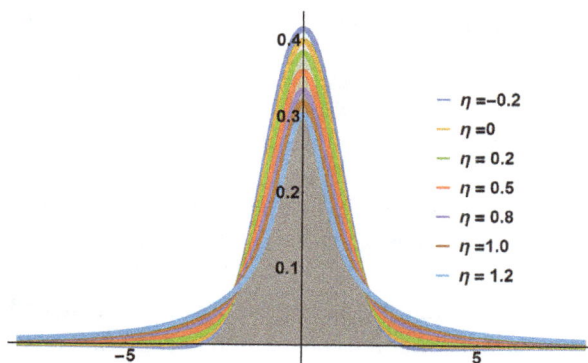

Figure 13.14: Normalized pseudo-Voigt functions located at $x_0 = 0$ with *fwhm* = 1 and $\eta = -0.2$, 0, .2, 0.5, 0.8, 1.0, 1.2.

In spite of its convenience and the fact that it fits experimental data well, there is little theoretical basis for physically relevant distributions to be of pseudo-Voigt shape. In particular, a convolution of one pseudo-Voigt function with a second is not a pseudo-Voigt, except for the trivial cases where $\eta_1 = \eta_2 = 0$ or $\eta_1 = \eta_2 = 1$. It is, however, possible to approximate a given PV by a Voigt function using eq. (13.89) and:

$$\eta = 13.36603 fwhm_L/fwhm - 0.47719 (fwhm_L/fwhm)^2 + 0.11116 (fwhm_L/fwhm)^3.$$

$$(13.94)$$

Then the convolution rules for Voigt functions can be approximated for pseudo-Voigt functions, at least if $0 \leq \eta \leq 1$ holds.

13.17.5 More width parameters

In addition to the *fwhm*, there are other important parameters quantifying widths of distributions, each with advantages and disadvantages. We consider these width parameters on the x scale as above.

The square root of the variance $var(x)^{1/2} = \sigma$ or the standard deviation $\langle (x - \langle x \rangle)^2 \rangle^{1/2}$ of a distribution described by $f(x)$ (which is normalized to an area of unity) is given as:

$$\langle (x - \langle x \rangle)^2 \rangle^{1/2} = \left(\int_{-\infty}^{\infty} x^2 f(x) dx \right)^{1/2}, \qquad (13.95)$$

with the average given as $\langle x \rangle = \int_{-\infty}^{\infty} x f(x) dx$, which agrees (in the case of a symmetric unimodal distribution) with the peak maximum x_0. The standard deviation of a Gaussian corresponds to:

$$\langle (x - \langle x \rangle)^2 \rangle^{1/2} = \frac{fwhm_G}{2\sqrt{2\ln 2}}, \qquad (13.96)$$

whereas the standard deviation of Lorentzian is not defined. The same holds for a pseudo-Voigt function in the case of $\eta > 0$. Voigt functions have a defined standard deviation only for sufficiently small values of $fwhm_L/fwhm$.

The integral breadth β of a function $f(x)$, which is normalized to an area of unity, is defined as:

$$\beta = \frac{1}{f(x_0)}, \qquad (13.97)$$

with the position of the peak maximum x_0. The integral breadth of a Gaussian is given by:

$$\beta_G = \sqrt{\frac{\pi}{4\ln2}}\, fwhm_G, \tag{13.98}$$

whereas that of a Lorentzian is given as:

$$\beta_L = \frac{\pi}{2} fwhm_L \tag{13.99}$$

Occasionally, a Voigt function is given in terms of its Gaussian and Lorentzian *fwhm* values in terms of:

$$fwhm_G = (1-\zeta)E$$
$$fwhm_L = \zeta E. \tag{13.100}$$

In this case, E is also some kind of width parameter related to the *fwhm* of the Voigt function in terms of:

$$fwhm = \Big((1-\zeta)^5 + 2.69269(1-\zeta)^4\zeta + 2.42843(1-\zeta)^3\zeta^2 + 4.47163(1-\zeta)^2\zeta^3$$

$$+\, 0.07842(1-\zeta)\zeta^4 + \zeta^5 \Big)^{\frac{1}{5}} E \tag{13.101}$$

Note that $fwhm = E$ holds for $\zeta = 0$ or 1. Note that calculation of η according to eq. (13.94) does not lead to the pseudo-Voigt approximating the corresponding Voigt to $\eta = \zeta$ except in the case of $\zeta = 0$ or 1.

13.17.6 Error function

When the results of a series of measurements are described by a normal distribution with standard deviation σ and expected value 0, then:

$$\mathrm{erf}\left(\frac{a}{\sigma\sqrt{2}}\right) \tag{13.102}$$

is the probability that the error of a single measurement lies between $-a$ and $+a$, for positive a, which is plotted in Figure 13.15. The error function is defined as:

$$\mathrm{erf}(x) = \frac{2}{\sqrt{\pi}} \int_0^x e^{t^2}\, \mathrm{d}t \tag{13.103}$$

The complementary error function is defined as:

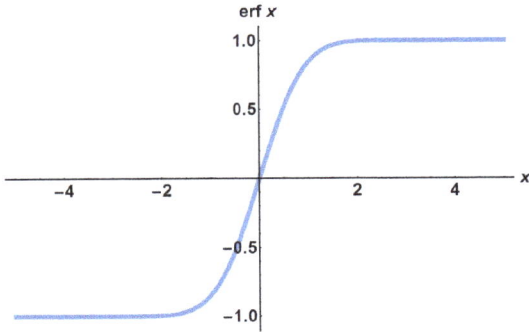

Figure 13.15: Plot of the error function given in eq. (13.104).

$$\text{erfc}(x) = 1 - \text{erf}(x) = \frac{2}{\sqrt{\pi}} \int_x^\infty e^{-t^2} dt.$$ (13.104)

Using complex-valued arguments z instead of x, one obtains the so-called Faddeeva function:

$$w(z) = e^{-z^2} \text{erfc}(-iz) = e^{-z^2} \left(1 + \frac{2i}{\sqrt{\pi}} \int_0^z e^{t^2} dt \right).$$ (13.105)

13.18 The method of least squares

The aim of the least-squares method is to adjust the parameters of a model function to best fit a data set with, for example, n data pairs (x_j, y_j) with the independent variable x_j. The dependent variable y_j, which is found by observation, can be treated as a vector:

$$\mathbf{y} = \begin{pmatrix} y_1 \\ y_2 \\ \vdots \\ y_n \end{pmatrix}$$ (13.106)

of dimension n. The model function of the form $f(x, \mathbf{p})$ contains adjustable parameters in form of the vector:

$$\mathbf{p} = \begin{pmatrix} p_1 \\ p_2 \\ \vdots \\ p_m \end{pmatrix} \tag{13.107}$$

of dimension m. The residual vector (observed – calc) of dimension n is then given as:

$$\mathbf{r} = \begin{pmatrix} y_1 - f(x_1, \mathbf{p}) \\ y_2 - f(x_2, \mathbf{p}) \\ \vdots \\ y_n - f(x_n, \mathbf{p}) \end{pmatrix}. \tag{13.108}$$

The vector $\mathbf{p}$ must be determined in a way that minimizes the weighted sum S of the squared residuals between the measured data and the model function at the positions x_j:

$$S \overset{Min}{\to} \sum_{j=1}^{n} r_j^2 = \sum_{j=1}^{n} \left(y_j - f(x_j, \mathbf{p}) \right)^2. \tag{13.109}$$

S is called the objective function. To minimize S, a set of m equations containing the partial derivatives of S must be calculated with respect to all parameters p_k with $k=1, \ldots, m$:

$$\frac{\partial S}{\partial p_k} = 2 \sum_{j=1}^{n} r_j \frac{\partial r_j}{\partial p_k} = -2 \sum_{j=1}^{n} r_j \frac{\partial f(x_j, \mathbf{p})}{\partial p_k}. \tag{13.110}$$

If the least squares estimate $\hat{\mathbf{p}}$ minimizes S, the partial derivatives must be zero:

$$\frac{\partial S}{\partial p_k} = 0. \tag{13.111}$$

Linear and *nonlinear* cases must be distinguished. A regression model is called *linear* when the model comprises a *linear combination* of the parameters, that is:

$$f(x, \mathbf{p}) = \sum_{k=1}^{m} p_k \, \Phi_k(x) \tag{13.112}$$

where the function $\Phi_j(x)$ is a function of x. The partial derivatives in this case are:

$$\frac{\partial f(x_j, \mathbf{p})}{\partial p_k} = \Phi_k(x_j) = X_{j,k} \tag{13.113}$$

or in matrix form

$$\mathbf{X} = \begin{pmatrix} X_{1,1} & \cdots & X_{1,n} \\ \vdots & \ddots & \vdots \\ X_{m,1} & \cdots & X_{m,n} \end{pmatrix}. \tag{13.114}$$

The partial derivatives of S can then be written as:

$$-2\sum_{j=1}^{n}\left(y_i - \sum_{k=1}^{m} X_{j,k}\hat{p}_k\right)X_{j,k} = 0, \tag{13.115}$$

which can be rearranged to obtain the m normal equations denoted by index i:

$$\sum_{j=1}^{n}\sum_{k=1}^{m} X_{j,i}X_{j,k}\hat{p}_k = \sum_{k=1}^{n} X_{j,i}y_i \ (i = 1, ..., m). \tag{13.116}$$

Eq. (13.116) can be conveniently written in matrix notation:

$$\overline{\mathbf{X}}\mathbf{X}\hat{\mathbf{p}} = \overline{\mathbf{X}}\mathbf{y}. \tag{13.117}$$

If there are weights applied to the observations with $w_j > 0$ as the weight of the jth observation (usually the reciprocal of the variance of the measured value) and:

$$\mathbf{W} = \begin{pmatrix} W_{1,1} & \cdots & 0 \\ \vdots & \ddots & \vdots \\ 0 & \cdots & W_{n,m} \end{pmatrix} \tag{13.118}$$

is the diagonal matrix of such weights, then the normal equations are written as:

$$(\overline{\mathbf{X}}\mathbf{W}\mathbf{X})\hat{\mathbf{p}} = \overline{\mathbf{X}}\mathbf{W}\mathbf{y}. \tag{13.119}$$

The algebraic solution is:

$$\hat{\mathbf{p}} = (\overline{\mathbf{X}}\mathbf{W}\mathbf{X})^{-1}\overline{\mathbf{X}}\mathbf{W}\mathbf{y}. \tag{13.120}$$

Many different ways exist for solving eq. (13.120). One possibility is the Cholesky decomposition where $\overline{\mathbf{X}}\mathbf{X}$ is decomposed into $\overline{\mathbf{R}}\mathbf{R}$ where $\mathbf{R}$ is an upper triangular matrix.

In practice, $f(x, \mathbf{p})$ is usually a nonlinear function. There is no closed-form solution to a nonlinear least squares problem. Instead, numerical algorithms are used to find the values of the parameters of $\mathbf{p}$ that minimize the objective. Most algorithms involve choosing initial values for the parameters. Then, the parameters are refined iteratively, that is, the values are obtained by successive approximation:

$$\mathbf{p} \approx \mathbf{p}^{t+1} = \mathbf{p}^t + \Delta\mathbf{p} \tag{13.121}$$

where t is an iteration number, and the vector of increments $\Delta\mathbf{p}$ is called the shift vector. In most algorithms, at each iteration the model may be linearized approximately by a first-order Taylor series expansion (see Section 13.2) around the current values of the parameters p^t (Figure 13.16) :

$$f(x_j, \mathbf{p}) \approx f(x_j, \mathbf{p}^t) + \sum_{k=1}^{m} \frac{\partial f(x_j, \mathbf{p}^t)}{\partial p_k} (p_k - p_k^t) = f(x_j, \mathbf{p}^t) + \sum_{k=1}^{m} J_{jk} \Delta p_k \tag{13.122}$$

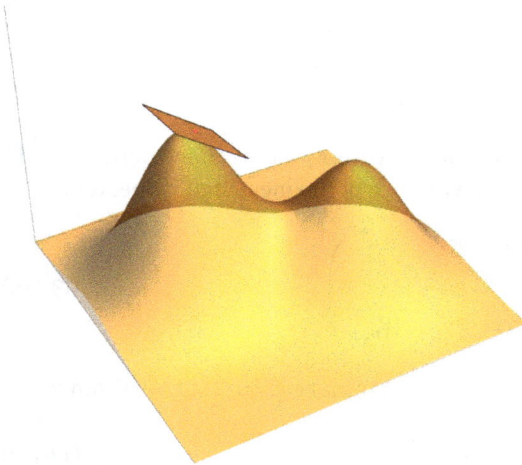

Figure 13.16: Taylor expansion of first order around the red point in a complex two-dimensional function.

with the residual vector of dimension n now defined as:

$$\Delta\mathbf{y} = \begin{pmatrix} y_1 - f(x_1, \mathbf{p}^t) \\ y_2 - f(x_2, \mathbf{p}^t) \\ \vdots \\ y_n - f(x_n, \mathbf{p}^t) \end{pmatrix}, \tag{13.123}$$

which in matrix notation can be then written as:

$$\bar{J}J\Delta p = \bar{J}\Delta y \tag{13.124}$$

or in case of weighted data:

$$(\bar{J}WJ)\Delta p = \bar{J}W\Delta y. \tag{13.125}$$

Methods for solving the system of equations comprise the Gauss–Newton algorithm, Levenberg–Marquardt algorithm, QR decomposition, gradient and direct search methods.

13.19 Spherical harmonics

The spherical harmonics $Y_{lm}(\theta, \varphi)$ are a complete and orthogonal set of solutions of the angular part of Laplace's equation in three dimensions:

$$Y_{lm}(\theta, \varphi) = \frac{1}{\sqrt{2\pi}} N_{lm} P_{lm}(\cos\theta) e^{im\varphi} \tag{13.126}$$

with scaling factors:

$$N_{lm} = \sqrt{\frac{2l+1}{2} \frac{(l-m)!}{(l+m)!}} \tag{13.127}$$

and the corresponding Legendre polynomials:

$$P_{lm}(x) = \frac{(-1)^m}{2^l l!} \left(1-x^2\right)^{m/2} \frac{d^{l+m}}{dx^{l+m}} \left(1-x^2\right)^l. \tag{13.128}$$

θ and φ are the coordinates of a spherical surface. They are similar to latitude and longitude except that θ goes from 0 to π and φ goes from 0 to 2π. The simplest spherical harmonic represents a sphere (Figure 13.17):

$$Y_{0,0}(\theta, \varphi) = \frac{1}{2\sqrt{\pi}}. \tag{13.129}$$

The spherical harmonics functions are orthogonal because if one integrates the product of any two different harmonics over the surface of the sphere, the result is zero. The spherical harmonics of second order are defined as:

$$Y_{2,0}(\theta, \varphi) = \frac{1}{4}\sqrt{\frac{5}{\pi}}(3\cos^2\theta - 1)$$

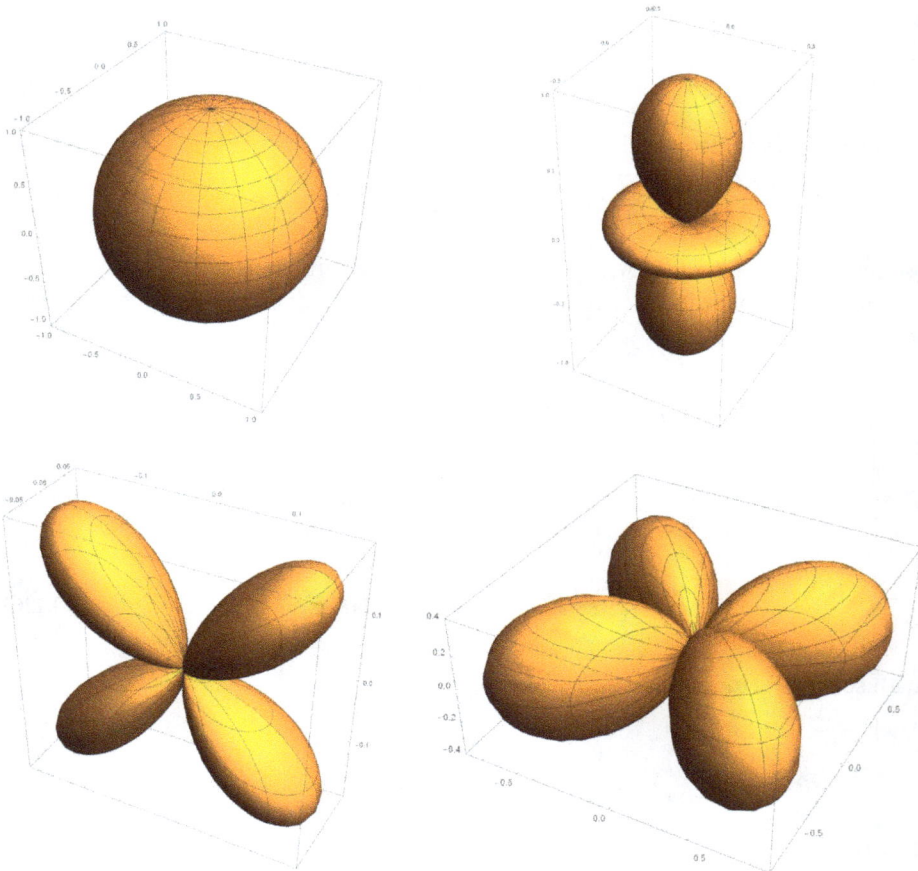

Figure 13.17: Graphical visualization of the normalized real components of second-order spherical harmonics $Y_{0,0}(\theta, \varphi)$, $Y_{2,0}(\theta, \varphi)$, $Y_{2,1}(\theta, \varphi)$ and $Y_{2,-2}(\theta, \varphi)$.

$$Y_{2,1}(\theta, \varphi) = -\tfrac{1}{2} e^{i\varphi} \sqrt{\tfrac{15}{2\pi}} \cos\theta \sin\theta$$

$$Y_{2,-1}(\theta, \varphi) = \tfrac{1}{2} e^{-i\varphi} \sqrt{\tfrac{15}{2\pi}} \cos\theta \sin\theta$$

$$Y_{2,2}(\theta, \varphi) = \tfrac{1}{4} e^{2i\varphi} \sqrt{\tfrac{15}{2\pi}} \sin^2\theta \qquad (13.130)$$

$$Y_{2,-2}(\theta, \varphi) = \tfrac{1}{4} e^{-2i\varphi} \sqrt{\tfrac{15}{2\pi}} \sin^2\theta$$

For powder diffraction the symmetrized and normalized real spherical harmonics of even order (due to the inversion center introduced by diffraction) are of most importance. The functions are normalized such that the maximum value of each component is 1 (Järvinen, 1993). As an example, the normalized real components of second-order spherical harmonics become:

$$Y_{2,0}(\theta, \varphi) = \tfrac{1}{2}(3\cos^2\theta - 1)$$
$$Y_{2,1}(\theta, \varphi) = 2\cos\varphi\cos\theta\sin\theta$$
$$Y_{2,-1}(\theta, \varphi) = 2\sin\varphi\cos\theta\sin\theta \qquad (13.131)$$
$$Y_{2,2}(\theta, \varphi) = \cos(2\varphi)\sin^2\theta$$
$$Y_{2,-2}(\theta, \varphi) = \sin(2\varphi)\sin^2\theta,$$

which are visualized in Figure 13.17.

The spherical harmonic functions can be expanded in a series to describe, in principle, any direction (θ, φ) dependent function.

References

Giacovazzo, C., Monaco, H.L., Artioli, G., Viterbo, D., Milaneso, M., Ferraris, G., Gilli, G., Gilli, P., Zanotti, G., Catti, M. (2011): *Fundamentals of crystallography*, Edited by C. Giacovazzo. IUCr Texts on Crystallography No. 15, 3rd editon, New York, IUCr/Oxford University Press, 842 pages.

Järvinen, M. (1993): *Application of symmetrized harmonics expansion to correction of the preferred orientation effect*. J. Appl. Cryst. 26, 525–531.

Müller, U. (2013): *Symmetry relationships between crystal structures*, Oxford University Press, New York, IUCr Texts on Crystallography No. 18, 332 pages.

Papula, L. (2014): *Mathematik für Ingenieure und Naturwissenschaftler" Band 1*, Springer Vieweg, 14th edition, 854 pages. (in German)

Thompson, P., Cox, D.E., Hastings, J.B. (1987): *Rietveld refinement of Debye-Scherrer synchrotron X-ray data from Al_2O_3*. J. Appl. Cryst. 20, 79–83.

Index

https://doi.org/10.1515/9783110461381-014